STOCHASTIC PROCESSES with APPLICATIONS to FINANCE

STOCHASTIC PROCESSES with APPLICATIONS to FINANCE

Masaaki Kijima

CHAPMAN & HALL/CRC

A CRC Press Company
Boca Raton   London   New York   Washington, D.C.

Library of Congress Cataloging-in-Publication Data

Kijima, Masaaki, 1957-
    Stochastic processes with applications to finance / Masaaki Kijima.
        p. cm.
    Includes bibliographical references and index.
    ISBN 1-58488-224-7
    1. Stochastic processes. 2. Business mathematics I. Title.

QA274 .K554 2002
519.2—dc21   2002067482

This book contains information obtained from authentic and highly regarded sources. Reprinted material is quoted with permission, and sources are indicated. A wide variety of references are listed. Reasonable efforts have been made to publish reliable data and information, but the author and the publisher cannot assume responsibility for the validity of all materials or for the consequences of their use.

Neither this book nor any part may be reproduced or transmitted in any form or by any means, electronic or mechanical, including photocopying, microfilming, and recording, or by any information storage or retrieval system, without prior permission in writing from the publisher.

The consent of CRC Press LLC does not extend to copying for general distribution, for promotion, for creating new works, or for resale. Specific permission must be obtained in writing from CRC Press LLC for such copying.

Direct all inquiries to CRC Press LLC, 2000 N.W. Corporate Blvd., Boca Raton, Florida 33431.

Trademark Notice: Product or corporate names may be trademarks or registered trademarks, and are used only for identification and explanation, without intent to infringe.

Visit the CRC Press Web site at www.crcpress.com

© 2003 by Chapman & Hall/CRC

No claim to original U.S. Government works
International Standard Book Number 1-58488-224-7
Library of Congress Card Number 2002067482
Printed in the United States of America   2 3 4 5 6 7 8 9 0
Printed on acid-free paper

Preface

This book presents the theory of stochastic processes and their applications to finance. In recent years, it has become increasingly important to use the idea of stochastic processes to describe financial uncertainty. For example, a stochastic differential equation is commonly used for the modeling of price fluctuations, a Poisson process is suitable for the description of default events, and a Markov chain is used to describe the dynamics of credit ratings of a corporate bond. However, the general theory of stochastic processes seems difficult to understand by anyone other than mathematical experts, because it requires a deep knowledge of various mathematical analyses. For example, although stochastic calculus is the basic tool for the pricing of derivative securities, it is based on a sophisticated theory of functional analyses.

In contrast, this book explains the idea of stochastic processes based on a simple class of discrete processes. In particular, a Brownian motion is a limit of random walks and a stochastic differential equation is a limit of stochastic *difference* equations. A random walk is a stochastic process with independent, identically distributed binomial random variables. Despite its strong assumptions, the random walk can serve as the basis for many stochastic processes.

A discrete process precludes pathological situations (so it makes the presentation simpler) and is easy to understand intuitively, while the actual process often fluctuates continuously. However, the continuous process is obtained from discrete processes by taking a limit under some regularity conditions. In other words, a discrete process is considered to be an approximation of the continuous counterpart. Hence, it seems important to start with discrete processes in order to understand such sophisticated continuous processes. This book presents important results in discrete processes and explains how to transfer those results to the continuous counterparts.

This book is suitable for the reader with a little knowledge of mathematics. It gives an elementary introduction to the areas of real analysis and probability. Ito's formula is derived as a result of the elementary Taylor expansion. Other important subjects such as the change of measure formula (the Girsanov theorem), the reflection principle and the Kolmogorov backward equation are explained by using random walks. Applications are taken from the pricing of derivative securities, and the Black-Scholes formula is obtained as a limit of binomial models. This book can serve as a text for courses on stochastic processes with special emphasis on finance.

The pricing of corporate bonds and credit derivatives is also a main theme of this book. The idea of this important new subject is explained in terms of discrete default models. In particular, a Poisson process is explained as a limit of random walks and a sophisticated Cox process is constructed in terms of a discrete hazard model. An absorbing Markov chain is another important topic, because it is used to describe the dynamics of credit ratings.

This book is organized as follows. In the next chapter, we begin with an introduction of some basic concepts of real calculus. A goal of this chapter is to derive Ito's formula from Taylor's expansion. The derivation is not mathematically rigorous, but the idea is helpful in many practical situations. Important results from elementary calculus are also presented for the reader's convenience.

Chapters 2 and 3 are concerned with the theory of basic probability and probability distributions useful for financial engineering. In particular, binomial and normal distributions are explained in some detail.

In Chapter 4, we explain derivative securities such as forward and futures contracts, options and swaps actually traded in financial markets. A *derivative security* is a security whose value depends on the values of other more basic underlying variables and, in recent years, has become more important than ever in financial markets.

Chapter 5 gives an overview of a general discrete-time model for the securities market in order to introduce various concepts important in financial engineering. In particular, self-financing and replicating portfolios play key roles for the pricing of derivative securities.

Chapter 6 provides a concise summary of the theory of random walks, the basis of the binomial model for the securities market. Detailed discussions of the binomial model are given in Chapter 7.

Chapter 8 considers a discrete model for the pricing of defaultable securities. A promising tool for this is the hazard-rate model. We demonstrate that a flexible model can be constructed in terms of random walks.

While Chapter 9 explains the importance of Markov chains in discrete time by showing various examples from finance, Chapter 10 is devoted to Monte Carlo simulation. As finance models become ever more complicated, practitioners want to use Monte Carlo simulation more and more. This chapter is intended to serve as an introduction to Monte Carlo simulation with emphasis on how to use it in financial engineering.

In Chapter 11, we introduce Brownian motions and Poisson processes in continuous time. The central limit theorem is invoked to derive the Black–Scholes formula from the discrete binomial counterpart as a continuous limit. A technique to transfer results in random walks to results in the corresponding continuous-time setting is discussed.

Chapter 12 introduces the stochastic processes necessary for the development of continuous-time securities market models. Namely, we discuss diffusion processes, martingales and stochastic integrals with respect to Brownian motions. A diffusion process is a natural extension of a Brownian motion and a solution to a stochastic differential equation, a key tool to describe the stochastic behavior of security prices in finance. Our arguments in this chapter are not rigorous, but work for almost all practical situations.

Finally, in Chapter 13, we describe a general continuous-time model for the securities market, parallel to the discrete-time counterpart discussed in Chapter 5. The Black–Scholes model is discussed in detail as an application of the no-arbitrage pricing theorem. The risk-neutral and forward-neutral pricing methods are also presented, together with their applications for the pricing of financial products.

I dedicate this book to all of my students for their continuous support. Without their warm encouragement over the years, I could not have completed this book. In particular, I wish to thank Katsumasa Nishide, Takashi Shibata and Satoshi Yamanaka for their technical contributions and helpful comments. The solutions of the exercises of this book listed in my web page (http://www.econ.kyoto-u.ac.jp/~kijima/) are made by them. Technical contributions by Yasuko Naito are also appreciated. Lastly, I would like to thank my wife Mayumi and my family for their patience.

Masaaki Kijima
Kyoto University

Contents

Preface		v
1	**Elementary Calculus: Towards Ito's Formula**	**1**
	1.1 Exponential and Logarithmic Functions	1
	1.2 Differentiation	4
	1.3 Taylor's Expansion	8
	1.4 Ito's Formula	10
	1.5 Integration	11
	1.6 Exercises	15
2	**Elements in Probability**	**19**
	2.1 The Sample Space and Probability	19
	2.2 Discrete Random Variables	21
	2.3 Continuous Random Variables	23
	2.4 Multivariate Random Variables	25
	2.5 Expectation	28
	2.6 Conditional Expectation	32
	2.7 Moment Generating Functions	35
	2.8 Exercises	37
3	**Useful Distributions in Finance**	**41**
	3.1 Binomial Distributions	41
	3.2 Other Discrete Distributions	43
	3.3 Normal and Log-Normal Distributions	46
	3.4 Other Continuous Distributions	50
	3.5 Multivariate Normal Distributions	53
	3.6 Exercises	57
4	**Derivative Securities**	**61**
	4.1 The Money-Market Account	61
	4.2 Various Interest Rates	62
	4.3 Forward and Futures Contracts	66
	4.4 Options	68
	4.5 Interest-Rate Derivatives	70
	4.6 Exercises	73

5 A Discrete-Time Model for Securities Market — 75
- 5.1 Price Processes — 75
- 5.2 The Portfolio Value and Stochastic Integral — 78
- 5.3 No-Arbitrage and Replicating Portfolios — 80
- 5.4 Martingales and the Asset Pricing Theorem — 84
- 5.5 American Options — 88
- 5.6 Change of Measure — 90
- 5.7 Exercises — 92

6 Random Walks — 95
- 6.1 The Mathematical Definition — 95
- 6.2 Transition Probabilities — 96
- 6.3 The Reflection Principle — 99
- 6.4 The Change of Measure Revisited — 102
- 6.5 The Binomial Securities Market Model — 105
- 6.6 Exercises — 108

7 The Binomial Model — 111
- 7.1 The Single-Period Model — 111
- 7.2 The Multi-Period Model — 114
- 7.3 The Binomial Model for American Options — 118
- 7.4 The Trinomial Model — 119
- 7.5 The Binomial Model for Interest-Rate Claims — 121
- 7.6 Exercises — 124

8 A Discrete-Time Model for Defaultable Securities — 127
- 8.1 The Hazard Rate — 127
- 8.2 A Discrete Hazard Model — 129
- 8.3 Pricing of Defaultable Securities — 131
- 8.4 Correlated Defaults — 135
- 8.5 Exercises — 138

9 Markov Chains — 141
- 9.1 Markov and Strong Markov Properties — 141
- 9.2 Transition Probabilities — 142
- 9.3 Absorbing Markov Chains — 145
- 9.4 Applications to Finance — 148
- 9.5 Exercises — 154

10 Monte Carlo Simulation — 157
- 10.1 Mathematical Backgrounds — 157
- 10.2 The Idea of Monte Carlo — 159
- 10.3 Generation of Random Numbers — 162
- 10.4 Some Examples from Financial Engineering — 165
- 10.5 Variance Reduction Methods — 169
- 10.6 Exercises — 172

11 From Discrete to Continuous: Towards the Black–Scholes   175
    11.1 Brownian Motions   175
    11.2 The Central Limit Theorem Revisited   178
    11.3 The Black–Scholes Formula   181
    11.4 More on Brownian Motions   183
    11.5 Poisson Processes   187
    11.6 Exercises   190

12 Basic Stochastic Processes in Continuous Time   193
    12.1 Diffusion Processes   193
    12.2 Sample Paths of Brownian Motions   197
    12.3 Martingales   199
    12.4 Stochastic Integrals   202
    12.5 Stochastic Differential Equations   205
    12.6 Ito's Formula Revisited   208
    12.7 Exercises   211

13 A Continuous-Time Model for Securities Market   215
    13.1 Self-Financing Portfolio and No-Arbitrage   215
    13.2 Price Process Models   217
    13.3 The Black–Scholes Model   222
    13.4 The Risk-Neutral Method   225
    13.5 The Forward-Neutral Method   231
    13.6 The Interest-Rate Term Structure   234
    13.7 Pricing of Interest-Rate Derivatives   241
    13.8 Pricing of Corporate Debts   245
    13.9 Exercises   253

References   261

Index   265

CHAPTER 1

Elementary Calculus: Towards Ito's Formula

Undoubtedly, one of the most useful formulae in financial engineering is Ito's formula. A goal of this chapter is to derive Ito's formula from Taylor's expansion. The derivation is not mathematically rigorous, but the idea is helpful in many practical situations of financial engineering. Important results from elementary calculus are also presented for the reader's convenience. See, e.g., Bartle (1976) for more details.

1.1 Exponential and Logarithmic Functions

Exponential and logarithmic functions naturally arise in the theory of finance when we consider a continuous-time model. This section summarizes important properties of these functions.

Consider the limit of sequence $\{a_n\}$ defined by

$$a_n = \left(1 + \frac{1}{n}\right)^n, \quad n = 1, 2, \ldots. \tag{1.1}$$

Note that the sequence $\{a_n\}$ is strictly increasing in n (Exercise 1.1). Associated with the sequence $\{a_n\}$ is the sequence $\{b_n\}$ defined by

$$b_n = \left(1 + \frac{1}{n}\right)^{n+1}, \quad n = 1, 2, \ldots. \tag{1.2}$$

It can be readily shown that the sequence $\{b_n\}$ is strictly decreasing in n (Exercise 1.1) and $a_n < b_n$ for all n. Since

$$\lim_{n \to \infty} \frac{b_n}{a_n} = \lim_{n \to \infty} \left(1 + \frac{1}{n}\right) = 1,$$

we conclude that the two sequences $\{a_n\}$ and $\{b_n\}$ converge to the same limit. The limit is usually called the *base of natural logarithm* and denoted by e (the reason for this will become apparent later). That is,

$$\mathrm{e} = \lim_{n \to \infty} \left(1 + \frac{1}{n}\right)^n. \tag{1.3}$$

The value is an irrational number (e = 2.718281828459 $\cdots$), i.e. it cannot be represented by a fraction of two integers. Note from the above observation

Table 1.1 *Convergence of the sequences to* e

n	1	2	3	4	5	$\cdots$	100	$\cdots$
a_n	2	2.25	2.37	2.44	2.49	$\cdots$	2.71	$\cdots$
b_n	4	3.38	3.16	3.05	2.99	$\cdots$	2.73	$\cdots$

that
$$\left(1+\frac{1}{n}\right)^n < \mathrm{e} < \left(1+\frac{1}{n}\right)^{n+1}, \quad n=1,2,\ldots,$$
which provides an approximation for e. However, the convergence of the sequences to e are rather slow (see Table 1.1).

For any positive number $x > 0$, consider the limit of sequence
$$\left(1+\frac{x}{n}\right)^n, \quad n=1,2,\ldots.$$
Let $N = n/x$. Since
$$\left(1+\frac{x}{n}\right)^n = \left[\left(1+\frac{1}{N}\right)^N\right]^x,$$
and since the function $g(y) = y^x$ is continuous in y for each x, it follows (see Proposition 1.3 below) that
$$\mathrm{e}^x = \lim_{n\to\infty}\left(1+\frac{x}{n}\right)^n, \quad x \geq 0,$$
where $\mathrm{e}^0 = 1$ by definition. The function e^x is called an *exponential function* and plays an essential role in the theory of continuous finance. In the following, we often denote the exponential function e^x by $\exp\{x\}$.

More generally, the exponential function is defined by
$$\mathrm{e}^x = \lim_{y\to\pm\infty}\left(1+\frac{x}{y}\right)^y, \quad x \in \mathbf{R}, \tag{1.4}$$
where $\mathbf{R} = (-\infty, \infty)$ is the real line. The proof is left in Exercise 1.8. Notice that, since x/y can be made arbitrarily small in the magnitude as $y \to \pm\infty$, the exponential function takes positive values only. Also, it is readily seen that e^x is strictly increasing in x. In fact, the exponential function e^x is positive and strictly increasing from 0 to ∞ as x runs from $-\infty$ to ∞.

The logarithm $\log_a x$ is the inverse function of $y = a^x$, where $a > 0$ and $a \neq 1$. That is, $x = \log_a y$ if and only if $y = a^x$. Hence, the *logarithmic function* $y = \log_a x$ (NOT $x = \log_a y$) is the mirror image of the function $y = a^x$ along the line $y = x$. The positive number a is called the *base* of logarithm. In particular, when $a = \mathrm{e}$, the inverse function of $y = \mathrm{e}^x$ is called the *natural logarithm* and denoted simply by $\log x$ or $\ln x$. The natural logarithmic function $y = \log x$ is the mirror image of the exponential function $y = \mathrm{e}^x$ along the line $y = x$

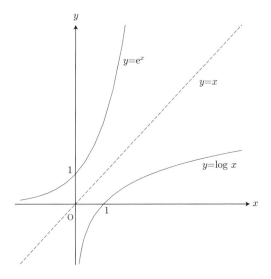

Figure 1.1 *The natural logarithmic and exponential functions*

(see Figure 1.1). The function $y = \log x$ is therefore increasing in $x > 0$ and ranges from $-\infty$ to ∞.

At this point, some basic properties of exponential functions are provided without proof.

Proposition 1.1 *Let $a > 0$ and $a \neq 1$. Then, for any real numbers x and y, we have the following.*

$$(1) \quad a^{x+y} = a^x a^y \qquad (2) \quad (a^x)^y = a^{xy}$$

The following basic properties of logarithmic functions can be obtained from Proposition 1.1. The proof is left in Exercise 1.3.

Proposition 1.2 *Let $a > 0$ and $a \neq 1$. Then, for any positive numbers x and y, we have the following.*

$$(1) \quad \log_a xy = \log_a x + \log_a y \qquad (2) \quad \log_a x^y = y \log_a x$$

We note that, since $a^{-y} = 1/a^y$, Proposition 1.1(1) implies

$$a^{x-y} = \frac{a^x}{a^y}.$$

Similarly, since $\log_a y^{-1} = -\log_a y$, Proposition 1.2(1) implies

$$\log_a \frac{x}{y} = \log_a x - \log_a y.$$

For positive numbers a and b such that $a, b \neq 1$, let

$$c = \log_b a.$$

Then, since $a = b^c$, we obtain from Proposition 1.1(2) that
$$a^x = b^{x \log_b a}, \quad x \in \mathbf{R}. \tag{1.5}$$
Since the logarithm is the inverse function of the exponential function, it follows from (1.5) that
$$\log_a x = \frac{\log_b x}{\log_b a}, \quad x > 0. \tag{1.6}$$
The formula (1.6) is called the *change of base* formula in logarithmic functions. By this reason, we usually take the most natural base, that is e, in logarithmic functions.

1.2 Differentiation

Let $f(x)$ be a real-valued function defined on an open interval $I \subset \mathbf{R}$. The function $f(x)$ is said to be *continuous* if, for any $x \in I$,
$$\lim_{h \to 0} f(x+h) = f(x). \tag{1.7}$$
Recall that the limit $\lim_{h \to 0} f(x+h)$ exists if and only if
$$\lim_{h \downarrow 0} f(x+h) = \lim_{h \uparrow 0} f(x+h),$$
i.e. both the right and the left limits exist and coincide with each other. Hence, a function $f(x)$ is continuous if the limit $\lim_{h \to 0} f(x+h)$ exists and coincides with the value $f(x)$ for all $x \in I$. If the interval I is not open, then the limit in (1.7) for each endpoint of the interval is understood to be an appropriate right or left limit.

One of the most important properties of continuous functions is the following.

Proposition 1.3 *Let I be an open interval, and suppose that a function $f(x)$ is continuous in $x \in I$. Then,*
$$\lim_{x \to c} f(g(x)) = f\left(\lim_{x \to c} g(x)\right),$$
provided that the limit $\lim_{x \to c} g(x) \in I$ exists.

For an open interval I, the real-valued function $f(x)$ defined on I is said to be *differentiable* at $x \in I$, if the limit
$$\lim_{h \to 0} \frac{f(x+h) - f(x)}{h}, \quad x \in I, \tag{1.8}$$
exists; in which case we denote the limit by $f'(x)$. The limit is often called the *derivative* in mathematics.* The function $f(x)$ is called *differentiable* in the open interval I if it is differentiable at all $x \in I$. Note from (1.7) and (1.8) that any differentiable function is continuous.

* In financial engineering, a derivative security is also called a derivative.

Let $y = f(x)$, and consider the changes in x and y. In (1.8), the denominator is the *difference* of x, denoted by Δx, i.e. x is incremented by $\Delta x = h$. On the other hand, the numerator in (1.8) is the difference of y, denoted by Δy, when x changes to $x + \Delta x$. Hence, the derivative is the limit of the fraction $\Delta y/\Delta x$ as $\Delta x \to 0$. When $\Delta x \to 0$, the differences Δx and Δy become infinitesimally small. An infinitesimally small difference is called a *differential*. The differential of x is denoted by $\mathrm{d}x$ and the differential of y is $\mathrm{d}y$. The derivative can then be viewed as the fraction of the differentials, i.e. $f'(x) = \mathrm{d}y/\mathrm{d}x$. However, in finance literature, it is common to write this as

$$\mathrm{d}y = f'(x)\mathrm{d}x, \tag{1.9}$$

provided that the function $f(x)$ is differentiable. The reason of this expression becomes apparent later.

Proposition 1.4 (Chain Rule) *For a composed function $y = f(g(x))$, let $u = g(x)$ and $y = f(u)$. If both $f(u)$ and $g(x)$ are differentiable, then*

$$\frac{\mathrm{d}y}{\mathrm{d}x} = \frac{\mathrm{d}y}{\mathrm{d}u}\frac{\mathrm{d}u}{\mathrm{d}x}.$$

In other words, we have $y' = f'(g(x))g'(x)$.

The *chain rule* in Proposition 1.4 for derivatives is easy to remember, since the derivative y' is the limit of the fraction $\Delta y/\Delta x$ as $\Delta x \to 0$, if it exists. The fraction is equal to

$$\frac{\Delta y}{\Delta x} = \frac{\Delta y}{\Delta u}\frac{\Delta u}{\Delta x}.$$

If both $f(u)$ and $g(x)$ are differentiable, then each fraction on the right-hand side converges to its derivative and the result follows.

The next result can be proved by a similar manner and the proof is left to the reader.

Proposition 1.5 *For a given function $f(x)$, let $g(x)$ denote its inverse function. That is, $y = f(x)$ if and only if $x = g(y)$. If $f(x)$ is differentiable, then $g(x)$ is also differentiable and*

$$\frac{\mathrm{d}y}{\mathrm{d}x} = \left(\frac{\mathrm{d}x}{\mathrm{d}y}\right)^{-1}.$$

Suppose that the derivative $f'(x)$ is further differentiable. Then, we can consider the derivative of $f'(x)$, which is denoted by $f''(x)$. The function $f(x)$ is said to be *twice differentiable* and $f''(x)$ the second-order derivative. A higher order derivative can also be considered in the same way. In particular, the nth order derivative is denoted by $f^{(n)}(x)$ with $f^{(0)}(x) = f(x)$. Alternatively, for $y = f(x)$, the nth order derivative is denoted by $\dfrac{\mathrm{d}^n y}{\mathrm{d}x^n}$, $\dfrac{\mathrm{d}^n}{\mathrm{d}x^n}f(x)$, etc.

Now, consider the exponential function $y = a^x$, where $a > 0$ and $a \neq 1$. Since, as in Exercise 1.2,

$$\lim_{k \to 0} \frac{\log(1+k)}{k} = 1,$$

the change of variable $k = e^h - 1$ yields

$$\lim_{h \to 0} \frac{e^h - 1}{h} = 1.$$

Hence, by the definition (1.8), we obtain

$$(a^x)' = \lim_{h \to 0} \frac{a^{x+h} - a^x}{h} = \lim_{h \to 0} \frac{e^{h \log a} - 1}{h} a^x = a^x \log a.$$

In particular, when $a = e$, one has

$$(e^x)' = e^x. \tag{1.10}$$

Thus, the base e is the most convenient (and natural) as far as its derivatives are concerned.

Equation (1.10) tells us more. That is, the exponential function e^x is differentiable infinitely many times and the derivative is the same as the original function. The logarithmic function is also differentiable infinitely many times, but the derivative is not the same as the original function. This can be easily seen from Proposition 1.5. Consider the exponential function $y = e^x$. The derivative is given by $dy/dx = e^x$, and so we obtain

$$\frac{dx}{dy} = \frac{1}{e^x} = \frac{1}{y}, \quad y > 0.$$

It follows that the derivative of the logarithmic function $y = \log x$ is given, by interchanging the roles of x and y, as $y' = 1/x$. The next proposition summarizes.

Proposition 1.6 *For any positive integer n, we have the following.*

(1) $\dfrac{d^n}{dx^n} e^x = e^x$ (2) $\dfrac{d^n}{dx^n} \log x = (-1)^{n-1} \dfrac{(n-1)!}{x^n}$

Here, the factorial $n!$ is defined by $n! = n(n-1)!$ with $0! = 1$.

Many important properties of differentiable functions are obtained from the next fundamental result (see Exercise 1.4).

Proposition 1.7 (Mean Value Theorem) *Suppose that $f(x)$ is continuous on a closed interval $[a, b]$ and differentiable in the open interval (a, b). Then, there exists a point c in (a, b) such that*

$$f(b) - f(a) = f'(c)(b - a).$$

The statement of the mean value theorem is easy to remember (and verify) by drawing an appropriate diagram as in Figure 1.2.

The next result is the familiar rule on the evaluation of 'indeterminant forms' and can be proved by means of a slight extension of the mean value theorem (see Exercises 1.5 and 1.6). Differentiable functions $f(x)$ and $g(x)$ are said to be of *indeterminant* form if $f(c) = g(c) = 0$ and $g(x)$ and $g'(x)$ do not vanish for $x \neq c$.

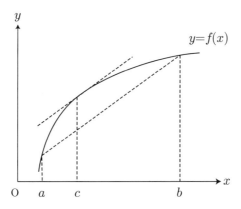

Figure 1.2 *The mean value theorem*

Proposition 1.8 (L'Hospital's Rule) *Let $f(x)$ and $g(x)$ be differentiable in an open interval $I \subset \mathbf{R}$, and suppose that $f(c) = g(c) = 0$ for $c \in I$. Then,*

$$\lim_{x \to c} \frac{f(x)}{g(x)} = \lim_{x \to c} \frac{f'(x)}{g'(x)}.$$

If the derivatives $f'(x)$ and $g'(x)$ are still of indeterminant form, L'Hospital's rule can be applied until the indeterminant is resolved.

The case where the functions become infinite at $x = c$, or where we have an 'indeterminant' of some other form, can often be treated by taking logarithms, exponentials, or some other manipulations (see Exercises 1.7(3) and 1.8).

Let $f(x, y)$ be a real-valued function of two variables x and y. Fix y, and take the derivative of $f(x, y)$ with respect to x. This is possible if $f(x, y)$ is differentiable in x, since $f(x, y)$ is a function of single variable x with y being fixed. Such a derivative is called a *partial derivative* and denoted by $f_x(x, y)$. Similarly, the partial derivative of $f(x, y)$ with respect to y is defined and denoted by $f_y(x, y)$. A higher order partial derivative is also considered in the same way. For example, $f_{xx}(x, y)$ is the partial derivative of $f_x(x, y)$ with respect to x. Note that $f_{xy}(x, y)$ is the partial derivative of $f_x(x, y)$ with respect to y while $f_{yx}(x, y)$ is the partial derivative of $f_y(x, y)$ with respect to x.[†] In general, they are different, but they coincide with each other if they are continuous.

Let $z = f(x, y)$ and consider a plane tangent to the function at (x, y). Such a plane is given by

$$g(u, v) = f(x, y) + \ell(u - x) + m(v - y), \quad (u, v) \in \mathbf{R}^2 \equiv \mathbf{R} \times \mathbf{R}, \quad (1.11)$$

for some ℓ and m. In the single-variable case, the derivative $f'(x)$ is the slope of the uniquely determined straight line tangent to the function at x. Similarly,

---

[†] Alternatively, we may write $\dfrac{\partial}{\partial x} f(x, y)$ for $f_x(x, y)$, $\dfrac{\partial^2}{\partial y \partial x} f(x, y)$ for $f_{xy}(x, y)$, etc.

we call the function $f(x,y)$ *differentiable* at point (x,y) if such a plane *uniquely* exists. Then, the function $f(u,v)$ can be approximated arbitrarily close by the plane (1.11) at a neighborhood of $(u,v) = (x,y)$. Formally, taking $u = x + \Delta x$ and $v = y$, i.e. $\Delta y = 0$, we have

$$f(x + \Delta x, y) \approx g(x + \Delta x, y) = f(x,y) + \ell \Delta x,$$

and this approximation can be made exact as $\Delta x \to 0$. It follows that the coefficient ℓ is equal to the partial derivative $f_x(x,y)$, since

$$\ell \approx \frac{f(x + \Delta x, y) - f(x,y)}{\Delta x}$$

and this approximation becomes exact as $\Delta x \to 0$. Similarly, we have $m = f_y(x,y)$, and so the difference of the function z, i.e. $\Delta z \equiv f(x+\Delta x, y+\Delta y) - f(x,y)$, is approximated by

$$\Delta z \approx \Delta g(x,y) = f_x(x,y)\Delta x + f_y(x,y)\Delta y. \quad (1.12)$$

Now, let $\Delta x \to 0$ and $\Delta y \to 0$ so that the approximation in (1.12) becomes exact. Then, the relation (1.12) can be represented in terms of the differentials. That is, we proved that if $z = f(x,y)$ is differentiable then

$$dz = f_x(x,y)dx + f_y(x,y)dy, \quad (1.13)$$

which should be compared with (1.9). Equation (1.13) is often called the *total derivative* of z (see Exercise 1.9). We remark that the total derivative (1.13) exists if the first-order partial derivatives $f_x(x,y)$ and $f_y(x,y)$ are continuous. See Bartle (1976) for details.

1.3 Taylor's Expansion

This section begins by introducing the notion of small order.

Definition 1.1 A *small order* $o(h)$ of h is a function of h satisfying

$$\lim_{h \to 0} \frac{o(h)}{h} = 0.$$

That is, $o(h)$ converges to zero faster than h as $h \to 0$.

The derivative defined in (1.8) can be stated in terms of the small order. Suppose that the function $y = f(x)$ is differentiable. Then, from (1.8), we obtain

$$\Delta y = f'(x)\Delta x + o(\Delta x), \quad (1.14)$$

where $\Delta y = f(x + \Delta x) - f(x)$, since $o(\Delta x)/\Delta x$ converges to zero as $\Delta x \to 0$. We note that, by the definition, a finite sum (more generally, a linear combination) of small orders of h is again a small order of h.

Suppose that a function $f(x)$ is differentiable $(n+1)$ times with derivatives $f^{(k)}(x)$, $k = 1, 2, \ldots, n+1$. We want to approximate $f(x)$ in terms of the polynomial

$$g(x) = \sum_{k=0}^{n} \gamma_k (x-a)^k. \quad (1.15)$$

TAYLOR'S EXPANSION

The coefficients γ_k are determined in such a way that
$$f^{(k)}(a) = g^{(k)}(a), \quad k = 0, 1, \ldots, n,$$
where $f^{(0)}(a) = f(a)$ and $g^{(0)}(a) = g(a)$.

Repeated applications of L'Hospital's rule yield
$$\lim_{x \to a} \frac{f(x) - g(x)}{(x-a)^n} = \lim_{x \to a} \frac{f'(x) - g'(x)}{n(x-a)^{n-1}}$$
$$\vdots$$
$$= \lim_{x \to a} \frac{f^{(n)}(x) - g^{(n)}(x)}{n!} = 0.$$

It follows that
$$f(x) = g(x) + o((x-a)^n),$$
where $o((x-a)^n)$ denotes a small order of $(x-a)^n$. On the other hand, from (1.15), we have
$$g^{(k)}(a) = \gamma_k k!, \quad k = 0, 1, \ldots, n.$$
Therefore, we obtain
$$\gamma_k = \frac{f^{(k)}(a)}{k!}, \quad k = 0, 1, \ldots, n.$$

Proposition 1.9 (Taylor's Expansion) *Any function $f(x)$ can be expanded as*
$$f(x) = f(a) + f'(a)(x-a) + \cdots + \frac{f^{(n)}(a)}{n!}(x-a)^n + R, \tag{1.16}$$
provided that the derivative $f^{(n+1)}(x)$ exists in an open interval (c,d), where $a \in (c,d)$ and $R = o((x-a)^n)$.

Equation (1.16) is called *Taylor's expansion of order n for $f(x)$ around $x = a$*, and the term R is called the *remainder*.

In particular, when $a = 0$, we have
$$f(x) = f(0) + f'(0)x + \frac{f''(0)}{2!}x^2 + \cdots + \frac{f^{(n)}(0)}{n!}x^n + R.$$

For example, the exponential function e^x is differentiable infinitely many times, and its Taylor's expansion around the origin is given by
$$e^x = 1 + x + \frac{x^2}{2!} + \cdots + \frac{x^n}{n!} + \cdots = \sum_{n=0}^{\infty} \frac{x^n}{n!}. \tag{1.17}$$

We remark that some textbooks use this equation as the definition of the exponential function e^x. See Exercise 1.10 for Taylor's expansion of a logarithmic function.

In financial engineering, it is often useful to consider differences of variables. That is, taking $x - a = \Delta x$ and $a = x$ in Taylor's expansion (1.16), we have
$$\Delta y = f'(x)\Delta x + \cdots + \frac{f^{(n)}(x)}{n!}(\Delta x)^n + o((\Delta x)^n). \tag{1.18}$$

Hence, Taylor's expansion is a higher extension of (1.14). When $\Delta x \to 0$, the difference Δx is replaced by the differential dx and, using the convention

$$(dx)^n = 0, \quad n > 1, \tag{1.19}$$

we obtain (1.9).

Consider next a function $z = f(x, y)$ of two variables, and suppose that it has partial derivatives of enough order. Taylor's expansion for the function $z = f(x, y)$ can be obtained in the same manner. We omit the detail. But, analogous to (1.18), it can be shown that the difference of $z = f(x, y)$ is expanded as

$$\Delta z = f_x \Delta x + f_y \Delta y + \frac{f_{xx}(\Delta x)^2 + 2f_{xy}\Delta x \Delta y + f_{yy}(\Delta y)^2}{2} + R, \tag{1.20}$$

where R denotes the remainder term. Here, we omit the variable (x, y) in the partial derivatives. Although we state the second-order Taylor expansion only, it will become evident that (1.20) is indeed enough for financial engineering practices.

1.4 Ito's Formula

In this section, we show that Ito's formula can be obtained by a standard application of Taylor's expansion (1.20).

Let x be a function of two variables, e.g. time t and some state variable z, and consider the difference $\Delta x(t, z) = x(t + \Delta t, z + \Delta z) - x(t, z)$. In order to derive Ito's formula, we need to assume that

$$\Delta x(t, z) = \mu \Delta t + \sigma \Delta z \tag{1.21}$$

and, moreover, that Δt and Δz satisfy the relation

$$\Delta t = (\Delta z)^2. \tag{1.22}$$

For any *smooth* function $y = f(t, x)$ with partial derivatives of enough order, we then have the following. From (1.21) and (1.22), we obtain

$$\Delta t \Delta x = \mu(\Delta t)^2 + \sigma(\Delta t)^{3/2} = o(\Delta t)$$

and

$$(\Delta x)^2 = \mu^2(\Delta t)^2 + 2\mu\sigma(\Delta t)^{3/2} + \sigma^2 \Delta t = \sigma^2 \Delta t + o(\Delta t),$$

where the variable (t, z) is omitted. Since the term $o(\Delta t)$ is negligible compared to Δt, it follows from (1.20) that

$$\Delta y = \left(f_t + \mu f_x + \frac{\sigma^2}{2} f_{xx}\right) \Delta t + \sigma f_x \Delta z + R,$$

where $R = o(\Delta t)$. The next result can be informally verified by replacing the differences with the corresponding differentials as $\Delta t \to 0$.[‡] See Exercises 1.12 and 1.13 for applications of Ito's formula.

[‡] As we shall see later, the variable satisfying (1.22) is not differentiable in the ordinary sense.

INTEGRATION

Theorem 1.1 (Ito's Formula) *Under the above conditions, we have*

$$dy = \left(f_t(t,x) + \mu f_x(t,x) + \frac{\sigma^2}{2} f_{xx}(t,x)\right) dt + \sigma f_x(t,x) dz.$$

In financial engineering, the state variable z is usually taken to be a standard Brownian motion (defined in Chapter 11), which is known to satisfy the relation (1.22) with the differences being replaced by the differentials. We shall explain this very important paradigm in Chapter 12.

If we assume another relation rather than (1.22), we obtain the other form of Ito's formula, as the next example reveals.

Example 1.1 As in the case of Poisson processes (to be defined in Chapter 11), suppose that Δz is the same order as Δt and

$$(\Delta z)^n = \Delta z, \quad n = 1, 2, \ldots. \tag{1.23}$$

Then, from (1.21), we have

$$(\Delta x)^n = \sigma^n \Delta z + o(\Delta t), \quad n = 2, 3, \ldots.$$

Hence, for a function $y = f(t,x)$ with partial derivatives of any order, we obtain from Taylor's expansion that

$$\begin{aligned}
\Delta y &= f_t(t,x)\Delta t + \sum_{n=1}^{\infty} \frac{\partial^n}{\partial x^n} f(t,x) \frac{(\Delta x)^n}{n!} \\
&= [f_t(t,x) + \mu f_x(t,x)]\Delta t + \left(\sum_{n=1}^{\infty} \frac{\partial^n}{\partial x^n} f(t,x) \frac{\sigma^n}{n!}\right) \Delta z,
\end{aligned}$$

where $\dfrac{\partial^n}{\partial x^n} f(t,x)$ denotes the nth order partial derivative of $f(t,x)$ with respect to x. But, from (1.16) with $x - a = \sigma$ and $a = x$, we have

$$f(x + \sigma) = f(x) + f'(x)\sigma + \cdots + \frac{f^{(n)}(x)}{n!}\sigma^n + \cdots.$$

It follows that

$$\sum_{n=1}^{\infty} \frac{\partial^n}{\partial x^n} f(t,x) \frac{\sigma^n}{n!} = f(t, x+\sigma) - f(t,x),$$

whence we obtain, as $\Delta t \to 0$,

$$dy = [f_t(t,x) + \mu f_x(t,x)]\, dt + [f(t, x+\sigma) - f(t,x)]\, dz. \tag{1.24}$$

We shall return to a similar model in Chapter 13 (see Example 13.3).

1.5 Integration

This section defines the Riemann–Stieltjes integral of a bounded function on a closed interval of $\mathbf{R}$, since the integral plays an important role in the theory of probability.

Recall that the area of a rectangle is given by a multiplication of two lengths.

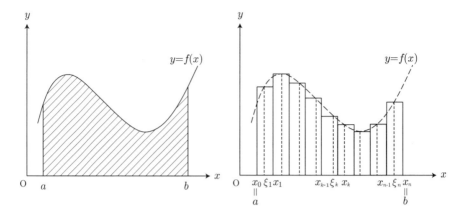

Figure 1.3 *The Riemann sum as a sum of areas*

Roughly speaking, the shadowed area depicted in Figure 1.3 is calculated as a limit of the area of the unions of rectangles. To be more specific, let $f(x)$ be a positive function defined on a closed interval $J = [a, b]$. A *partition* of J is a finite collection of non-overlapping subintervals whose union is equal to J. That is, $Q = (x_0, x_1, \ldots, x_n)$ is a partition of J if

$$a = x_0 < x_1 < \cdots < x_n = b.$$

The interval $[x_{k-1}, x_k]$, $k = 1, 2, \ldots, n$, is called a *subinterval* in Q. Given the partition Q, a *Riemann sum* of $f(x)$ is a real number $S(f; Q)$ of the form

$$S(f; Q) = \sum_{k=1}^{n} f(\xi_k)(x_k - x_{k-1}), \tag{1.25}$$

where the numbers ξ_k are chosen arbitrary such that $x_{k-1} \leq \xi_k \leq x_k$. Note that the Riemann sum provides an approximation of the shadowed area depicted in Figure 1.3.

The Riemann sum (1.25) can also be defined for a negative function by considering negative areas. In fact, it is defined for any bounded function. To see this, let us denote the *positive part* of a function $f(x)$ by $f_+(x)$, i.e.

$$f_+(x) = \max\{f(x), 0\}.$$

The *negative part*, denoted by $f_-(x)$, is defined by

$$f_-(x) = f_+(x) - f(x) \geq 0.$$

If $f(x)$ is bounded, then the positive and negative parts are also bounded, and the Riemann sum of $f(x)$ is given by the difference of the Riemann sum of $f_+(x)$ and that of $f_-(x)$. In the rest of this section, we assume unless stated otherwise that the functions under consideration are bounded.

The next step is to consider a limit of Riemann sums. To this end, we need to define an appropriate sequence of partitions. That is, if Q_1 and Q_2 are partitions of J, we say that Q_2 is a *refinement* of Q_1 in the case that every subinterval in Q_2 is contained in some subinterval in Q_1. The maximal length of subintervals in partition Q_n is denoted by d_n.

Now, consider a sequence of partitions, $Q_1, Q_2, \ldots$, such that Q_{n+1} is a refinement of Q_n and that $d_n \to 0$ as $n \to \infty$. If the sequence of Riemann sums $S(f; Q_n)$ converges to a real number for any such partitions as $n \to \infty$, then the limit is called the *Riemann integral* and denoted by $\int_a^b f(x)\mathrm{d}x$. That is, we define

$$\int_a^b f(x)\mathrm{d}x = \lim_{n \to \infty} S(f; Q_n), \tag{1.26}$$

where the limit is taken over all possible sequences of partitions of $J = [a, b]$ such that Q_{n+1} is a refinement of Q_n and $d_n \to 0$ as $n \to \infty$.

Suppose that the integral $\int_a^b f(x)\mathrm{d}x$ exists, so that the integral is approximated by a Riemann sum as

$$\int_a^b f(x)\mathrm{d}x \approx \sum_{k=1}^n f(\xi_k)(x_k - x_{k-1}).$$

Note that, under the above conditions, the difference $x_k - x_{k-1}$ tends to the differential $\mathrm{d}x$ and $f(\xi_k)$ to $f(x)$ as $n \to \infty$. Hence, the meaning of the notation $\int_a^b f(x)\mathrm{d}x$ should be clear.

Next, instead of considering the length $x_k - x_{k-1}$ in (1.25), we consider some other measure of magnitude of the subinterval $[x_{k-1}, x_k]$. Namely, suppose that the (possibly negative) total mass on the interval $[a, x]$ is represented by a bounded function $g(x)$. Then, the difference $\Delta g(x_k) \equiv g(x_k) - g(x_{k-1})$ denotes the (possibly negative) mass on the subinterval $[x_{k-1}, x_k]$. Similar to (1.25), a *Riemann–Stieltjes sum* of $f(x)$ with respect to $g(x)$ and corresponding to Q is defined as a real number $S(f, g; Q)$ of the form

$$S(f, g; Q) = \sum_{k=1}^n f(\xi_k) \Delta g(x_k), \tag{1.27}$$

where the numbers ξ_k are chosen arbitrary such that $x_{k-1} \leq \xi_k \leq x_k$. The *Riemann–Stieltjes integral* of $f(x)$ with respect to $g(x)$ is a limit of $S(f, g; Q_n)$ for any appropriate sequence of partitions and denoted by

$$\int_a^b f(x)\mathrm{d}g(x) = \lim_{n \to \infty} S(f, g; Q_n).$$

The meaning of the differential $\mathrm{d}g$ in the integral should be clear. The function $f(x)$ is called the *integrand* and $g(x)$ the *integrator*. Sometimes, we say that $f(x)$ is g-integrable if the Riemann–Stieltjes integral exists. In Chapter 12, we shall define a stochastic integral as an integral of a process with respect to a Brownian motion.

Some easily verifiable properties of Riemann–Stieltjes integrals are in order. We assume that

$$\int_a^b f(x)\mathrm{d}g(x) = -\int_b^a f(x)\mathrm{d}g(x).$$

In the next proposition, we suppose that the Riemann–Stieltjes integrals under consideration exist. The proof is left to the reader.

Proposition 1.10 *Let α_1 and α_2 be any real numbers.*

(1) Suppose $f(x) = \alpha_1 f_1(x) + \alpha_2 f_2(x)$. Then,

$$\int_a^b f(x)\mathrm{d}g(x) = \alpha_1 \int_a^b f_1(x)\mathrm{d}g(x) + \alpha_2 \int_a^b f_2(x)\mathrm{d}g(x).$$

(2) Suppose $g(x) = \alpha_1 g_1(x) + \alpha_2 g_2(x)$. Then,

$$\int_a^b f(x)\mathrm{d}g(x) = \alpha_1 \int_a^b f(x)\mathrm{d}g_1(x) + \alpha_2 \int_a^b f(x)\mathrm{d}g_2(x).$$

(3) For any real number $c \in [a, b]$, we have

$$\int_a^b f(x)\mathrm{d}g(x) = \int_a^c f(x)\mathrm{d}g(x) + \int_c^b f(x)\mathrm{d}g(x).$$

We note that, if the integrator $g(x)$ is differentiable, i.e. $\mathrm{d}g(x) = g'(x)\mathrm{d}x$, then the Riemann–Stieltjes integral is reduced to the ordinary Riemann integral. That is,

$$\int_a^b f(x)\mathrm{d}g(x) = \int_a^b f(x)g'(x)\mathrm{d}x. \tag{1.28}$$

On the other hand, if $g(x)$ is a *simple* function in the sense that

$$g(x_0) = \alpha_0; \quad g(x) = \alpha_k, \quad x_{k-1} < x \leq x_k, \tag{1.29}$$

where $Q = (x_0, x_1, \ldots, x_n)$ is a partition of the interval $[a, b]$, and if $f(x)$ is continuous at each x_k, then $f(x)$ is g-integrable and

$$\int_a^b f(x)\mathrm{d}g(x) = \sum_{k=0}^{n-1} (\alpha_{k+1} - \alpha_k) f(x_k). \tag{1.30}$$

Exercise 1.14 contains some calculation of Riemann–Stieltjes integrals.

As is well known, the integral is the inverse operator of the differential. To confirm this, suppose that a function $f(x)$ is continuous on $[a, b]$ and that $g(x)$ has a continuous derivative $g'(x)$. Define

$$F(x) = \int_a^x f(y)\mathrm{d}g(y), \quad x \in (a, b). \tag{1.31}$$

Let $h > 0$, and consider the difference of $F(x)$. Then, by the mean value theorem (Proposition 1.7), there exists a point c in $(x, x+h)$ such that

$$\frac{F(x+h) - F(x)}{h} = \frac{1}{h} \int_x^{x+h} f(y)\mathrm{d}g(y) = f(c)g'(c).$$

A similar result holds for the case $h < 0$. Since $c \to x$ as $h \to 0$ and since $f(x)$ and $g'(x)$ are continuous, we have from (1.8) that $F'(x) = f(x)g'(x)$, as desired. This result can be expressed as

$$F(x) = \int_a^x f(y)\mathrm{d}g(y) \iff \mathrm{d}F(x) = f(x)\mathrm{d}g(x), \tag{1.32}$$

provided that the required differentiability and integrability are satisfied.

Now, if $(\Delta x)^2$ is negligible as in the ordinary differentiation, we obtain

$$(f(x)g(x))' = f'(x)g(x) + f(x)g'(x). \tag{1.33}$$

The proof is left to the reader. Since integral is the inverse operator of the differential, we have

$$f(b)g(b) - f(a)g(a) = \int_a^b f'(x)g(x)\mathrm{d}x + \int_a^b f(x)g'(x)\mathrm{d}x,$$

provided that $f(x)$ and $g(x)$ are differentiable. In general, from (1.28), we expect the following important result:

$$\int_a^b f(x)\mathrm{d}g(x) + \int_a^b g(x)\mathrm{d}f(x) = f(b)g(b) - f(a)g(a). \tag{1.34}$$

The identity (1.34) is called the *integration by parts* formula (cf. Exercise 1.16). See Exercise 1.15 for another important formula for integrals.

Finally, we present two important results for integration without proof.

Proposition 1.11 (Dominated Convergence) *Let* $J = [a, b]$, *and suppose that* $g(x)$ *is non-decreasing in* $x \in J$. *Suppose further that* $h(x)$ *is g-integrable on J and, for each n,*

$$|f_n(x)| \leq h(x), \quad x \in J.$$

If $f_n(x)$ are g-integrable and $f_n(x)$ converge to a g-integrable function as $n \to \infty$ for each $x \in J$, then we have

$$\lim_{n \to \infty} \int_a^b f_n(x)\mathrm{d}g(x) = \int_a^b \lim_{n \to \infty} f_n(x)\mathrm{d}g(x).$$

Proposition 1.12 (Fubini's Theorem) *For a function $f(x, t)$ of two variables, suppose that either of the following converges:*

$$\int_c^d \left\{ \int_a^b |f(x,t)|\mathrm{d}x \right\} \mathrm{d}t, \quad \int_a^b \left\{ \int_c^d |f(x,t)|\mathrm{d}t \right\} \mathrm{d}x.$$

Then, the other integral also converges, and we have

$$\int_c^d \left\{ \int_a^b f(x,t)\mathrm{d}x \right\} \mathrm{d}t = \int_a^b \left\{ \int_c^d f(x,t)\mathrm{d}t \right\} \mathrm{d}x.$$

Moreover, if $f(x,t) \geq 0$, then the equation always holds, including the possibility that both sides are infinite.

1.6 Exercises

Exercise 1.1 Show that the sequence $\{a_n\}$ defined by (1.1) is strictly increasing in n while $\{b_n\}$ defined by (1.2) is strictly decreasing. *Hint*: Use the inequalities $(1-x)^n > 1 - nx$, $0 < x < 1$, and $(1+x)^n > 1 + nx$, $x > 0$, for $n \geq 2$.

Exercise 1.2 Using (1.4), prove the following.

(1) $e = \lim_{h \to 0} (1+h)^{1/h}$ \quad (2) $\lim_{h \to 0} \dfrac{\log(1+h)}{h} = 1$

Note: Do not use L'Hospital's rule.

Exercise 1.3 The logarithm $y = \log_a x$ is the inverse function of $y = a^x$. That is, $y = a^x$ if and only if $x = \log_a y$. Using this relation and Proposition 1.1, prove that Proposition 1.2 holds.

Exercise 1.4 Suppose that a function $f(x)$ is continuous on $[a,b]$ and differentiable in (a,b). Prove the following.
(1) If $f'(x) = 0$ for $a < x < b$ then $f(x)$ is constant.
(2) If $f'(x) = g'(x)$ for $a < x < b$ then $f(x)$ and $g(x)$ differ by a constant, where $g(x)$ is assumed to be continuous on $[a,b]$ and differentiable in (a,b).
(3) If $f'(x) \geq 0$ for $a < x < b$ then $f(x_1) \leq f(x_2)$ for $x_1 < x_2$.
(4) If $f'(x) > 0$ for $a < x < b$ then $f(x_1) < f(x_2)$ for $x_1 < x_2$.
(5) If $|f'(x)| \leq M$ for $a < x < b$ then $f(x)$ satisfies the *Lipschitz condition*:
$$|f(x_1) - f(x_2)| \leq M|x_1 - x_2|, \quad a < x_1, x_2 < b.$$

Exercise 1.5 (Cauchy Mean Value Theorem) Let $f(x)$ and $g(x)$ be continuous on $[a,b]$ and differentiable in (a,b). Prove that there exists a point c in (a,b) such that
$$f'(c)[g(b) - g(a)] = g'(c)[f(b) - f(a)].$$
Hint: For $g(b) \neq g(a)$, consider the function
$$\phi(x) = f(x) - f(a) - \frac{f(b) - f(a)}{g(b) - g(a)}[g(x) - g(a)].$$

Exercise 1.6 Prove L'Hospital's rule (Proposition 1.8) by using the Cauchy mean value theorem.

Exercise 1.7 Using L'Hospital's rule, calculate the following limits.

(1) $\lim_{x \to 1} \dfrac{x^2 - 1}{x - 1}$ \quad (2) $\lim_{x \to \infty} \dfrac{x^2}{e^x}$ \quad (3) $\lim_{x \downarrow 0} x^x$

Hint: For (3), consider $y = \log x^x$.

Exercise 1.8 Prove (1.4) by applying L'Hospital's rule.

Exercise 1.9 Obtain the total derivatives of the following functions whenever possible.

(1) $z = xy$ \quad (2) $z = \exp\{x^2 + xy + y^2\}$ \quad (3) $z = \dfrac{y}{\sqrt{x^2 - y^2}}$

Exercise 1.10 Prove the following.
$$\log(1+x) = \sum_{n=1}^{\infty} \frac{(-1)^{n-1}}{n} x^n, \quad |x| < 1.$$

EXERCISES

Exercise 1.11 Obtain the second-order Taylor expansions of the following functions.

(1) $z = xy$ (2) $z = \exp\{x^2 + xy\}$ (3) $z = \dfrac{x}{y}$ $(y \neq 0)$

Exercise 1.12 Suppose that the variable $x = x(t, z)$ at time t satisfies
$$\mathrm{d}x = x(\mu \mathrm{d}t + \sigma \mathrm{d}z),$$
where $(\mathrm{d}z)^2 = \mathrm{d}t$. Applying Ito's formula, prove that
$$\mathrm{d}\log x = \left(\mu - \frac{\sigma^2}{2}\right)\mathrm{d}t + \sigma \mathrm{d}z.$$

Hint: Let $f(t, x) = \log x$.

Exercise 1.13 Suppose that the variable $x = x(t, z)$ at time t satisfies
$$\mathrm{d}x = \mu \mathrm{d}t + \sigma \mathrm{d}z,$$
where $(\mathrm{d}z)^2 = \mathrm{d}t$. Let $y = e^{x - \mu t}$. Applying Ito's formula, obtain the equation for $\mathrm{d}y$.

Exercise 1.14 Calculate the following Riemann–Stieltjes integrals.

(1) $\displaystyle\int_0^1 x \, \mathrm{d}(x^2)$ (2) $\displaystyle\int_0^4 x^2 \, \mathrm{d}[x]$ (3) $\displaystyle\int_0^1 x \, \mathrm{d}(e^x)$

Here, $[x]$ denotes the largest integer less than or equal to x.

Exercise 1.15 (Change of Variable Formula) Let $\phi(x)$ be defined on an interval $[\alpha, \beta]$ with a continuous derivative, and suppose that $a = \phi(\alpha)$ and $b = \phi(\beta)$. Show that, if $f(x)$ is continuous on the range of $\phi(x)$, then
$$\int_a^b f(x) \mathrm{d}x = \int_\alpha^\beta f(\phi(t))\phi'(t)\mathrm{d}t.$$

Hint: Define $F(\xi) = \int_a^\xi f(x)\mathrm{d}x$ and consider $H(t) = F(\phi(t))$.

Exercise 1.16 As in Exercise 1.13, suppose that the variables $x = x(t, z)$ and $y = y(t, z)$ satisfy
$$\mathrm{d}x = \mu_x \mathrm{d}t + \sigma_x \mathrm{d}z, \quad \mathrm{d}y = \mu_y \mathrm{d}t + \sigma_y \mathrm{d}z,$$
respectively, where $(\mathrm{d}z)^2 = \mathrm{d}t$. Using the result in Exercise 1.11(1), show that
$$\mathrm{d}(xy) = x\mathrm{d}y + y\mathrm{d}x + \sigma_x \sigma_y \mathrm{d}t.$$

Compare this with (1.33). The integration by parts formula (1.34) does not hold in this setting.

CHAPTER 2

Elements in Probability

This chapter provides a concise summary of the theory of basic probability. In particular, a finite probability space is of interest because, in a discrete securities market model, the time horizon is finite and possible outcomes in each time is also finite. Probability distributions useful in financial engineering such as normal distributions are given in the next chapter. See, e.g., Çinlar (1975), Karlin and Taylor (1975), Neuts (1973), and Ross (1983) for more detailed discussions of basic probability theory.

2.1 The Sample Space and Probability

In probability theory, the set of possible outcomes is called the *sample space* and usually denoted by Ω. Each outcome ω belonging to the sample space Ω is called an *elementary event*, while a subset A of Ω is called an *event*. In the terminology of set theory, Ω is the universal set, $\omega \in \Omega$ is an element, and $A \subset \Omega$ is a set.

In order to make the probability model precise, we need to declare what family of events will be considered. Let $\mathcal{F}$ denote such a family of events. The family $\mathcal{F}$ of events must satisfy the following properties:

[F1] $\Omega \in \mathcal{F}$,

[F2] If $A \subset \Omega$ is in $\mathcal{F}$ then $A^c \in \mathcal{F}$, and

[F3] If $A_n \subset \Omega$, $n = 1, 2, \ldots$, are in $\mathcal{F}$ then $\cup_{n=1}^{\infty} A_n \in \mathcal{F}$.

The family of events, $\mathcal{F}$, satisfying these properties is called a σ-*field*.

For each event $A \in \mathcal{F}$, the *probability* of A is denoted by $P(A)$. In modern probability theory, probability is understood to be a set function, defined on $\mathcal{F}$, satisfying the following properties:

[PA1] $P(\Omega) = 1$,

[PA2] $0 \leq P(A) \leq 1$ for any event $A \in \mathcal{F}$, and

[PA3] For mutually exclusive events $A_n \in \mathcal{F}$, $n = 1, 2, \ldots$, i.e. $A_i \cap A_j = \emptyset$ for $i \neq j$, we have

$$P(A_1 \cup A_2 \cup \cdots) = \sum_{n=1}^{\infty} P(A_n). \qquad (2.1)$$

Note that the physical meaning of probability is not important in this definition. The property (2.1) is called the σ-*additivity*.

Definition 2.1 Given a sample space Ω and a σ-field $\mathcal{F}$, if a set function P defined on $\mathcal{F}$ satisfies the above properties, we call P a *probability measure* (or probability for short). The triplet $(\Omega, \mathcal{F}, P)$ is called a *probability space*.

In the rest of this chapter, we assume that the σ-field $\mathcal{F}$ is rich enough, meaning that all the events under consideration belong to $\mathcal{F}$. That is, we consider a probability of event without mentioning that the event indeed belongs to the σ-field. See Exercise 2.1 for an example of probability space.

From Axioms [PA1] and [PA2], probability takes values between 0 and 1, and the probability that something happens, i.e. the event Ω occurs, is equal to unity. Axiom [PA3] plays an elementary role for the calculation of probability. For example, since an event A and its complement A^c are disjoint, we obtain

$$P(A) + P(A^c) = P(\Omega) = 1.$$

Hence, the *probability of complementary event* is given by

$$P(A^c) = 1 - P(A). \tag{2.2}$$

In particular, since $\emptyset = \Omega^c$, we have $P(\emptyset) = 0$ for the empty set $\emptyset$. Of course, if $A = B$ then $P(A) = P(B)$.

The role of Axiom [PA3] will be highlighted by the following example. See Exercise 2.2 for another example.

Example 2.1 For any events A and B, we have

$$A \cup B = (A \cap B^c) \cup (A \cap B) \cup (B \cap A^c).$$

What is crucial here is the fact that the events in the right-hand side are mutually exclusive even when A and B are *not* disjoint. Hence, we can apply (2.1) to obtain

$$P(A \cup B) = P(A \cap B^c) + P(A \cap B) + P(B \cap A^c).$$

Moreover, in the same spirit, we have

$$P(A) = P(A \cap B^c) + P(A \cap B)$$

and

$$P(B) = P(B \cap A^c) + P(A \cap B).$$

It follows that

$$P(A \cup B) = P(A) + P(B) - P(A \cap B) \tag{2.3}$$

for any events A and B (see Figure 2.1). See Exercise 2.3 for a more general result.

The probability $P(A \cap B)$ is called the *joint* probability of events A and B. Now, for events A and B such that $P(B) > 0$, the *conditional probability* of A under B is defined by

$$P(A|B) = \frac{P(A \cap B)}{P(B)}. \tag{2.4}$$

That is, the conditional probability $P(A|B)$ is the ratio of the joint probability

DISCRETE RANDOM VARIABLES

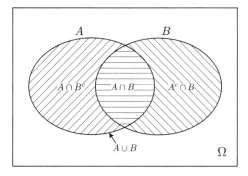

Figure 2.1 *An interpretation of $P(A \cup B)$*

$P(A \cap B)$ to the probability $P(B)$. In this regard, the event B plays the role of the sample space, and the meaning of the definition (2.4) should be clear. Moreover, from (2.4), we have

$$P(A \cap B) = P(B)P(A|B) = P(A)P(B|A). \quad (2.5)$$

The joint probability of events A and B is given by the probability of B (A, respectively) times the conditional probability of A (B) under B (A). See Exercise 2.4 for a more general result.

Suppose that the sample space Ω is divided into mutually exclusive events $A_1, A_2, \ldots$, i.e. $A_i \cap A_j = \emptyset$ for $i \neq j$ and $\Omega = \cup_{i=1}^{\infty} A_i$. Such a family of events is called a *partition* of Ω. Let B be another event, and consider

$$B = B \cap \Omega = \bigcup_{i=1}^{\infty}(B \cap A_i).$$

The family of events $\{B \cap A_i\}$ is a partition of B, since $(B \cap A_i) \cap (B \cap A_j) = \emptyset$ for $i \neq j$. It follows from (2.1) and (2.5) that

$$P(B) = \sum_{i=1}^{\infty} P(A_i)P(B|A_i). \quad (2.6)$$

The formula (2.6) is called the *law of total probability* and plays a fundamental role in probability calculations. See Exercise 2.5 for a related result.

2.2 Discrete Random Variables

Suppose that the sample space Ω consists of a finite number of elementary events, $\{\omega_1, \omega_2, \ldots, \omega_n\}$ say, and consider a set function defined by

$$P(\{\omega_i\}) = p_i, \quad i = 1, 2, \ldots, n,$$

where $p_i > 0$ and $\sum_{i=1}^n p_i = 1$. Here, we implicitly assume that $\{\omega_i\} \in \mathcal{F}$, i.e. $\mathcal{F}$ contains all of the subsets of Ω. It is easy to see that

$$P(A) = \sum_{i:\, \omega_i \in A} p_i$$

for any event $A \in \mathcal{F}$. The set function P satisfies the condition (2.1), whence it defines a probability.

As an example, consider an experiment of rolling a die. In this case, the sample space is given by $\Omega = \{\omega_1, \omega_2, \ldots, \omega_6\}$, where ω_i represents the outcome that the die lands with number i. If the die is fair, the plausible probability is to assign $P(\{\omega_i\}) = 1/6$ to each outcome ω_i. However, this is not the only case (see Exercise 2.1). Even if the die is actually fair, you may assign another probability to each outcome. This causes no problem as far as the probability satisfies the requirement (2.1). Such a probability is often called the *subjective* probability.

The elementary events may not be real numbers. In the above example, ω_i itself does not represent a number. If you are interested in the number i, you must consider the mapping $\omega_i \to i$. But, if the problem is whether the outcome is an even number or not, the following mapping X may suffice:

$$X(\omega) = \begin{cases} 0, & \omega = \omega_i \text{ and } i \text{ is odd}, \\ 1, & \omega = \omega_i \text{ and } i \text{ is even}. \end{cases}$$

What is crucial here is that the probability of each outcome in X can be calculated based on the given probability $\{p_1, p_2, \ldots, p_6\}$, where $p_i = P(\{\omega_i\})$. In this case, we obtain

$$P\{X = 1\} = p_2 + p_4 + p_6.$$

Here, it should be noted that we consider the event

$$A = \{\omega : X(\omega) = 1\},$$

and the probability $P(A)$ is denoted by $P\{X = 1\}$. Similarly, we have

$$P\{X = 0\} = p_1 + p_3 + p_5 = 1 - P\{X = 1\}.$$

Mathematically speaking, a mapping X from Ω to a subset of $\boldsymbol{R}$ is called a *random variable* if the probability of each outcome of X can be calculated based on the given probability P. More formally, a mapping X is said to be a random variable if $P(\{\omega : a < X(\omega) \leq b\})$ can be calculated for every $a < b$. See Definition 2.2 below for the formal definition of random variables.

In what follows, we denote the probability $P(\{\omega : X(\omega) \leq r\})$ by $P\{X \leq r\}$, which reads 'the probability that X is less than or equal to r.' The curry bracket indicates that this is a simplified notation, and $\{X \leq r\}$ represents the event $\{\omega : X(\omega) \leq r\}$. The probabilities $P\{X > r\}$, $P\{a < X \leq b\}$, etc. are understood similarly. Note that $X(\omega)$ is a real number for each $\omega \in \Omega$. We will call $X(\omega)$ a *realization* of X.

When the sample space Ω is finite, there are only finitely many realizations of X, $\Omega_X \equiv \{x_1, x_2, \ldots, x_m\}$ say, where $m \leq n$ and n is the number of

CONTINUOUS RANDOM VARIABLES

elementary events. The probability of $B \subset \Omega_X$ can be calculated based on the probabilities
$$p_j^X \equiv P\{X = x_j\}, \quad j = 1, 2, \ldots, m.$$
Moreover, $\sum_{j=1}^m p_j^X = 1$. This is so, since X is a function defined on Ω and the events $\{X = x_j\}$ are mutually exclusive. The probability $\{p_1^X, p_2^X, \ldots, p_m^X\}$ is often called an *induced* probability; however, we call it simply a probability by abusing the terminology. The set Ω_X is also called the *sample space* of X.

Conversely, suppose that we are given sets of real numbers $\{x_1, x_2, \ldots, x_m\}$ and $\{p_1, p_2, \ldots, p_m\}$, where $p_j > 0$, $j = 1, 2, \ldots, m$, and $\sum_{j=1}^m p_j = 1$. Following the above arguments backwards, we conclude that there exists a random variable X defined on Ω such that $P\{X = x_j\} = p_j$. Therefore, the set of real numbers $\{x_1, x_2, \ldots, x_m\}$ can be considered as a sample space of X and $\{p_1, p_2, \ldots, p_m\}$ a probability. The pair of sample space and probability is called a *probability distribution* or distribution for short. This is the reason why we often start from a probability distribution without constructing a probability space explicitly. It should be noted that there may exist another random variable Y that has the same distribution as X. If this is the case, we denote $X \stackrel{\mathrm{d}}{=} Y$, and X and Y are said to be *equal in law*.

Let $h(x)$ be a function defined on $\mathbf{R}$, and consider $h(X)$ for a random variable X. If the sample space is finite, it is easily seen that $h(X)$ is also a random variable, since $h(X)$ can be seen as a composed mapping from Ω to $\mathbf{R}$, and the distribution of $h(X)$ is obtained from the distribution of X.

Example 2.2 Suppose that the set of realizations of a random variable X is $\{-2, -1, 0, 1, 2\}$, and consider $Y = X^2$. The sample space of Y is $\{0, 1, 4\}$ and the probability of Y is obtained as $P\{Y = 0\} = P\{X = 0\}$ and
$$P\{Y = 1\} = P\{X = 1\} + P\{X = -1\}.$$
The probability $P\{Y = 4\}$ can be obtained similarly. It is readily seen that $\sum_i P\{Y = i\} = 1$ and so, Y is a random variable.

2.3 Continuous Random Variables

A discrete probability model precludes pathological situations (so it makes the presentation simpler) and is easy to understand intuitively. However, there are many instances whose outcomes cannot be described by a discrete model. Also, a continuous model is often easier to calculate. This section provides a concise summary of continuous random variables.

Before proceeding, we mention that a continuous probability model requires a careful mathematical treatment. To see this, consider a fictitious experiment that you drop a pin on the interval $(0, 1)$. First, the interval is divided into n subintervals with equal lengths. If your throw is completely random, the probability that the pin drops on some particular subinterval is one over n. This is the setting of finite probability model. Now, consider the continuous model, i.e. we let n tend to infinity. Then, the probability that the pin drops on any point is zero, although the pin will be somewhere on the interval! There must be wrong in this argument. The flaw occurs because we take the same argument for the continuous case as for the discrete case.

In order to avoid such a flaw, we make the role of the σ-field more explicit. In the above example, we let $\mathcal{F}$ be the σ-field that contains all the open subintervals of $(0, 1)$, not just all the points in $(0, 1)$ as above. Because your throw is completely random, the probability that the pin drops on a subinterval $(a, b) \subset (0, 1)$ will be $b - a$. The probability space constructed in this way has no ambiguity. The problem of how to construct the probability space becomes important for the continuous case.

We give the formal definition of random variables.

Definition 2.2 Given a probability space $(\Omega, \mathcal{F}, P)$, let X be a mapping from Ω to an interval I of the real line $\mathbf{R}$. The mapping X is said to be a *random variable* if, for any $a < b$, $\{\omega : a < X(\omega) \leq b\} \in \mathcal{F}$.

In other words, since the probability P is a set function defined on $\mathcal{F}$, X is a random variable if the probability $P(\{\omega : a < X(\omega) \leq b\})$ for any $a < b$ is known. That is, if X is a random variable, then we know in principle the probability that a realization of X is in the interval $(a, b]$. For notational simplicity, we write $P\{a < X \leq b\}$ as before.

A random variable X is said to be *continuous*, if the set of realizations of X is an interval I of the real line and there exists a non-negative function $f(x)$ such that

$$P\{a \leq X \leq b\} = \int_a^b f(x) \mathrm{d}x \tag{2.7}$$

for any $[a, b] \subset I$. The function $f(x)$ defined on I is called the *density function* of X.* Conversely, the density function $f(x)$, $x \in I$, defines the continuous random variable X. Also, from (2.7), if we define

$$F(x) = \int_{-\infty}^x f(y) \mathrm{d}y = P\{X \leq x\}, \quad x \in I,$$

where $f(x) = 0$ for $x \notin I$, we obtain

$$f(x) = \frac{\mathrm{d}}{\mathrm{d}x} F(x), \quad x \in I. \tag{2.8}$$

The function $F(x)$ is called the *distribution function* of X.

It is readily seen that the density function satisfies

$$f(x) \geq 0, \quad \int_I f(x) \mathrm{d}x = 1, \tag{2.9}$$

and is characterized by these properties. Also, for sufficiently small $h > 0$, the mean value theorem (Proposition 1.7) asserts that there exists some c such that

$$\frac{1}{h} P\{x < X \leq x + h\} = f(c), \quad x < c < x + h.$$

---

* The density function is not uniquely defined. However, it is common to take the most 'natural' one as the density function, e.g. the one being continuous, if it exists.

Hence, the density function can be thought of as the *intensity* that the random variable X takes values on a neighborhood of x. It should be noted that the density function itself does not represent a probability. So, $f(x)$ can be greater than unity for some x as far as it satisfies the properties given in (2.9). Finally, for any $x \in \mathbf{R}$, the continuous random variable satisfies

$$P\{X = x\} = 0.$$

Hence, e.g., we have $P\{a < X < b\} = P\{a \le X \le b\}$ for continuous random variable X.

Example 2.3 Since the density function $f(x)$ is non-negative, the distribution function $F(x)$ is non-decreasing in x. Also, it satisfies $F(-\infty) = 0$ and $F(\infty) = 1$.[†] Conversely, a distribution function is characterized by these properties. For example, the function defined by

$$F(x) = \frac{\log(1 + \beta x)}{\log(1 + \beta)}, \quad 0 \le x \le 1,$$

is a distribution function for any $\beta > 0$. The density function is obtained from (2.8) and given by

$$f(x) = \frac{\beta}{\log(1 + \beta)} \frac{1}{1 + \beta x}, \quad 0 \le x \le 1.$$

This continuous distribution is called the *Bradford distribution*. See Exercise 2.6 for other examples.

2.4 Multivariate Random Variables

Given a probability space $(\Omega, \mathcal{F}, P)$, consider a two-dimensional mapping (X, Y) from Ω to $\mathbf{R}^2$. As for the univariate case, the mapping is said to be a (bivariate) random variable if the probability

$$P(\{\omega : x_1 < X(\omega) \le x_2, \, y_1 < Y(\omega) \le y_2\})$$

can be calculated for any $(x_1, x_2) \times (y_1, y_2) \subset \mathbf{R}^2$. As before, we denote the probability by $P\{x_1 < X \le x_2, y_1 < Y \le y_2\}$, which reads the probability that X is in the interval $(x_1, x_2]$ and Y is in the interval $(y_1, y_2]$. A higher-variate random variable can be defined in the same manner.

When the sample space Ω is finite, there are only finitely many realizations of (X, Y), $\Omega_{XY} \equiv \{(x_i, y_j); i = 1, 2, \ldots, m, \, j = 1, 2, \ldots, n\}$ say, and the probability of event in Ω_{XY} can be calculated based on the *joint probability*

$$p_{ij} \equiv P\{X = x_i, Y = y_j\}, \quad i = 1, 2, \ldots, m, \, j = 1, 2, \ldots, n.$$

It is noted that the joint probability satisfies the properties

$$p_{ij} \ge 0, \quad \sum_{i=1}^{m} \sum_{j=1}^{n} p_{ij} = 1$$

[†] Moreover, the distribution function $F(x)$ is right-continuous in the sense that $F(x) = \lim_{h \downarrow 0} F(x + h)$.

and is characterized by these properties.

The *marginal* distribution of X can be obtained from the joint distribution. That is,

$$P\{X = x_i\} = \sum_{j=1}^{n} P\{X = x_i, Y = y_j\}, \quad i = 1, 2, \ldots, m. \qquad (2.10)$$

The marginal distribution of Y is obtained similarly. Also, from (2.4), the *conditional* distribution of X under $\{Y = y_j\}$ is defined as

$$P\{X = x_i | Y = y_j\} = \frac{P\{X = x_i, Y = y_j\}}{P\{Y = y_j\}}, \quad i = 1, 2, \ldots, m, \qquad (2.11)$$

provided $P\{Y = y_j\} > 0$. The conditional distribution of Y under $\{X = x_i\}$ is defined similarly.

From (2.11) and (2.10), we have another form of the *law of total probability*

$$P\{X = x_i\} = \sum_{j=1}^{n} P\{Y = y_j\} P\{X = x_i | Y = y_j\}, \quad i = 1, 2, \ldots, m, \qquad (2.12)$$

for discrete random variables X and Y; cf. (2.6).

In general, the marginal distribution can be calculated from the joint distribution; see (2.10), but not vice versa. The converse is true only under independence, which we formally define next.

Definition 2.3 Two discrete random variables X and Y are said to be *independent* if, for any realizations x_i and y_j,

$$P\{X = x_i, Y = y_j\} = P\{X = x_i\} P\{Y = y_j\}. \qquad (2.13)$$

Independence of more than two discrete random variables is defined similarly.

Suppose that discrete random variables X and Y are independent. Then, from (2.11), we have

$$P\{X = x_i | Y = y_j\} = P\{X = x_i\},$$

where $P\{Y = y_j\} > 0$. That is, the information $\{Y = y_j\}$ does not affect the probability of X. Similarly, the information $\{X = x_i\}$ does not affect the probability of Y, since

$$P\{Y = y_j | X = x_i\} = P\{Y = y_j\}.$$

Independence is intuitively interpreted by this observation. More generally, if random variables $X_1, X_2, \ldots, X_n$ are independent, then

$$P\{X_n \in A_n | X_1 \in A_1, \ldots, X_{n-1} \in A_{n-1}\} = P\{X_n \in A_n\},$$

provided that $P\{X_1 \in A_1, \ldots, X_{n-1} \in A_{n-1}\} > 0$. That is, the probability distribution of X_n is unaffected by knowing the information about the other random variables.

Example 2.4 Independence property is merely a result of Equation (2.13) and can be counter-intuitive. For example, consider a random variable (X,Y) with the joint probability

$X\backslash Y$	-1	0	1
-1	3/32	5/32	1/32
0	5/32	8/32	3/32
1	3/32	3/32	1/32

Since $P\{X = 1\} = 7/32$, $P\{Y = 1\} = 5/32$, and $P\{X = 1, Y = 1\} = 1/32$, X and Y are not independent. However, it is readily seen that X^2 and Y^2 are independent. The proof is left to the reader. See Exercise 2.8 for another example.

We now turn to the continuous case. For a bivariate random variable (X,Y) defined on $\mathbf{R}^2$, if there exists a two-dimensional function $f(x,y)$ such that

$$P\{a \leq X \leq b, c \leq Y \leq d\} = \int_c^d \int_a^b f(x,y) dx dy$$

for any $[a,b] \times [c,d] \subset \mathbf{R}^2$, then (X,Y) is called *continuous*. The function $f(x,y)$ is called the *joint* density function of (X,Y). The joint density function satisfies

$$f(x,y) \geq 0, \quad \int_{-\infty}^{\infty} \int_{-\infty}^{\infty} f(x,y) dx dy = 1,$$

and is characterized by these properties. See Exercise 2.9 for an example.

As for the discrete case, the *marginal* density of X is obtained by

$$f_X(x) = \int_{-\infty}^{\infty} f(x,y) dy, \quad x \in \mathbf{R}. \tag{2.14}$$

The marginal density of Y is obtained similarly. The *conditional* density of X given $\{Y = y\}$ is defined by

$$f(x|y) = \frac{f(x,y)}{f_Y(y)}, \quad x \in \mathbf{R}, \tag{2.15}$$

provided that $f_Y(y) > 0$. The conditional density of Y given $\{X = x\}$ is defined similarly.

From (2.14) and (2.15), the law of total probability is given by

$$f_X(x) = \int_{-\infty}^{\infty} f_Y(y) f(x|y) dy, \quad x \in \mathbf{R}, \tag{2.16}$$

in the continuous case; cf. (2.12).

Example 2.5 Suppose that, given a parameter θ, a (possibly multivariate) random variable X has a density function $f(x|\theta)$, but the parameter θ itself is a random variable defined on I. In this case, X is said to follow a *mixture distribution*. Suppose that the density function of θ is $g(\theta)$, $\theta \in I$. Then, from (2.16), we have

$$f(x) = \int_I f(x|\theta) g(\theta) d\theta.$$

On the other hand, given a realization x, the function defined by

$$h(\theta|x) = \frac{f(x|\theta)g(\theta)}{\int_I f(x|\theta)g(\theta)d\theta}, \quad \theta \in I, \tag{2.17}$$

satisfies the properties in (2.9), so $h(\theta|x)$ is a density function of θ for the given x. Compare this result with the Bayes formula given in Exercise 2.5. The function $h(\theta|x)$ is called the *posterior* distribution of θ given the realization x, while the density $g(\theta)$ is called the *prior* distribution.

Definition 2.4 Two continuous random variables X and Y are said to be *independent* if, for all x and y,

$$f(x,y) = f_X(x)f_Y(y). \tag{2.18}$$

Independence of more than two continuous random variables is defined similarly.

Finally, we formally define a family of independent and identically distributed (abbreviated IID) random variables. The IID assumption leads to such classic limit theorems as the strong law of large numbers and the central limit theorem (see Chapter 10).

Definition 2.5 A family of random variables $X_1, X_2, \ldots$ is said to be *independent and identically distributed* (IID) if any finite collection of them, $(X_{k_1}, X_{k_2}, \ldots, X_{k_n})$ say, are independent and the distribution of each random variable is the same.

2.5 Expectation

Let X be a discrete random variable with probability distribution

$$p_i = P\{X = x_i\}, \quad i = 1, 2, \ldots.$$

For a real-valued function $h(x)$, the *expectation* is defined by

$$E[h(X)] = \sum_{i=1}^{\infty} h(x_i)p_i, \tag{2.19}$$

provided that

$$E[|h(X)|] = \sum_{i=1}^{\infty} |h(x_i)|p_i < \infty.$$

If $E[|h(X)|] = \infty$, we say that the expectation does not exist. The expectation plays a central role in the theory of probability.

In particular, when $h(x) = x^n$, $n = 1, 2, \ldots$, we have

$$E[X^n] = \sum_{i=1}^{\infty} x_i^n p_i, \tag{2.20}$$

called the *nth moment* of X, if it exists. The case that $n = 1$ is usually called

EXPECTATION

the *mean* of X. The *variance* of X is defined, if it exists, by

$$V[X] \equiv E\left[(X - E[X])^2\right] = \sum_{i=1}^{\infty}(x_i - E[X])^2 p_i. \qquad (2.21)$$

The square root of the variance, $\sigma_X = \sqrt{V[X]}$, is called the *standard deviation* of X. The standard deviation (or variance) is used as a measure of spread or dispersion of the distribution around the mean. The bigger the standard deviation, the wider the spread, and there is more chance of having a value far below (and also above) the mean.

From (2.21), the variance (and also the standard deviation) has the following property. We note that the next result holds in general.

Proposition 2.1 *For any random variable X, the variance $V[X]$ is non-negative, and $V[X] = 0$ if and only if X is constant, i.e. $P\{X = x\} = 1$ for some x.*

The expected value of a continuous random variable is defined in terms of the density function. Let $f(x)$ be the density function of random variable X. Then, for any real-valued function $h(x)$, the expectation is defined by

$$E[h(X)] = \int_{-\infty}^{\infty} h(x)f(x)\mathrm{d}x, \qquad (2.22)$$

provided that $\int_{-\infty}^{\infty} |h(x)|f(x)\mathrm{d}x < \infty$; otherwise, the expectation does not exist. The mean of X, if it exists, is

$$E[X] = \int_{-\infty}^{\infty} xf(x)\mathrm{d}x,$$

while the variance of X is defined by

$$V[X] = \int_{-\infty}^{\infty} (x - E[X])^2 f(x)\mathrm{d}x.$$

See Exercise 2.10 for another expression of the mean $E[X]$.

It should be noted that the expectation, if it exists, can be expressed in terms of the Riemann–Stieltjes integral. That is, in general,

$$E[h(X)] = \int_{-\infty}^{\infty} h(x)\mathrm{d}F(x),$$

where $F(x)$ denotes the distribution function of X.

Expectation and probability are related through the indicator function. For any event A, the *indicator function* 1_A means $1_A = 1$ if A is true and $1_A = 0$ otherwise. See Exercise 2.11 for an application of the next result.

Proposition 2.2 *For any random variable X and an interval $I \subset \mathbf{R}$, we have*

$$E\left[1_{\{X \in I\}}\right] = P\{X \in I\}.$$

Proof. From (2.19), we obtain

$$E\left[1_{\{X \in I\}}\right] = 1 \times P\{1_{\{X \in I\}} = 1\} + 0 \times P\{1_{\{X \in I\}} = 0\}.$$

But, the event $\{1_{\{X \in I\}} = 1\}$ is equivalent to the event $\{X \in I\}$. □

Let (X, Y) be a bivariate random variable with joint distribution function $F(x, y) = P\{X \le x, Y \le y\}$ defined on $\mathbb{R}^2$. For a two-dimensional function $h(x, y)$, the expectation is defined by

$$E[h(X, Y)] = \int_{-\infty}^{\infty} \int_{-\infty}^{\infty} h(x, y) \mathrm{d}F(x, y), \qquad (2.23)$$

provided that $\int_{-\infty}^{\infty} \int_{-\infty}^{\infty} |h(x, y)| \mathrm{d}F(x, y) < \infty$. That is, in the discrete case, the expectation is given by

$$E[h(X, Y)] = \sum_{i=1}^{\infty} \sum_{j=1}^{\infty} h(x_i, y_j) P\{X = x_i, Y = y_j\},$$

while we have

$$E[h(X, Y)] = \int_{-\infty}^{\infty} \int_{-\infty}^{\infty} h(x, y) f(x, y) \mathrm{d}x \mathrm{d}y$$

in the continuous case, where $f(x, y)$ is the joint density function of (X, Y).

In particular, when $h(x, y) = ax + by$, we have

$$\begin{aligned} E[aX + bY] &= \int_{-\infty}^{\infty} \int_{-\infty}^{\infty} (ax + by) \mathrm{d}F(x, y) \\ &= a \int_{-\infty}^{\infty} x \int_{y=-\infty}^{\infty} \mathrm{d}F(x, y) + b \int_{-\infty}^{\infty} y \int_{x=-\infty}^{\infty} \mathrm{d}F(x, y) \\ &= a \int_{-\infty}^{\infty} x \, \mathrm{d}F_X(x) + b \int_{-\infty}^{\infty} y \, \mathrm{d}F_Y(y), \end{aligned}$$

so that

$$E[aX + bY] = aE[X] + bE[Y], \qquad (2.24)$$

provided that all the expectations exist. Here, $F_X(x)$ and $F_Y(x)$ denote the marginal distribution functions of X and Y, respectively.

In general, we have the following important result. The next result is referred to as the *linearity of expectation*.

Proposition 2.3 *Let $X_1, X_2, \ldots, X_n$ be any random variables. For any real numbers $a_1, a_2, \ldots, a_n$, we have*

$$E\left[a_1 X_1 + a_2 X_2 + \cdots + a_n X_n\right] = \sum_{i=1}^{n} a_i E[X_i],$$

provided that the expectations exist.

The *covariance* between two random variables X and Y is defined by

$$\begin{aligned} C[X, Y] &\equiv E[(X - E[X])(Y - E[Y])] \qquad (2.25) \\ &= \int_{-\infty}^{\infty} \int_{-\infty}^{\infty} (x - E[X])(y - E[Y]) \mathrm{d}F(x, y). \end{aligned}$$

The *correlation coefficient* between X and Y is defined by

$$\rho[X,Y] = \frac{C[X,Y]}{\sigma_X \sigma_Y}. \tag{2.26}$$

From Schwarz's inequality (see Exerise 2.15), we have $-1 \le \rho[X,Y] \le 1$, and $\rho[X,Y] = \pm 1$ if and only if $X = \pm aY + b$ with $a > 0$.

Note that, if there is a tendency that $(X - E[X])$ and $(Y - E[Y])$ have the same sign, then the covariance is likely to be positive; in this case, X and Y are said to have a *positive correlation*. If there is a tendency that $(X - E[X])$ and $(Y - E[Y])$ have the opposite sign, then the covariance is likely to be negative; in this case, they are said to have a *negative correlation*. In the case that $C[X,Y] = 0$, they are called *uncorrelated*.

The next result can be obtained from the linearity of expectation. The proof is left in Exercise 2.12.

Proposition 2.4 *For any random variables $X_1, X_2, \ldots, X_n$, we have*

$$V[X_1 + X_2 + \cdots + X_n] = \sum_{i=1}^n V[X_i] + 2\sum_{i<j} C[X_i, X_j].$$

In particular, if they are mutually uncorrelated, i.e. $C[X_i, X_j] = 0$ for $i \ne j$, then

$$V[X_1 + X_2 + \cdots + X_n] = V[X_1] + V[X_2] + \cdots + V[X_n].$$

The covariance exhibits the *bilinearity* as shown in Part (4) of the next result. The proof of the next proposition is left to the reader. Note that $C[X,X] = V[X]$ by the definition.

Proposition 2.5 *Let a and b be any real numbers. For any random variables X, Y, and Z, we have*

(1) $C[X,Y] = E[XY] - E[X]E[Y]$,

(2) $C[X+a, Y+b] = C[X,Y]$,

(3) $C[aX, bY] = abC[X,Y]$, *and*

(4) $C[X, aY + bZ] = aC[X,Y] + bC[X,Z]$.

Independence of random variables can be characterized in terms of the expectations. The proof of the next result is beyond the scope of this book and omitted.

Proposition 2.6 *Suppose that X and Y are independent. Then, for any functions $f(x)$ and $g(x)$, we have*

$$E[f(X)g(Y)] = E[f(X)]E[g(Y)], \tag{2.27}$$

for which the expectations exist. Conversely, if (2.27) holds for any bounded functions $f(x)$ and $g(x)$, then two random variables X and Y are independent.

Taking $f(x) = g(x) = x$ in (2.27), we conclude that if X and Y are independent then they are uncorrelated, but not vice versa.

2.6 Conditional Expectation

In this section, we define the concept of conditional expectations, one of the most important subjects in financial engineering.

We have already defined the conditional probability (2.11) and the conditional density (2.15). Since they define a probability distribution, the *conditional expectation* of X under the event $\{Y = y\}$ is given by

$$E[X|Y=y] = \int_{-\infty}^{\infty} x\,\mathrm{d}F(x|y), \quad y \in \mathbf{R}, \tag{2.28}$$

provided that $\int_{-\infty}^{\infty} |x|\mathrm{d}F(x|y) < \infty$, where $F(x|y) = P\{X \leq x|Y = y\}$ denotes the conditional distribution function of X under $\{Y = y\}$. That is, in the discrete case, the conditional expectation is given by

$$E[X|Y=y] = \sum_{i=1}^{\infty} x_i P\{X = x_i | Y = y\},$$

while we have

$$E[X|Y=y] = \int_{-\infty}^{\infty} x f(x|y)\mathrm{d}x$$

in the continuous case.

Notice that $E[X|Y=y]$ is a function of y. Since $Y(\omega) = y$ for some $\omega \in \Omega$, the conditional expectation can be thought of as a composed function of ω, whence $E[X|Y(\omega)]$ is a random variable. It is common to denote this random variable by $E[X|Y]$. Recall that, since $E[X|Y]$ is the expectation with respect to X, it is a random variable in terms of Y.[‡] We shall give a more general notion of conditional expectations in Chapter 5.

Two random variables X and Y are said to be equal *almost surely* (abbreviated *a.s.*) if $P\{X = Y\} = 1$; in which case, we write $X = Y$, *a.s.* Similarly, $X \geq Y$, *a.s.* means $P\{X \geq Y\} = 1$. However, the notation *a.s.* will be suppressed if the meaning is clear.

In the next three propositions, we provide important properties of conditional expectations. All the conditional expectations under consideration are assumed to exist.

Proposition 2.7 *For any random variables X, Y, and Z and any real numbers a and b, we have the following.*

(1) *If X and Z are independent, then $E[X|Z] = E[X]$.*

(2) $E[XZ|Y,Z] = ZE[X|Y,Z].$

(3) $E[aX + bY|Z] = aE[X|Z] + bE[Y|Z].$

Note that any random variable is independent of constants. Hence, taking Z to be a constant, the properties in Proposition 2.7 also hold for the ordinary expectation. We remark that the converse statement of (1) is not true. That is, even if $E[X|Z] = E[X]$ holds, the random variables X and Z may not

---

[‡] In order to make this point clearer, we often denote it by $E_X[X|Y]$, meaning that the expectation is taken with respect to X.

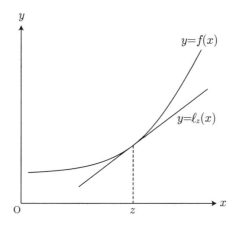

Figure 2.2 *A straight line tangent to the convex function*

be independent in general (see Exercise 2.7). The proof of Part (2) is left in Exercise 2.13. The other proofs are simple and omitted. Property (3) is called the *linearity* of conditional expectation.

Proposition 2.8 *Let X and Y be any random variables.*

(1) *If $f_1(x) \leq f_2(x)$ for all x, then $E[f_1(X)|Y] \leq E[f_2(X)|Y]$.*

(2) *For any convex function, we have $f(E[X|Y]) \leq E[f(X)|Y]$.*

Proof. Part (1) follows, since $E[Z|Y] \geq 0$ for any $Z \geq 0$ by the definition (2.28) of the conditional expectation. To prove Part (2), let y be arbitrary and let $z = E[X|Y = y]$. Since $f(x)$ is convex, there exists a linear function $\ell_z(x)$ such that $f(z) = \ell_z(z)$ and $f(x) \geq \ell_z(x)$ for all x (see Figure 2.2). That is, we have

$$f(E[X|Y = y]) = \ell_z(E[X|Y = y])$$

and $f(X) \geq \ell_z(X)$. But, since $\ell_z(x)$ is linear, the linearity of conditional expectation (Proposition 2.7(3)) implies that

$$\ell_z(E[X|Y = y]) = E[\ell_z(X)|Y = y].$$

It follows from Part (1) that

$$f(E[X|Y = y]) \leq E[f(X)|Y = y].$$

Since y is arbitrary, Part (2) follows. □

Part (1) of Proposition 2.8 is called the monotonicity of (conditional) expectation, while Part (2) is called *Jensen's inequality*. In particular, when $f(x) = |x|$, we have

$$|E[X|Y]| \leq E[|X||Y]. \tag{2.29}$$

See Exercises 2.14 and 2.15 for other important inequalities in probability theory.

Proposition 2.9 Let $h(x)$ be a real-valued function. For random variables X, Y, and Z, we have
$$E[E[h(X)|Y,Z]|Z] = E[h(X)|Z].$$

Proof. We prove the proposition for the continuous case only. The proof for the discrete case is left to the reader. Let $f(x,y,z)$ be the joint density function of (X,Y,Z). The marginal densities are denoted by, e.g., $f_X(x)$ and $f_{YZ}(y,z)$. Also, the conditional density of Y under $\{Z = z\}$ is denoted by $f_{Y|Z}(y|z)$. The other densities are denoted in an obvious manner. Now, given $\{Z = z\}$, we have from (2.28) that

$$E[h(X)|Z = z] = \int_{-\infty}^{\infty} h(x) f_{X|Z}(x|z) \mathrm{d}x = \frac{\int_{-\infty}^{\infty} h(x) f_{XZ}(x,z) \mathrm{d}x}{f_Z(z)}.$$

Similarly, given $\{Y = y, Z = z\}$, we obtain

$$E[h(X)|Y = y, Z = z] = \frac{\int_{-\infty}^{\infty} h(x) f(x,y,z) \mathrm{d}x}{f_{YZ}(y,z)},$$

which we denote by $g(y,z)$, since it is a function of (y,z). Then,

$$E[E[h(X)|Y,Z]|Z = z] = E[g(Y,Z)|Z = z]$$

and so

$$\begin{aligned} E[g(Y,z)|Z = z] &= \int_{-\infty}^{\infty} g(y,z) f_{Y|Z}(y|z) \mathrm{d}y \\ &= \int_{-\infty}^{\infty} \frac{\int_{-\infty}^{\infty} h(x) f(x,y,z) \mathrm{d}x}{f_{YZ}(y,z)} f_{Y|Z}(y|z) \mathrm{d}y. \end{aligned}$$

From the definition (2.15) of the conditional density, we have

$$\frac{f_{Y|Z}(y|z)}{f_{YZ}(y,z)} = \frac{1}{f_Z(z)}.$$

It follows that

$$\begin{aligned} E[g(Y,z)|Z = z] &= \frac{\int_{-\infty}^{\infty} \int_{-\infty}^{\infty} h(x) f(x,y,z) \mathrm{d}x \mathrm{d}y}{f_Z(z)} \\ &= \frac{\int_{-\infty}^{\infty} h(x) f_{XZ}(x,z) \mathrm{d}x}{f_Z(z)} \\ &= E[h(X)|Z = z]. \end{aligned}$$

The result is now established, because

$$E[E[h(X)|Y,Z]|Z = z] = E[h(X)|Z = z]$$

holds for all z, completing the proof. □

In Proposition 2.9, if we take Z to be a constant, then

$$E[E[h(X)|Y]] = E[h(X)]. \qquad (2.30)$$

Equation (2.30) is another form of the *law of total probability* and very useful in applied probability. In particular, as in Proposition 2.2, let $h(x) = 1_{\{x \in A\}}$. Then, from (2.30), we obtain

$$E[P\{X \in A|Y\}] = P\{X \in A\}. \tag{2.31}$$

See Exercise 2.16 for an application of the law of total probability (2.30).

2.7 Moment Generating Functions

Let X be a random variable, and suppose that the expectation $E\left[e^{tX}\right]$ exists at a neighborhood of the origin, i.e. for $|t| < \varepsilon$ with some $\varepsilon > 0$. The function

$$m_X(t) = E\left[e^{tX}\right], \quad |t| < \varepsilon, \tag{2.32}$$

with respect to t is called the *moment generating function* (MGF for short) of X. The MGF, if it exists, generates the moments of X. Note that the integral (expectation) and the derivative are interchangeable if the MGF exists at a neighborhood of the origin. It follows that

$$\left.\frac{\mathrm{d}}{\mathrm{d}t} m_X(t)\right|_{t=0} = E\left[\left.\frac{\mathrm{d}}{\mathrm{d}t} e^{tX}\right|_{t=0}\right] = E[X].$$

The next result convinces the term 'moment generating' function.

Proposition 2.10 *Let X be a random variable, and suppose that the MGF $m_X(t)$ exists at a neighborhood of the origin. Then, for any positive integer n, the nth moment $E[X^n]$ exists and is given by*

$$E[X^n] = m_X^{(n)}(0), \quad n = 1, 2, \ldots,$$

where $m_X^{(n)}(t)$ denotes the nth derivative of $m_X(t)$.

It should be noted that the converse statement of Proposition 2.10 is not true. That is, even if a random variable has the moments of any order, the random variable may not have the MGF. A famous and important example in financial engineering is log-normally distributed random variables. We shall return to this topic later (see Chapter 3). In this section, we provide important properties of MGFs necessary for later developments.

Importance of MGFs is largely due to the following two results. In particular, the classic central limit theorem can be proved using these results (see Chapter 10). We refer to Feller (1971) for the proofs.

Proposition 2.11 *The MGF determines the distribution of a random variable uniquely. That is, for two random variables X and Y, we have*

$$m_X(t) = m_Y(t) \iff X \stackrel{\mathrm{d}}{=} Y,$$

provided that the MGFs exist at a neighborhood of the origin. Here, $\stackrel{\mathrm{d}}{=}$ stands for equality in law.

In the next proposition, a sequence of random variables X_n is said to *converge in law* to a random variable X if the distribution function $F_n(x)$ of X_n

converges to a distribution function $F(x)$ of X as $n \to \infty$ for every point of continuity of $F(x)$.

Proposition 2.12 *Let $\{X_n\}$ be a sequence of random variables, and suppose that the MGF $m_n(t)$ of each X_n exists at a neighborhood of the origin. Let X be another random variable with the MGF $m(t)$. Then, X_n converges in law to X as $n \to \infty$ if and only if $m_n(t)$ converges to $m(t)$ as $n \to \infty$ at a neighborhood of the origin.*

We note that if X_n converges to X in law then

$$\lim_{n \to \infty} E[f(X_n)] = E[f(X)] \tag{2.33}$$

for any bounded, continuous function f, for which the expectations exist.

Finally, we provide a novel application of MGFs to construct an Archimedean copula. The reader is referred to Marshall and Olkin (1988) for more detailed discussions. See Nelsen (1999) for an introduction of copulas.

Example 2.6 (Archimedean Copula) For any positive random variable R, suppose that it has the MGF at a neighborhood of the origin. Let

$$\psi(s) = E\left[e^{-sR}\right], \quad s \geq 0. \tag{2.34}$$

It is readily seen that

$$(-1)^n \psi^{(n)}(s) > 0, \quad s > 0,$$

where $\psi^{(n)}(s)$ denotes the nth derivative of $\psi(s)$. Such a function is called *completely monotone*. We denote the inverse by $\psi^{-1}(s)$.

Suppose further that $\psi(0) = 1$ and $\psi(\infty) = 0$. Then, the inverse function is decreasing in s with $\psi^{-1}(0) = \infty$ and $\psi^{-1}(1) = 0$. Now, for any $(x_1, \ldots, x_n) \in (0, 1]^n$, let

$$C(x_1, \ldots, x_n) = \psi(\psi^{-1}(x_1) + \cdots + \psi^{-1}(x_n)). \tag{2.35}$$

The function $C(x_1, \ldots, x_n)$, $0 < x_i \leq 1$, is called a *copula* if it defines an n-variate probability distribution.

To see that the function $C(x_1, \ldots, x_n)$ indeed defines a probability distribution, consider a family of IID random variables $Y_1, \ldots, Y_n$ whose survival function[§] is given by

$$P\{Y_i > y\} = e^{-y}, \quad y \geq 0. \tag{2.36}$$

It is assumed that R and Y_i are independent. Define

$$X_i = \psi(Y_i/R), \quad i = 1, \ldots, n,$$

and consider the n-variate random variable $(X_1, \ldots, X_n)$. Note that the range of X_i is $(0, 1]$. Since $\psi^{-1}(x)$ is decreasing in x, its distribution function is given by

$$P\{X_1 \leq x_1, \ldots, X_n \leq x_n\} = P\{Y_1 \geq \psi^{-1}(x_1)R, \ldots, Y_n \geq \psi^{-1}(x_n)R\}.$$

[§] The probability $P\{Y > y\}$ is called the *survival function* of Y.

On the other hand, by the law of total probability (2.31), we have
$$P\{X_1 \leq x_1, \ldots, X_n \leq x_n\}$$
$$= E\left[P\{Y_1 \geq \psi^{-1}(x_1)R, \ldots, Y_n \geq \psi^{-1}(x_n)R|R\}\right].$$
But, since R and Y_i are independent, we obtain
$$P\{Y_1 \geq \psi^{-1}(x_1)R, \ldots, Y_n \geq \psi^{-1}(x_n)R|R = r\}$$
$$= P\{Y_1 \geq \psi^{-1}(x_1)r, \ldots, Y_n \geq \psi^{-1}(x_n)r\}$$
$$= \prod_{i=1}^n P\{Y_i \geq \psi^{-1}(x_i)r\}$$
$$= e^{-(\psi^{-1}(x_1)+\cdots+\psi^{-1}(x_n))r},$$
where we have used (2.36) for the third equality. It follows from (2.34) that
$$P\{X_1 \leq x_1, \ldots, X_n \leq x_n\} = E\left[e^{-(\psi^{-1}(x_1)+\cdots+\psi^{-1}(x_n))R}\right]$$
$$= \psi\left(\psi^{-1}(x_1) + \cdots + \psi^{-1}(x_n)\right),$$
as desired.

2.8 Exercises

Exercise 2.1 Consider an experiment of rolling a die. Show that $(\Omega, \mathcal{F}, P)$ defined below is a probability space:
$$\Omega = \{\omega_1, \omega_2, \ldots, \omega_6\}, \quad \mathcal{F} = \{\Omega, \emptyset, A, B, C, A^c, B^c, C^c\},$$
where $A = \{\omega_1, \omega_2\}$, $B = \{\omega_3, \omega_4\}$, $C = \{\omega_5, \omega_6\}$, and A^c denotes the complement of A, and $P(A) = 1/2$ and $P(B) = P(C) = 1/4$.

Exercise 2.2 For any events A and B such that $A \subset B$, prove that $P(A) \leq P(B)$. This property is called the *monotonicity of probability*.

Exercise 2.3 For any events $A_1, A_2, \ldots, A_n$, prove by induction on n that
$$P(A_1 \cup A_2 \cup \cdots \cup A_n) = \sum_{i=1}^n P(A_i) - \sum_{i<j} P(A_i \cap A_j)$$
$$+ \cdots + (-1)^{n-1} P(A_1 \cap A_2 \cap \cdots \cap A_n),$$
where the general term is given by
$$(-1)^{m-1} \sum_{i_1 < \cdots < i_m} P(A_{i_1} \cap \cdots \cap A_{i_m}).$$
In particular, for events A, B, and C, we have
$$P(A \cup B \cup C) = P(A) + P(B) + P(C) + P(A \cap B \cap C)$$
$$- P(A \cap B) - P(A \cap C) - P(B \cap C).$$

Exercise 2.4 (Chain Rule) For any events $A_1, A_2, \ldots, A_n$, suppose that $P(A_1 \cap \cdots \cap A_{n-1}) > 0$. Prove that

$$P(A_1 \cap \cdots \cap A_n) = P(A_1)P(A_2|A_1) \cdots P(A_n|A_1 \cap \cdots \cap A_{n-1}).$$

Exercise 2.5 (Bayes Formula) In the same notation as in (2.6), prove that

$$P(A_k|B) = \frac{P(A_k)P(B|A_k)}{\sum_{i=1}^{\infty} P(A_i)P(B|A_i)}.$$

Exercise 2.6 Suppose that $p_j = Kq^j$, $j = 0, 1, \ldots, N$, for some $q > 0$. Determine K so that $\{p_j\}$ is a probability distribution. Also, suppose that $f(x) = Kx^2$, $0 \le x \le 4$. When does $f(x)$ become a density function?

Exercise 2.7 For any events A, B, and C with $P(C) > 0$, suppose that

$$P(A \cap B|C) = P(A|C)P(B|C).$$

Then, the events A and B are said to be *conditionally independent* given the event C. Suppose, in addition, that $P(C) = 0.5$, $P(A|C) = P(B|C) = 0.9$, $P(A|C^c) = 0.2$ and $P(B|C^c) = 0.1$. Prove that A and B are not independent, i.e. $P(A \cap B) \ne P(A)P(B)$. In general, the conditional independence does not imply independence and vice versa.

Exercise 2.8 Suppose that a random variable (X, Y, Z) takes the values $(1, 0, 0)$, $(0, 1, 0)$, $(0, 0, 1)$, or $(1, 1, 1)$ equally likely. Prove that they are pairwise independent but not independent.

Exercise 2.9 Show that the function defined by

$$f(x, y) = \frac{1}{\pi}, \quad x^2 + y^2 \le 1,$$

is a joint density function. Also, obtain its marginal density functions.

Exercise 2.10 Let X be a continuous random variable taking values on $\mathbf{R}$, and suppose that it has a finite mean. Prove that

$$E[X] = \int_0^{\infty} (1 - F(x)) \mathrm{d}x - \int_{-\infty}^0 F(x) \mathrm{d}x,$$

where $F(x)$ denotes the distribution function of X.

Exercise 2.11 (Markov's Inequality) Let X be any random variable. For any smooth function $h(x)$, show that

$$P\{|h(X)| > a\} \le \frac{E[|h(X)|]}{a}, \quad a > 0,$$

provided that the expectation exists. *Hint*: Take $A = \{x : |h(x)| > a\}$ and apply Proposition 2.2.

Exercise 2.12 For any random variables X and Y, prove that

$$V[X + Y] = V[X] + V[Y] + 2C[X, Y].$$

Also, prove Proposition 2.4 by a mathematical induction.

EXERCISES

Exercise 2.13 By mimicking the proof of Proposition 2.9, prove the second statement of Proposition 2.7.

Exercise 2.14 (Chebyshev's Inequality) Suppose that a random variable X has the mean $\mu = E[X]$ and the variance $\sigma^2 = V[X]$. For any $\varepsilon > 0$, prove that
$$P\{|X - \mu| \geq \varepsilon\} \leq \frac{\sigma^2}{\varepsilon^2}.$$
Hint: Use Markov's inequality with $h(x) = (x - \mu)^2$ and $a = \varepsilon^2$.

Exercise 2.15 (Schwarz's Inequality) For any random variables X and Y with finite variances σ_X^2 and σ_Y^2, respectively, prove that
$$|C[X, Y]| \leq \sigma_X \sigma_Y,$$
and that an equality holds only if $Y = aX + b$ for some constants $a \neq 0$ and b. Because of this and (2.26), we have $-1 \leq \rho[X, Y] \leq 1$ and $\rho[X, Y] = \pm 1$ if and only if $X = \pm aY + b$ with $a > 0$.

Exercise 2.16 Let $X_1, X_2, \ldots$ be a sequence of IID random variables, and let N be an integer-valued random variable which is independent of $\{X_n\}$. Define $Y = \sum_{i=1}^{N} X_i$, where $Y = 0$ if $N = 0$. Prove that the moment generating function (MGF) of Y, if it exists, is given by
$$m_Y(t) = g_N(m_X(t)),$$
where $g_N(z)$ is the *generating function* of N, i.e.
$$g_N(z) \equiv \sum_{n=0}^{\infty} P\{N = n\} z^n,$$
provided that $g_N(z)$ exists at a neighborhood of $z = 1$.

Exercise 2.17 Let Y be defined in Exercise 2.16 and assume that the MGF exists at a neighborhood of the origin. Prove that
$$E[Y] = E[X]E[N]$$
and
$$V[Y] = E[N]V[X] + E^2[X]V[N],$$
where $E^2[X] = (E[X])^2$.

CHAPTER 3

Useful Distributions in Finance

This chapter provides some useful distributions in financial engineering. In particular, binomial and normal distributions are of particular importance. The reader is referred to Johnson and Kotz (1970) and Johnson, Kotz, and Kemp (1992) for details. See also Tong (1990) for detailed discussions of normal distributions.

3.1 Binomial Distributions

A discrete random variable X is said to follow a *Bernoulli* distribution with parameter p (denoted by $X \sim Be(p)$ for short) if its realization is either 1 or 0 and the probability is given by

$$P\{X = 1\} = 1 - P\{X = 0\} = p, \quad 0 < p < 1. \tag{3.1}$$

The realization 1 (0, respectively) is often interpreted as a success (failure), and so the parameter p represents the probability of success.

A discrete random variable X is said to follow a *binomial* distribution with parameter (n, p) (denoted by $X \sim B(n, p)$ for short) if the set of realizations is $\{0, 1, 2, \ldots, n\}$ and the probability is given by

$$P\{X = k\} = {}_nC_k\, p^k (1-p)^{n-k}, \quad k = 0, 1, \ldots, n, \tag{3.2}$$

where ${}_nC_k = n!/[k!(n-k)!]$ denotes the combination of k different objects from n objects.

In order to interpret binomial distributions, consider a sequence of trials that are performed independently. If each trial follows the same Bernoulli distribution, then the sequence of trials is called *Bernoulli trials*. The random variable $X \sim B(n, p)$ is interpreted as the number of successes in n Bernoulli trials with success probability p.

More specifically, let $X_1, X_2, \ldots$ be a sequence of IID random variables. The random variable X_k represents the realization of the kth Bernoulli trial with success probability p. Let

$$X = X_1 + X_2 + \cdots + X_n. \tag{3.3}$$

The discrete convolution property then shows that $X \sim B(n, p)$. The proof of this result is left in Exercise 3.1.

Example 3.1 (Binomial Model) Let S_n denote the time n price of a financial security that pays no dividends, and assume that $S_0 = S$. Formally, the *binomial model* is defined by

$$S_n = S \prod_{i=1}^{n} Y_i, \quad n = 0, 1, 2, \ldots,$$

where Y_i are IID random variables whose probability distribution is given by

$$P\{Y_i = u\} = 1 - P\{Y_i = d\} = p, \quad 0 < p < 1.$$

Here, $0 < d < 1 < u$, and we have used the convention that $\prod_{i=1}^{0} = 1$. Now, let X be a binomial random variable defined by (3.3). It is easily seen that S_n and $\hat{S}_n$ defined by

$$\hat{S}_n = S u^X d^{n-X}, \quad n = 0, 1, 2, \ldots,$$

have the same distribution. In fact, we have

$$P\{S_n = S u^k d^{n-k}\} = b_k(n, p), \quad k = 0, 1, 2, \cdots, n,$$

where $b_k(n, p)$ denotes the binomial probability (3.2). We shall discuss details of the binomial model in Chapters 6 and 7.

For later uses, we discuss more about the binomial distribution (3.2). For notational convenience, we continue to use the notation $b_k(n, p)$, $k = 0, 1, \ldots, n$, for the probability (3.2). Then, denoting

$$r_k \equiv \frac{b_{k+1}(n, p)}{b_k(n, p)}, \quad k = 0, 1, \ldots, n - 1,$$

it is readily seen that

$$r_k = \frac{n - k}{k + 1} \frac{p}{1 - p}, \quad k = 0, 1, \ldots, n - 1.$$

Since r_k is decreasing in k, if $r_0 < 1$, i.e. $p < 1/(n + 1)$, then $b_k(n, p)$ is decreasing in k, while it is increasing for the case that $r_{n-1} > 1$, i.e. $p > n/(n + 1)$. Otherwise, for the case that

$$\frac{1}{n+1} \leq p \leq \frac{n}{n+1},$$

$b_k(n, p)$ is *unimodal* in k in the sense that $b_k(n, p)$ is non-decreasing first until some k^* and then non-increasing after k^*. The k^* is called the *mode* and given by some integer k satisfying $r_{k-1} \geq 1 \geq r_k$. The probability $b_k(n, p)$ can be easily calculated by the recursive relation

$$b_{k+1}(n, p) = b_k(n, p) \frac{n - k}{k + 1} \frac{p}{1 - p}, \quad k = 0, 1, \ldots, n - 1.$$

In what follows, the *cumulative* probability and the *survival* probability of the binomial distribution $B(n, p)$ are denoted, respectively, by

$$B_k(n, p) \equiv \sum_{j=0}^{k} b_j(n, p), \quad \overline{B}_k(n, p) \equiv \sum_{j=k}^{n} b_j(n, p); \quad k = 0, 1, \ldots, n.$$

Since the binomial probability is symmetric in the sense that
$$b_k(n,p) = b_{n-k}(n, 1-p),$$
we obtain the symmetry result
$$B_k(n,p) = \overline{B}_{n-k}(n, 1-p), \quad k = 0, 1, \ldots, n. \tag{3.4}$$
In particular, because of this relationship, $b_k(n, 1/2)$ is symmetric around $k = n/2$. The following approximation due to Peizer and Pratt (1968) is useful in practice:
$$B_k(n,p) \approx \Phi(v) \equiv \frac{1}{\sqrt{2\pi}} \int_{-\infty}^{v} e^{-x^2/2} dx,$$
where
$$v = \frac{k' - np + (1-2p)/6}{|k' - np|}$$
$$\times \sqrt{\frac{2}{1 + 1/(6n)} \left\{ k' \log \frac{k'}{np} + (n-k') \log \frac{n-k'}{n(1-p)} \right\}}$$
and $k' = k + 0.5$. Here, $\Phi(x)$ is the distribution function of the standard normal distribution (to be defined later). The survival probability $\overline{B}_k(n,p)$ can be approximated using the symmetry (3.4).

Let X be a random variable that follows a Bernoulli distribution with parameter p, i.e. $X \sim Be(p)$. Then, the MGF of X exists for any $t \in \mathbf{R}$ and is given by
$$m(t) = pe^t + (1-p), \quad t \in \mathbf{R}. \tag{3.5}$$
The mean and variance are given, from Proposition 2.10, as
$$E[X] = m'(0) = p$$
and
$$V[X] = m''(0) - E^2[X] = p(1-p),$$
respectively, where $E^2[X] = (E[X])^2$. Using the interpretation (3.3) for $X \sim B(n,p)$ and Proposition 2.11, the MGF of $X \sim B(n,p)$ is given by
$$m(t) = \left[pe^t + (1-p)\right]^n, \quad t \in \mathbf{R}. \tag{3.6}$$
The mean and variance of $X \sim B(n,p)$ are obtained, respectively, as
$$E[X] = np, \quad V[X] = np(1-p). \tag{3.7}$$
Note that the variance is maximized at $p = 1/2$. See Exercise 3.2 for a related problem.

3.2 Other Discrete Distributions

In this section, we introduce some useful, discrete distributions in finance that are obtained from Bernoulli trials. See Exercises 3.4–3.6 for other important discrete distributions.

3.2.1 Poisson Distributions

A random variable X is said to follow a *Poisson* distribution with parameter λ (denoted by $X \sim Poi(\lambda)$ for short) if the set of realizations is $\{0, 1, 2, \ldots\}$ and the probability is given by

$$P\{X = n\} = \frac{\lambda^n}{n!} e^{-\lambda}, \quad n = 0, 1, 2, \ldots, \tag{3.8}$$

where $\lambda > 0$. From Taylor's expansion (1.17), i.e.

$$e^x = \sum_{n=0}^{\infty} \frac{x^n}{n!}, \quad x \in \mathbf{R},$$

we obtain $\sum_{n=0}^{\infty} P\{X = n\} = 1$, whence (3.8) defines a probability distribution. The MGF of $X \sim Poi(\lambda)$ is given by

$$m(t) = \exp\left\{\lambda(e^t - 1)\right\}, \quad t \in \mathbf{R}. \tag{3.9}$$

The mean and variance of $X \sim Poi(\lambda)$ are obtained as

$$E[X] = V[X] = \lambda.$$

The proof is left to the reader.

The Poisson distribution is characterized as a limit of binomial distributions. To see this, consider a binomial random variable $X_n \sim B(n, \lambda/n)$, where $\lambda > 0$ and n is sufficiently large. From (3.6), the MGF of X_n is given by

$$m_n(t) = \left[1 + (e^t - 1)\frac{\lambda}{n}\right]^n, \quad t \in \mathbf{R}.$$

Using (1.3), we then have

$$\lim_{n \to \infty} m_n(t) = \exp\left\{\lambda(e^t - 1)\right\}, \quad t \in \mathbf{R},$$

which is the MGF (3.9) of a Poisson random variable $X \sim Poi(\lambda)$. It follows from Proposition 2.12 that X_n converges in law to the Poisson random variable $X \sim Poi(\lambda)$.

Let $X_1 \sim Poi(\lambda_1)$ and $X_2 \sim Poi(\lambda_2)$, and suppose that they are independent. Then, from (2.11) and the independence, we obtain

$$P\{X_1 = k | X_1 + X_2 = n\} = \frac{P\{X_1 = k\} P\{X_2 = n - k\}}{P\{X_1 + X_2 = n\}}, \quad k = 0, 1, \ldots, n.$$

Since $X_1 + X_2 \sim Poi(\lambda_1 + \lambda_2)$ from the result of Exercise 3.3, simple algebra shows that

$$P\{X_1 = k | X_1 + X_2 = n\} = {}_nC_k \left(\frac{\lambda_1}{\lambda_1 + \lambda_2}\right)^k \left(\frac{\lambda_2}{\lambda_1 + \lambda_2}\right)^{n-k}.$$

Hence, the random variable X_1 conditional on the event $\{X_1 + X_2 = n\}$ follows the binomial distribution $B(n, \lambda_1/(\lambda_1 + \lambda_2))$.

3.2.2 Geometric Distributions

A discrete random variable X is said to follow a *geometric* distribution with parameter p (denoted by $X \sim Geo(p)$ for short) if the set of realizations is $\{0, 1, 2, \ldots\}$ and the probability is given by

$$P\{X = n\} = p(1-p)^n, \quad n = 0, 1, 2, \ldots. \tag{3.10}$$

The geometric distribution is interpreted as the number of failures until the first success, i.e. the *waiting time* until the first success, in Bernoulli trials with success probability p. It is readily seen that

$$E[X] = \frac{1-p}{p}, \quad V[X] = \frac{1-p}{p^2}.$$

The proof is left to the reader.

Geometric distributions satisfy the *memoryless property*. That is, let $X \sim Geo(p)$. Then,

$$P\{X - n \geq m | X \geq n\} = P\{X \geq m\}, \quad m, n = 0, 1, 2, \ldots. \tag{3.11}$$

To see this, observe that the survival probability is given by

$$P\{X \geq m\} = (1-p)^m, \quad m = 0, 1, 2, \ldots.$$

The left-hand side in (3.11) is equal to

$$P\{X \geq n + m | X \geq n\} = \frac{P\{X \geq n + m\}}{P\{X \geq n\}},$$

whence the result. The memoryless property (3.11) states that, given $X \geq n$, the survival probability is independent of the past. This property is intuitively obvious because of the independence in the Bernoulli trials. That is, the event $\{X \geq n\}$ in the Bernoulli trials is equivalent to saying that there is no success until the $(n-1)$th trial. But, since each trial is performed independently, the first success after the nth trial should be independent of the past. It is noted that geometric distributions are the only discrete distributions that possess the memoryless property.

Example 3.2 In this example, we construct a default model of a corporate firm using Bernoulli trials. A general default model in the discrete-time setting will be discussed in Chapter 8. Let $Y_1, Y_2, \ldots$ be a sequence of IID Bernoulli random variables with parameter p. That is,

$$P\{Y_n = 1\} = 1 - P\{Y_n = 0\} = p, \quad n = 1, 2, \ldots.$$

Using these random variables, we define

$$N = \min\{n : Y_n = 0\},$$

and interpret N as the default time epoch of the firm (see Figure 3.1). Note that N is an integer-valued random variable and satisfies the relation

$$\{N > t\} = \{Y_n = 1 \text{ for all } n \leq t\}, \quad t = 1, 2, \ldots.$$

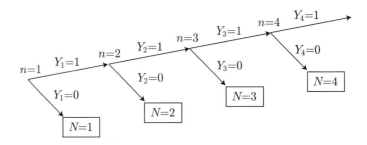

Figure 3.1 *The construction of default time epoch N*

It follows from the independence of Y_n that
$$P\{N > t\} = p^t, \quad t = 1, 2, \ldots,$$
whence $(N-1)$ follows a geometric distribution $Geo(p)$. The default time epoch constructed in this way has the memoryless property, which is obviously counter-intuitive in practice. However, by allowing the random variables Y_n to be dependent in some ways, we can construct a more realistic default model. The details will be described in Chapter 8.

3.3 Normal and Log-Normal Distributions

A central role in finance literature is played by normal distributions. This section is devoted to describe a concise summary of normal distributions with emphasis on applications to finance.

A continuous random variable X is said to be *normally distributed* with mean μ and variance σ^2 if the density function of X is given by
$$f(x) = \frac{1}{\sqrt{2\pi}\sigma} \exp\left\{-\frac{(x-\mu)^2}{2\sigma^2}\right\}, \quad x \in \mathbf{R}. \tag{3.12}$$

If this is the case, we denote $X \sim N(\mu, \sigma^2)$. In particular, $N(0,1)$ is called the *standard* normal distribution, and its density function is given by
$$\phi(x) = \frac{1}{\sqrt{2\pi}} e^{-x^2/2}, \quad x \in \mathbf{R}. \tag{3.13}$$

A significance of the normal density is the symmetry about the mean and its bell-like shape. The density functions of normal distributions with various variances are depicted in Figure 3.2. The distribution function of the standard normal distribution is given by
$$\Phi(x) = \int_{-\infty}^{x} \frac{1}{\sqrt{2\pi}} e^{-t^2/2} dt, \quad x \in \mathbf{R}. \tag{3.14}$$

Note that the distribution function $\Phi(x)$ does *not* have any analytical expression.

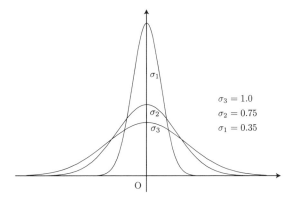

Figure 3.2 The density functions of normal distribution ($\sigma_1 < \sigma_2 < \sigma_3$)

For any random variable X with mean μ and variance σ^2, the transformation given by
$$Y = \frac{X - \mu}{\sigma} \tag{3.15}$$
is called the *standardization* of X. If, in particular, $X \sim N(\mu, \sigma^2)$, the random variable Y follows the standard normal distribution $N(0,1)$. Conversely, if $Y \sim N(0,1)$ then
$$X = \mu + \sigma Y \sim N(\mu, \sigma^2). \tag{3.16}$$
The proof is left in Exercise 3.7.

In what follows, we mainly work with standard normal random variables, and a particular normal variable is obtained from the transformation (3.16). Important properties of the standard normal density and distribution functions are summarized next. The proof is left to the reader.

Proposition 3.1 *The standard normal density $\phi(x)$ is symmetric about 0, i.e. $\phi(x) = \phi(-x)$, $x \in \mathbf{R}$, and satisfies*
$$\phi'(x) = -x\phi(x), \quad x \in \mathbf{R}.$$
Accordingly, we have $\Phi(x) = 1 - \Phi(-x)$ for all $x \in \mathbf{R}$.

The MGF of $X \sim N(\mu, \sigma^2)$ is given by
$$m(t) = \exp\left\{\mu t + \frac{\sigma^2 t^2}{2}\right\}, \quad t \in \mathbf{R}. \tag{3.17}$$
To see this, we only consider the case $N(0,1)$. The general case follows from the transformation (3.16). Suppose $X \sim N(0,1)$. Then,
$$m(t) = \int_{-\infty}^{\infty} e^{tx}\phi(x)\mathrm{d}x = \frac{1}{\sqrt{2\pi}}\int_{-\infty}^{\infty} e^{tx - x^2/2}\mathrm{d}x.$$
Since $tx - x^2/2 = [t^2 - (x-t)^2]/2$, we obtain
$$m(t) = e^{t^2/2}\int_{-\infty}^{\infty} \frac{1}{\sqrt{2\pi}} e^{-(x-t)^2/2}\mathrm{d}x.$$

But, since the integrand is the density function of the normal distribution $N(t,1)$, its integral over $\mathbf{R}$ is just unity. Hence, we obtain

$$m(t) = e^{t^2/2}, \quad t \in \mathbf{R}.$$

Recall from Proposition 2.11 that the MGF given by (3.17) determines the normal distribution $N(\mu, \sigma^2)$ uniquely. See Exercise 3.8 for the calculation of the moments of normal distributions.

Suppose $X \sim N(\mu, \sigma^2)$ and consider the random variable $Y = e^X$. The distribution of Y is called the *log-normal distribution*, since $\log Y$ is normally distributed. To obtain the density function of Y, we first consider its distribution function

$$F(y) \equiv P\{Y \leq y\} = P\{X \leq \log y\}, \quad y > 0.$$

But, since $X \sim N(\mu, \sigma^2)$, the standardization (3.15) yields

$$F(y) = \Phi\left(\frac{\log y - \mu}{\sigma}\right), \quad y > 0.$$

It follows that the density function is given by

$$f(y) = \frac{1}{\sqrt{2\pi}\sigma y} \exp\left\{-\frac{(\log y - \mu)^2}{2\sigma^2}\right\}, \quad y > 0. \tag{3.18}$$

The proof is left to the reader.

The nth moment of Y can be obtained from (3.17), since

$$Y^n = e^{nX}, \quad n = 1, 2, \ldots.$$

It follows that

$$E[Y^n] = \exp\left\{n\mu + \frac{n^2\sigma^2}{2}\right\}, \quad n = 1, 2, \ldots.$$

It is noted that, although Y has the moments of any order, the MGF of a log-normal distribution does not exist. This is so, because for any $h > 0$

$$E\left[e^{hY}\right] = E\left[\sum_{n=0}^{\infty} \frac{h^n Y^n}{n!}\right] = \sum_{n=0}^{\infty} \frac{h^n E[Y^n]}{n!} = \infty. \tag{3.19}$$

The proof is left in Exercise 3.11.

Let X be a random variable defined on $\mathbf{R}$. We assume that it has the MGF $E\left[e^{\theta X}\right]$ for some $\theta \in \mathbf{R}$. Consider the set function P^* defined by

$$P^*(A) = \frac{E\left[e^{\theta X} 1_A\right]}{E\left[e^{\theta X}\right]}, \quad A \in \mathcal{F}, \tag{3.20}$$

where 1_A is the indicator function of event A, and $\mathcal{F}$ denotes the σ-field. The function P^* defines a probability measure, since $P^*(\Omega) = 1$ and $P^*(A) \geq 0$. The σ-additivity (2.1) of P^* follows from that of the indicator function. That is, for mutually exclusive events A_i, $i = 1, 2, \ldots$, we have

$$1_{\sum_{i=1}^{\infty} A_i} = \sum_{i=1}^{\infty} 1_{A_i}.$$

Note that the right-hand side of (3.20) corresponds to the original probability P, and the transformation defines a new probability measure P^*. Such a transformation is called the *change of measure*. In particular, this change of measure is called the *Esscher transform* in mathematical insurance.

Suppose that the payoff of an insurance is given by $h(X)$. The premium of the insurance is calculated from (3.20) as

$$E^*[h(X)] = \frac{E\left[e^{\theta X} h(X)\right]}{E\left[e^{\theta X}\right]} = \frac{1}{E\left[e^{\theta X}\right]} \int_{-\infty}^{\infty} e^{\theta x} h(x) \mathrm{d}F(x),$$

where $F(x)$ is the distribution function of X. Define

$$F^*(x) = \frac{1}{E\left[e^{\theta X}\right]} \int_{-\infty}^{x} e^{\theta y} \mathrm{d}F(y), \quad x \in \mathbf{R}. \tag{3.21}$$

It is easy to see that $F^*(x)$ defines a distribution function on $\mathbf{R}$, since $F^*(x)$ is continuous and non-decreasing in x, $F(-\infty) = 0$ and $F(\infty) = 1$. Also,

$$E^*[h(X)] = \int_{-\infty}^{\infty} h(x) \mathrm{d}F^*(x).$$

Hence, $F^*(x)$ is a distribution function of X under P^*, and the change of measure (3.20) corresponds to the change of distribution function from $F(x)$ to $F^*(x)$ through (3.21).

Now, suppose that $X \sim N(\mu, \sigma^2)$, and let $\theta \neq 0$ be such that

$$E\left[e^{\theta X}\right] = 1.$$

It is easily seen that $\theta = -2\mu/\sigma^2$ in this particular case. From (3.21), the transformed density function is given by

$$f^*(x) = e^{\theta x} f(x), \quad x \in \mathbf{R}. \tag{3.22}$$

Since the MGF associated with $f^*(x)$ is obtained as

$$m^*(t) \equiv \int_{-\infty}^{\infty} e^{tx} f^*(x) \mathrm{d}x = \int_{-\infty}^{\infty} e^{(\theta+t)x} f(x) \mathrm{d}x = m(\theta + t),$$

we have from (3.17) that

$$m^*(t) = e^{-\mu t + \sigma^2 t^2 / 2}, \quad t \in \mathbf{R},$$

which is the MGF of the normal distribution $N(-\mu, \sigma^2)$. Since the transformation (3.22) is useful in the Black–Scholes formula (see, e.g., Gerber and Shiu (1994) for details), we summarize the result as a theorem.

Theorem 3.1 *Suppose that $X \sim N(-\psi^2/2, \psi^2)$, and denote its density function by $f(x)$. Then, the transformation*

$$f^*(x) = e^x f(x), \quad x \in \mathbf{R},$$

defines the density function of the normal distribution $N(\psi^2/2, \psi^2)$.

3.4 Other Continuous Distributions

This section introduces some useful continuous distributions in financial engineering other than normal and log-normal distributions. See Exercises 3.13–3.15 for other important continuous distributions.

3.4.1 Chi-Square Distributions

A continuous random variable X is said to follow a *chi-square* distribution with n degrees of freedom (denoted by $X \sim \chi_n^2$ for short) if its realization is in $(0, \infty)$ and the density function is given by

$$\psi_n(x) = \frac{1}{2^{n/2}\Gamma(n/2)} x^{\frac{n}{2}-1} e^{-x/2}, \quad x > 0,$$

where $\Gamma(x)$ denotes the *gamma function* defined by

$$\Gamma(x) = \int_0^\infty u^{x-1} e^{-u} du, \quad x > 0. \tag{3.23}$$

We note that $\Gamma(1) = 1$, $\Gamma(1/2) = \sqrt{\pi}$, and $\Gamma(\alpha) = (\alpha - 1)\Gamma(\alpha - 1)$. Hence, $\Gamma(n) = (n-1)!$ for positive integer n. The proof is left to the reader.

The chi-square distribution χ_n^2 is interpreted as follows. Let $X_i \sim N(0,1)$, $i = 1, 2, \ldots, n$, and suppose that they are independent. Then, it can be shown that

$$X_1^2 + X_2^2 + \cdots + X_n^2 \sim \chi_n^2. \tag{3.24}$$

Hence, if $X \sim \chi_n^2$, then $E[X] = n$ by the linearity of expectation (see Proposition 2.3). Also, we have $V[X] = 2n$ since, from Proposition 2.4,

$$V[X] = V[X_1^2] + \cdots + V[X_n^2] = n\left(E[X_1^4] - E^2[X_1^2]\right)$$

and $E[X_1^4] = 3$ from the result in Exercise 3.8. Note from the interpretation (3.24) that if $X \sim \chi_n^2$, $Y \sim \chi_m^2$ and they are independent, then we have $X + Y \sim \chi_{n+m}^2$.

Importance of chi-square distributions is apparent in statistical inference. Let $X_1, X_2, \ldots, X_n$ denote independent samples from a normal population $N(\mu, \sigma^2)$. The sample mean is defined as

$$\bar{X} = \frac{X_1 + X_2 + \cdots + X_n}{n},$$

while the sample variance is given by

$$S^2 = \frac{(X_1 - \bar{X})^2 + (X_2 - \bar{X})^2 + \cdots + (X_n - \bar{X})^2}{n-1}.$$

Then, it is well known that

$$\bar{X} \sim N\left(\mu, \frac{\sigma^2}{n}\right), \quad \frac{(n-1)S^2}{\sigma^2} \sim \chi_{n-1}^2,$$

and they are independent.

OTHER CONTINUOUS DISTRIBUTIONS

For independent random variables $X_i \sim N(\mu, \sigma^2)$, $i = 1, 2, \ldots, n$, define
$$X = \sum_{i=1}^{n} \frac{(X_i - \mu + \delta)^2}{\sigma^2}.$$
The density function of X is given by
$$\phi(x) = e^{-\beta^2/2} \sum_{k=0}^{\infty} \frac{(\beta^2/2)^k}{k!} \psi_{n+2k}(x), \quad \beta^2 = \frac{n\delta^2}{\sigma^2}, \qquad (3.25)$$
where $\psi_n(x)$ is the density function of χ_n^2. The distribution of X is called the *non-central, chi-square* distribution with n degrees of freedom and non-centrality δ. When $\delta = 0$, it coincides with the ordinary chi-square distribution. The distribution is a Poisson mixture of chi-square distributions. The degree of freedom n can be extended to a positive number.

3.4.2 Exponential Distributions

A continuous random variable X is said to follow an *exponential distribution* with parameter $\lambda > 0$ (denoted by $X \sim Exp(\lambda)$ for short) if its realization is on $[0, \infty)$, and the density function is given by
$$f(x) = \lambda e^{-\lambda x}, \quad x \geq 0. \qquad (3.26)$$
As was used in Example 2.6, the survival probability is given by
$$P\{X > x\} = e^{-\lambda x}, \quad x \geq 0.$$
Note that $E[X] = 1/\lambda$ and $V[X] = 1/\lambda^2$ (see Exercise 3.13). Also, the MGF is given by
$$E\left[e^{tX}\right] = \frac{\lambda}{\lambda - t}, \quad t < \lambda.$$
It is easily seen that the exponential distribution is a continuous limit of geometric distributions. Hence, exponential distributions share similar properties with geometric distributions. In particular, exponential distributions possess the memoryless property
$$P\{X > x + y | X > y\} = P\{X > x\}, \quad x, y \geq 0,$$
and are characterized by this property.

Example 3.3 This example is parallel to Example 3.2 and constructs a default time epoch in the continuous-time case. For a given positive number λ, define $H(x) = \lambda x$. Let $S \sim Exp(1)$ and define
$$N = \sup\{t : S > H(t)\}.$$
Since $H(x)$ is increasing in x, N is well defined (see Figure 3.3). As before, we interpret N as the default time epoch. Note that N is a positive random variable and satisfies the relation
$$\{N > t\} = \{S > H(t)\}, \quad t \geq 0.$$

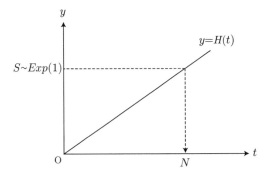

Figure 3.3 *The construction of default time epoch N in continuous time*

It follows that
$$P\{N > t\} = P\{S > H(t)\} = e^{-\lambda t}, \quad t \geq 0,$$
whence N is exponentially distributed with mean $1/\lambda$. We shall return to a more realistic model using a general non-decreasing function $H(x)$ (see Chapter 13 below).

For IID random variables $X_i \sim Exp(\lambda)$, $i = 1, 2, \ldots$, define
$$S_k = X_1 + X_2 + \cdots + X_k, \quad k = 1, 2, \ldots.$$
The density function of S_k is given by
$$f(x) = \frac{\lambda^k}{(k-1)!} x^{k-1} e^{-\lambda x}, \quad x \geq 0, \qquad (3.27)$$
which can be proved by the convolution formula (see Exercise 3.9). This distribution is called an *Erlang distribution* of order k with parameter $\lambda > 0$ (denoted by $S_k \sim E_k(\lambda)$ for short). The distribution function of $E_k(\lambda)$ is given by
$$F(x) = 1 - e^{-\lambda x} \sum_{n=0}^{k-1} \frac{(\lambda x)^n}{n!}, \quad x \geq 0. \qquad (3.28)$$
The Erlang distribution is related to a Poisson distribution through
$$P\{S_k > \lambda\} = P\{Y < k\},$$
where $S_k \sim E_k(1)$ and $Y \sim Poi(\lambda)$. The proof is left to the reader.

Next, let $S_k = X_1 + \cdots + X_k$ as above, and define
$$N(t) = \max\{k : S_k \leq t\} \quad t \geq 0. \qquad (3.29)$$
If X_i denotes the lifetime of the ith system, $N(t)$ is interpreted as the number of failed systems until time t. Since the event $\{N(t) \geq k\}$ is equivalent to the event $\{S_k \leq t\}$, it follows from (3.28) that
$$P\{N(t) = k\} = \frac{(\lambda t)^k}{k!} e^{-\lambda t}, \quad k = 0, 1, 2, \ldots. \qquad (3.30)$$

MULTIVARIATE NORMAL DISTRIBUTIONS

That is, $N(t)$ follows a Poisson distribution with parameter λt; see (3.8). The process $\{N(t), t \geq 0\}$ counting the number of failures is called a *Poisson process*, which we shall discuss later (see Chapter 11).

3.4.3 Uniform Distributions

The simplest continuous distributions are uniform distributions. A random variable X is said to follow a *uniform distribution* with interval (a, b) (denoted by $X \sim U(a, b)$ for short) if the density function is given by

$$f(x) = \frac{1}{b-a}, \quad a < x < b. \tag{3.31}$$

The distribution function of $U(a, b)$ is

$$F(x) = \frac{x-a}{b-a}, \quad a < x < b.$$

It is readily seen that

$$E[X] = \frac{a+b}{2}, \quad V[X] = \frac{(b-a)^2}{12}. \tag{3.32}$$

In particular, $U(0, 1)$ is called the *standard* uniform distribution, and if $Y \sim U(0, 1)$ then $(b-a)Y + a \sim U(a, b)$.

Importance of standard uniform distributions is due to the following fact. Let X be any random variable with distribution function $F(x)$. For simplicity, we assume that $F(x)$ has the inverse function $F^{-1}(x)$. Consider now a random variable $U \sim U(0, 1)$. Then,

$$X \stackrel{\mathrm{d}}{=} F^{-1}(U), \tag{3.33}$$

where $\stackrel{\mathrm{d}}{=}$ stands for equality in law. To see this, we calculate the distribution function of $F^{-1}(U)$ to obtain

$$P\{F^{-1}(U) \leq x\} = P\{U \leq F(x)\}, \quad x \in \mathbf{R},$$

since $F(x)$ is increasing in x. But, since $U \sim U(0, 1)$ so that $P\{U \leq x\} = x$, it follows that

$$P\{F^{-1}(U) \leq x\} = F(x) = P\{X \leq x\}, \quad x \in \mathbf{R},$$

as desired. For example, since $y = -\log(1-x)$ is the inverse function of $y = 1 - e^{-x}$, we conclude that $X = -\log(1-U)$ follows an exponential distribution $Exp(1)$.

3.5 Multivariate Normal Distributions

Let $(X_1, X_2, \ldots, X_n)$ be an n-variate random variable. In general, the probabilistic property of such a random variable is determined by the joint distribution. However, in the case of normal distributions, treatments become extremely simple. That is, suppose that $(X_1, X_2, \ldots, X_n)$ follows an n-variate normal distribution. Then, the joint distribution is determined by the means $\mu_i = E[X_i]$ and the covariances (variances in the case that $i = j$)

$$\sigma_{ij} = E[X_i X_j] - E[X_i]E[X_j], \quad i, j = 1, 2, \ldots, n.$$

To be more precise, we define the *mean vector* and the *covariance matrix*, respectively, as

$$\boldsymbol{\mu} = \begin{pmatrix} \mu_1 \\ \mu_2 \\ \vdots \\ \mu_n \end{pmatrix}, \quad \boldsymbol{\Sigma} = \begin{pmatrix} \sigma_{11} & \sigma_{12} & \cdots & \sigma_{1n} \\ \sigma_{21} & \sigma_{22} & \cdots & \sigma_{2n} \\ \vdots & \vdots & \ddots & \vdots \\ \sigma_{n1} & \sigma_{n2} & \cdots & \sigma_{nn} \end{pmatrix}.$$

The symmetry of covariances implies that $\sigma_{ij} = \sigma_{ji}$, so that the covariance matrix $\boldsymbol{\Sigma}$ is symmetric (in fact, it is positive semi-definite as proved below). The joint density function is given by

$$f(\mathbf{x}) = \frac{1}{(2\pi)^{n/2}\sqrt{\det(\boldsymbol{\Sigma})}} \exp\left\{-\frac{(\mathbf{x} - \boldsymbol{\mu})^\top \boldsymbol{\Sigma}^{-1}(\mathbf{x} - \boldsymbol{\mu})}{2}\right\}, \quad (3.34)$$

provided that $\boldsymbol{\Sigma}$ is *positive definite*, where $\mathbf{x} = (x_1, x_2, \ldots, x_n)^\top$ and $\top$ denotes the transpose. Note that the marginal distributions are normal and $X_i \sim N(\mu_i, \sigma_{ii})$. If the covariance matrix is not positive definite, the random variables are linearly dependent and some special treatments are required. See, e.g., Tong (1990) for details.

Definition 3.1 A symmetric matrix $\mathbf{A}$ is said to be *positive definite* if, for any non-zero vector $\mathbf{c}$, we have $\mathbf{c}^\top \mathbf{A} \mathbf{c} > 0$. It is called *positive semi-definite* if, for any vector $\mathbf{c}$, we have $\mathbf{c}^\top \mathbf{A} \mathbf{c} \geq 0$.

Recall that, even if X and Y are uncorrelated, they can be dependent on each other in general. However, in the case of normal distributions, uncorrelated random variables are necessarily independent, which is one of the most striking properties of normal distributions. To prove this, suppose that the covariance matrix is diagonal, i.e. $\sigma_{ij} = 0$ for $i \neq j$. Then, from (3.34), the joint density function becomes

$$f(\mathbf{x}) = \prod_{i=1}^{n} \frac{1}{\sqrt{2\pi\sigma_{ii}}} \exp\left\{-\frac{(x_i - \mu_i)^2}{2\sigma_{ii}}\right\}, \quad x_i \in \mathbf{R},$$

so that they are independent (see Definition 2.4). In particular, if $X_i \sim N(0, 1)$, $i = 1, 2, \ldots, n$, and they are independent, then the joint density function is given by

$$f(\mathbf{x}) = \prod_{i=1}^{n} \frac{1}{\sqrt{2\pi}} e^{-x_i^2/2}, \quad x_i \in \mathbf{R}. \quad (3.35)$$

Such a random variable $(X_1, \ldots, X_n)$ is said to follow the *standard n-variate normal distribution*.

For real numbers $c_1, c_2, \ldots, c_n$, consider the *linear combination* of normally

MULTIVARIATE NORMAL DISTRIBUTIONS

distributed random variables $X_1, X_2, \ldots, X_n$, i.e.

$$R = c_1 X_1 + c_2 X_2 + \cdots + c_n X_n. \tag{3.36}$$

A significant property of normal distributions is that the linear combination R is also normally distributed. Since a normal distribution is determined by the mean and the variance, it suffices to obtain the mean

$$E[R] = \sum_{i=1}^{n} c_i \mu_i \tag{3.37}$$

and the variance

$$V[R] = \sum_{i=1}^{n} \sum_{j=1}^{n} c_i \sigma_{ij} c_j. \tag{3.38}$$

Note that the variance $V[R]$ must be non-negative, whence the covariance matrix $\Sigma = (\sigma_{ij})$ is positive semi-definite by Definition 3.1. If the variance $V[R]$ is positive for any non-zero, real numbers c_i, then the covariance matrix Σ is positive definite. The positive definiteness implies that there are no redundant random variables. In other words, if the covariance matrix Σ is not positive definite, then there exists a linear combination such that R defined in (3.36) has zero variance, i.e. R is a constant.

Example 3.4 (Value at Risk) Consider a portfolio consisting of n securities, and denote the rate of return of security i by X_i. Let c_i denote the *weight* of security i in the portfolio. Then, the rate of return of the portfolio is given by the linear combination (3.36). To see this, let x_i be the current market value of security i, and let Y_i be its future value. The rate of return of security i is given by $X_i = (Y_i - x_i)/x_i$. Let $Q_0 = \sum_{i=1}^{n} \xi_i x_i$, where ξ_i denotes the number of shares (not weight) of security i, and let Q denote the random variable that represents the future portfolio value. Assuming that no rebalance will be made, the weight c_i in the portfolio is given by $c_i = \xi_i x_i / Q_0$. It follows that the rate of return, $R = (Q - Q_0)/Q_0$, of the portfolio is obtained as

$$R = \frac{\sum_{i=1}^{n} \xi_i (Y_i - x_i)}{Q_0} = \sum_{i=1}^{n} c_i X_i, \tag{3.39}$$

as claimed. The mean $E[R]$ and the variance $V[R]$ are given by (3.37) and (3.38), respectively, no matter what the random variables X_i are.

Now, of interest is the potential for *significant loss* in the portfolio. Namely, given a probability level α, what would be the maximum loss in the portfolio? The maximum loss is called the *value at risk* (VaR) with confidence level $100\alpha\%$, and VaR is a prominent tool for risk management. More specifically, VaR with confidence level $100\alpha\%$ is defined by $z_\alpha > 0$ that satisfies

$$P\{Q - Q_0 \geq -z_\alpha\} = \alpha,$$

and VaR's popularity is based on aggregation of many components of market risk into the *single* number z_α. Figure 3.4 depicts the meaning of this equation.

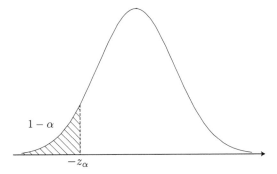

Figure 3.4 *The density function of $Q - Q_0$ and VaR*

Since $Q - Q_0 = Q_0 R$, we obtain

$$P\{Q_0 R \leq -z_\alpha\} = 1 - \alpha. \qquad (3.40)$$

Note that the definition of VaR does not require normality. However, calculation of VaR becomes considerably simpler, if we assume that $(X_1, \ldots, X_n)$ follows an n-variate normal distribution. Then, the rate of return R in (3.39) is normally distributed with mean (3.37) and variance (3.38). The value z_α satisfying (3.40) can be obtained using a table of the standard normal distribution. In many standard textbooks of statistics, a table of the survival probability

$$L(x) = \int_x^\infty \frac{1}{\sqrt{2\pi}} e^{-y^2/2} dy, \quad x > 0,$$

is given. Then, using the standardization (3.15) and symmetry of the density function (3.13) about 0, we can obtain the value z_α with ease. Namely, letting $r_\alpha = -z_\alpha/Q_0$, it follows from (3.40) that

$$1 - \alpha = P\left\{\frac{R - \mu}{\sigma} \leq \frac{r_\alpha - \mu}{\sigma}\right\},$$

where $\mu = E[R]$ and $\sigma = \sqrt{V[R]}$, whence

$$1 - \alpha = L\left(-\frac{r_\alpha - \mu}{\sigma}\right).$$

Therefore, letting x_α be such that $L(x_\alpha) = 1 - \alpha$, we obtain

$$z_\alpha = Q_0(x_\alpha \sigma - \mu),$$

as the VaR with confidence level $100\alpha\%$.* The value x_α is the $100(1 - \alpha)$-percentile of the standard normal distribution (see Table 3.1). For example, if $\mu = 0$, then the 99% VaR is given by $2.326\sigma Q_0$. See Exercise 3.17 for an example of the VaR calculation.

---

* Since the risk horizon is very short, e.g. one day or one week, in market risk management, the mean rate of return μ is often set to be zero. In this case, VaR with confidence level $100\alpha\%$ is given by $x_\alpha \sigma Q_0$.

Table 3.1 *Percentiles of the standard normal distribution*

$100(1-\alpha)$	10	5.0	1.0	0.5	0.1
x_α	1.282	1.645	2.326	2.576	3.090

3.6 Exercises

Exercise 3.1 (Discrete Convolution) Let X and Y be non-negative, discrete random variables with probability distributions $p_i^X = P\{X = i\}$ and $p_i^Y = P\{Y = i\}$, $i = 0, 1, 2, \ldots$, respectively, and suppose that they are independent. Prove that

$$P\{X + Y = k\} = \sum_{i=0}^{k} p_i^X p_{k-i}^Y, \quad k = 0, 1, 2, \ldots.$$

Using this formula, prove that if $X \sim B(n, p)$, $Y \sim Be(p)$ and they are independent then $X + Y \sim B(n+1, p)$.

Exercise 3.2 Suppose that $X \sim B(n, p)$, $Y \sim B(m, p)$ and they are independent. Using the MGF of binomial distributions, show that $X + Y \sim B(n+m, p)$. Interpret this result in terms of Bernoulli trials.

Exercise 3.3 Suppose that $X \sim Poi(\lambda)$, $Y \sim Poi(\mu)$ and they are independent. Using the MGF of Poisson distributions, show that $X+Y \sim Poi(\lambda+\mu)$. Interpret this result in terms of Bernoulli trials.

Exercise 3.4 (Logarithmic Distribution) Suppose that a discrete random variable X has the probability distribution

$$P\{X = n\} = a\frac{(1-p)^n}{n}, \quad n = 1, 2, \ldots,$$

where $0 < p < 1$ and $a = -1/\log p$. Show that

$$E\left[e^{tX}\right] = \frac{\log[1 - (1-p)e^t]}{\log p}$$

and $E[X] = a(1-p)/p$. This distribution represents the situation that the failure probability decreases as the number of trials increases.

Exercise 3.5 (Pascal Distribution) In Bernoulli trials with success probability p, let T represent the time epoch of the mth success. That is, denoting the time interval between the $(i-1)$th success and the ith success by T_i, we have $T = T_1 + T_2 + \cdots + T_m$. Prove that the probability function of T is given by

$$P\{T = k\} = {}_{k-1}C_{k-m}\, p^m(1-p)^{k-m}, \quad k = m, m+1, \ldots.$$

Exercise 3.6 (Negative Binomial Distribution) For the random variable T defined in Exercise 3.5, let $X = T - m$. Show that

$$P\{X = k\} = {}_{m+k-1}C_k\, p^m(1-p)^k, \quad k = 0, 1, \ldots.$$

Also, let $p = 1 - \lambda/n$ for some $\lambda > 0$. Prove that X converges in law to a Poisson random variable as $n \to \infty$.

Exercise 3.7 Suppose $X \sim N(0,1)$. Then, using the MGF (3.17), show that $\mu + \sigma X \sim N(\mu, \sigma^2)$.

Exercise 3.8 Using the MGF (3.17), verify that $E[Y] = \mu$ and $V[Y] = \sigma^2$ for $Y \sim N(\mu, \sigma^2)$. Also, show that $E[X^3] = 0$ and $E[X^4] = 3$ for $X \sim N(0,1)$.

Exercise 3.9 (Convolution Formula) Let X and Y be continuous random variables with density functions $f_X(x)$ and $f_Y(x)$, respectively, and suppose that they are independent. Prove that the density function of $X + Y$ is given by

$$f_{X+Y}(x) = \int_{-\infty}^{\infty} f_X(y) f_Y(x-y) dy, \quad x \in \mathbf{R}.$$

Using this, prove that if $X \sim N(\mu_X, \sigma_X^2)$, $Y \sim N(\mu_Y, \sigma_Y^2)$ and they are independent then $X + Y \sim N(\mu_X + \mu_Y, \sigma_X^2 + \sigma_Y^2)$. Also, prove (3.27) by an induction on k.

Exercise 3.10 In contrast to Exercise 3.9, suppose that $X \sim N(0, \sigma^2)$, but Y depends on the realization of X. Namely, suppose that $Y \sim N(\mu(x), \sigma^2(x))$ given $X = x$, $x \in \mathbf{R}$, for some functions $\mu(x)$ and $\sigma^2(x) > 0$. Obtain the density function of $Z = X + Y$.

Exercise 3.11 Equation (3.19) can be proved as follows.

(1) Show that $\int_1^{n+1} \log y \, dy \geq \log n!$ for any $n = 1, 2, \ldots$.

(2) Let $h > 0$. For all sufficiently large n, show that

$$n \log h + n\mu + n^2 \sigma^2 \geq \int_1^{n+1} \log y \, dy.$$

(3) Combining these inequalities, show that

$$\sum_{n=0}^{\infty} \frac{e^{n \log h + n\mu + n^2 \sigma^2}}{n!} = \infty.$$

Exercise 3.12 For a positive random variable X with mean μ and variance σ^2, the quantity $\eta = \sigma/\mu$ is called the *coefficient of variation* of X. Show that $\eta = 1$ for any exponential distribution, while $\eta < 1$ for Erlang distributions given by (3.27). Also, prove that $\eta > 1$ for the distribution with the density function

$$f(x) = \sum_{i=1}^{k} p_i \lambda_i e^{-\lambda_i x}, \quad x \geq 0,$$

where $p_i, \lambda_i > 0$ and $\sum_{i=1}^{k} p_i = 1$. The mixture of exponential distributions is called a *hyper-exponential distribution*.

Exercise 3.13 (Gamma Distribution) Suppose that a continuous random variable has the density function

$$f(x) = \frac{\lambda^\alpha}{\Gamma(\alpha)} x^{\alpha-1} e^{-\lambda x}, \quad x > 0,$$

where $\Gamma(x)$ is the gamma function (3.23). Prove that the MGF is given by

$$m(t) = \left(\frac{\lambda}{\lambda - t}\right)^\alpha, \quad t < \lambda.$$

Using this, show that $E[X] = \alpha/\lambda$ and $V[X] = \alpha/\lambda^2$. Note: An exponential distribution is a special case with $\alpha = 1$, while the case with $\alpha = n/2$ and $\lambda = 1/2$ corresponds to a chi-square distribution with n degrees of freedom.

Exercise 3.14 (Inverse Gamma Distribution) Suppose that X follows a Gamma distribution given in Exercise 3.13. Let $Y = X^{-1}$. Prove that the density function of Y is given by

$$f(x) = \frac{\lambda^\alpha}{\Gamma(\alpha)} x^{-\alpha - 1} e^{-\lambda/x}, \quad x > 0.$$

Exercise 3.15 (Extreme-Value Distribution) Consider a sequence of IID random variables $X_i \sim Exp(1/\theta)$, $i = 1, 2, \ldots$, and define

$$Y_n = \max\{X_1, X_2, \ldots, X_n\} - \theta \log n.$$

The distribution function of Y_n is denoted by $F_n(x)$. Prove that

$$F_n(x) = \left[1 - e^{-(x + \theta \log n)/\theta}\right]^n = \left[1 - \frac{e^{-x/\theta}}{n}\right]^n,$$

so that

$$\lim_{n \to \infty} F_n(x) = \exp\left\{-e^{-x/\theta}\right\}.$$

This distribution is called a *double exponential* distribution and is known as one of the extreme-value distributions.

Exercise 3.16 Let (X, Y) be any bivariate normal random variable. For any function $f(x)$ for which the following expectations exist, show that

$$E\left[f(X) e^{-Y}\right] = E\left[e^{-Y}\right] E[f(X - C[X, Y])],$$

where $C[X, Y]$ denotes the covariance between X and Y.

Exercise 3.17 Let X_i, $i = 1, 2$, denote the annual rate of return of security i, and suppose that (X_1, X_2) follows the bivariate normal distribution with means $\mu_1 = 0.1$, $\mu_2 = 0.15$, variances $\sigma_1 = 0.12$, $\sigma_2 = 0.18$ and correlation coefficient $\rho = -0.4$. Calculate the 95% VaR for 5 days of the portfolio $R = 0.4 X_1 + 0.6 X_2$, where the current market value of the portfolio is 1 million dollars.

CHAPTER 4

Derivative Securities

The aim of this chapter is to explain typical derivative securities actually traded in the market. A *derivative security* (or simply a *derivative*) is a security whose value depends on the values of other more basic underlying variables. In recent years, derivative securities have become more important than ever in financial markets. Futures and options are actively traded in exchange markets and swaps and many exotic options are traded outside of exchanges, called the over-the-counter (OTC) markets, by financial institutions and their corporate clients. See, e.g., Cox and Rubinstein (1985) and Hull (2000) for more detailed discussions of financial products.

4.1 The Money-Market Account

Consider a bank deposit with initial principal $F = 1$. The amount of the deposit after t periods is denoted by B_t. The interest paid for period t is equal to $B_{t+1} - B_t$. If the interest is proportional to the amount B_t, it is called the *compounded interest*. That is, the compounded interest is such that the amounts of deposit satisfy the relation

$$B_{t+1} - B_t = rB_t, \quad t = 0, 1, 2, \ldots,$$

where the multiple $r > 0$ is called the *interest rate*. Since $B_{t+1} = (1+r)B_t$, it follows that

$$B_t = (1+r)^t, \quad t = 0, 1, 2, \ldots. \quad (4.1)$$

The deposit B_t is often called the *money-market account*.

Suppose now that the annual interest rate is r and interest is paid n times each year. We divide one year into n equally spaced subperiods, so that the interest rate for each period is given by r/n. Following the same arguments as above, it is readily seen that the amount of deposit after m periods is given by

$$B_m = \left(1 + \frac{r}{n}\right)^m, \quad m = 0, 1, 2, \ldots. \quad (4.2)$$

For example, when the interest is semiannual compounding, the amount of deposit after 2 years is given by $B_4 = (1 + r/2)^4$.

Suppose $t = m/n$ for some integers m and n, and let $B(t)$ denote the amount of deposit at time t for the compounding interest with annual interest rate r. From (4.2), we have

$$B(t) = \left(1 + \frac{r}{n}\right)^{nt}.$$

Now, what if we let n tend to infinity? That is, what if the interest is *continuous* compounding? This is easy because, from (1.3), we have

$$B(t) = \left[\lim_{n\to\infty}\left(1+\frac{r}{n}\right)^n\right]^t = e^{rt}, \quad t \geq 0. \tag{4.3}$$

Here, we have employed Proposition 1.3, since the function $f(x) = x^t$ is continuous.

Consider next the case that the interest rates vary in time. For simplicity, we first assume that the interest rates are a step function of time. That is, the interest rate at time t is given by

$$r(t) = r_i \quad \text{if } t_{i-1} \leq t < t_i, \quad i = 1, 2, \ldots,$$

where $t_0 = 0$. Then, from (4.3), we know that $B(t_1) = e^{r_1 t_1}$, $B(t_2)/B(t_1) = e^{r_2(t_2-t_1)}$, etc. Hence, for time t such that $t_{n-1} \leq t < t_n$, we obtain

$$B(t) = \exp\left\{\sum_{k=1}^{n-1} r_k \delta_k + r_n(t - t_{n-1})\right\}, \quad \delta_k \equiv t_k - t_{k-1}.$$

Recall that the integral of any (Riemann) integrable function $r(t)$ is the limit of the sum, i.e.

$$\int_0^t r(u)\mathrm{d}u = \lim_{n\to\infty}\left[\sum_{k=1}^{n-1} r_k \delta_k + r_n(t - t_{n-1})\right],$$

where the limit is taken over all possible sequences of partitions of the interval $[0, t]$. The next result summarizes.

Theorem 4.1 *Suppose that the instantaneous interest rate at time t is $r(t)$. If the interest is continuous compounding, then the time t money-market account is given by*

$$B(t) = \exp\left\{\int_0^t r(u)\mathrm{d}u\right\}, \quad t \geq 0, \tag{4.4}$$

provided that the integral exists.

We note that, even though the interest rates $r(t)$ are random, the money-market account $B(t)$ is given by (4.4); in which case, $B(t)$ is also a random variable.

4.2 Various Interest Rates

In order to value derivative securities, we need to *discount* future cashflows with respect to the default-free interest rates. Hence, all derivative securities are exposed to interest-rate risk, and the term structure models of interest rates are of importance in financial engineering. In this section, we introduce various interest rates needed for later discussions in this book.

VARIOUS INTEREST RATES

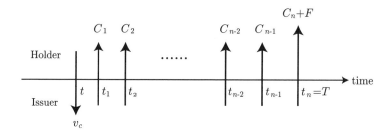

Figure 4.1 *The cashflow of a coupon-bearing bond (v_C denotes the time t price of the coupon-bearing bond)*

4.2.1 Discount Bonds and Coupon-Bearing Bonds

A financial security that promises to pay a single cashflow of magnitude F at a future time, called the *maturity*, is called a *discount bond*, and the amount F is referred to as the *face value*. By taking F as the unit of money, we can assume without loss of generality that $F = 1$.

A bond more commonly traded in practice is a security that promises to pay a stream of certain payments, called *coupons*, at future times as well as the face value at the maturity. Such a bond is called a *coupon-bearing bond* and its cashflow, if it pays coupon C_i at time t_i, is depicted in Figure 4.1. Since each cashflow C_i is equivalent to a cashflow of a discount bond with face value C_i, a coupon-bearing bond can be thought of as a *portfolio* of discount bonds with face values C_i (see Exercise 4.1).

There are two types of bonds, default-free bonds and corporate bonds. A *corporate bond* issued by a firm promises to pay a stream of payments; but there is a possibility that the bond defaults before the maturity and cannot meet its debt obligations. That is, corporate bonds are exposed to *credit risk*. On the other hand, a *default-free bond* issued by a government such as U.S. or Japan involves no such risk.

Suppose that the prices of default-free discount bonds for all maturities are observed in (or can be imputed from) the market. Since the prices of default-free discount bonds with the same maturity are the same, the discount bonds can be used as a benchmark of future values. Namely, let $v(t, T)$, $t \leq T$, be the time t price of the discount bond that pays 1 dollar for sure at maturity T. Then, the present value of the certain cashflow B at future time T is given by $Bv(t, T)$, where t denotes the current time. Of course, $v(t, t) = 1$. The function $v(t, T)$ with respect to T is called the *term structure* of the default-free discount bonds. It is also called the *discount function*, since the present value of 1 dollar paid at future time T is given by $v(t, T)$.

4.2.2 Yield-to-Maturity

Suppose that, at time t, an investor purchases a security for $S(t)$ that pays $S(T)$ dollars for sure at maturity T. The *rate of return* per unit of time, $R(t,T)$ say, from this investment is defined by

$$R(t,T) = \frac{S(T) - S(t)}{(T-t)S(t)}, \quad t \leq T. \tag{4.5}$$

It follows that
$$S(T) = S(t)\left[1 + (T-t)R(t,T)\right].$$

Next, suppose that the rate of return per unit of time is computed in the sense of compounded interests. From (4.2), we then have

$$S(T) = S(t)\left[1 + \frac{(T-t)R_n(t,T)}{n}\right]^n, \quad n = 1, 2, \ldots, \tag{4.6}$$

where the subscript n in $R_n(t,T)$ means that the interests are compounded n times each year. Denoting the rate of return per unit of time in the sense of the continuous compounding by $Y(t,T) = \lim_{n\to\infty} R_n(t,T)$, it follows from (4.3) that

$$S(T) = S(t)\,\mathrm{e}^{(T-t)Y(t,T)}, \quad t \leq T,$$

or, equivalently,

$$Y(t,T) = \frac{1}{T-t} \log \frac{S(T)}{S(t)}, \quad t \leq T.$$

The rate of return per unit of time in the continuous compounding is called the *yield-to-maturity* (or simply the *yield*). In particular, if the security is the default-free discount bond with maturity T, i.e. $S(t) = v(t,T)$ and $S(T) = 1$, then we obtain

$$Y(t,T) = -\frac{\log v(t,T)}{T-t}, \quad t \leq T. \tag{4.7}$$

In what follows, we use Equation (4.7) as the definition of the yield of the discount bond.

Example 4.1 Consider a coupon-bearing bond with coupons C and face value 1. The coupons are paid semiannually, and the maturity is 5 years later. Suppose that a coupon has just been paid at time 0, and the current market price of the bond is q. Let r be the rate of return per unit of time of this coupon-bearing bond in the semiannual compounding. In practice, it is common to define the rate of return of the coupon-bearing bond as a solution of the equation

$$q = \frac{C}{1+r/2} + \frac{C}{(1+r/2)^2} + \cdots + \frac{1+C}{(1+r/2)^{10}}. \tag{4.8}$$

The function in the right-hand side of (4.8) is strictly decreasing in r, $q = 1 + 10C$ for $r = 0$, and $q \to 0$ as $r \to \infty$. Hence, as far as the current bond price q is less than $1 + 10C$, there exists a unique solution that satisfies (4.8). See Exercise 4.2 for a related problem. The rate of return, r, can be calculated numerically by the bisection method or the standard Newton method.

VARIOUS INTEREST RATES

4.2.3 Spot Rates

Given the yield curve $Y(t,T)$, $t < T$, of the default-free discount bonds, the *instantaneous interest rate* (or the *spot rate* for short) at time t is defined by the limit

$$r(t) = \lim_{T \to t} Y(t,T).$$

That is, the spot rate is the yield of the default-free discount bond with infinitesimal maturity. Of course, no such discount bonds exist in the real market and, hence, the spot rates are conceptual interest rates.* Also, from (4.7), we have

$$r(t) = -\lim_{T \to t} \frac{\log v(t,T)}{T-t} = -\frac{\partial}{\partial T} \log v(t,T)\bigg|_{T=t}. \quad (4.9)$$

Hence, as easily seen from (4.9), the discount bond price $v(t,T)$ cannot be recovered from the spot rates $r(t)$ alone. Nevertheless, the spot rates play a crucial role in the valuation of derivative securities. For example, as in (4.4), the money-market account $B(t)$ is given in terms of the spot rates $r(t)$.

4.2.4 Forward Yields and Forward Rates

The time t yield, $f(t,T,\tau)$, of the default-free discount bond over the future time interval $[T,\tau]$ is defined by

$$\frac{v(t,\tau)}{v(t,T)} = e^{-(\tau-T)f(t,T,\tau)}, \quad t < T < \tau.$$

The yield $f(t,T,\tau)$ is called the *forward yield*. Comparing this with (4.7), the forward yield can be thought of as the yield of $v(t,\tau)/v(t,T)$, i.e.

$$f(t,T,\tau) = -\frac{\log \dfrac{v(t,\tau)}{v(t,T)}}{\tau - T}, \quad t \leq T < \tau.$$

As we will see soon, the value $v(t,\tau)/v(t,T)$ is the *forward price* of the discount bond; see (4.17) below.

The instantaneous forward yield,

$$\begin{aligned}
f(t,T) &= \lim_{\tau \to T} f(t,T,\tau) \\
&= -\lim_{h \to 0} \frac{\log v(t,T+h) - \log v(t,T)}{h} \\
&= -\frac{\partial}{\partial T} \log v(t,T), \quad t \leq T, \quad (4.10)
\end{aligned}$$

is called the *forward rate*. The forward rates cannot be observed in the real

---

* In the empirical finance literature, the yield of discount bonds with short maturity is used as an approximation of the instantaneous spot rate. For example, Chan, Karolyi, Longstaff and Sanders (1992) used the one-month treasury bill yield, while Ait-Sahalia (1996) used the 7-day Euro dollar deposit rate as the spot rate.

market either. However, from (4.10), we obtain

$$v(t,T) = \exp\left\{-\int_t^T f(t,s)\mathrm{d}s\right\}, \quad t \leq T. \tag{4.11}$$

Hence, the discount bond prices are recovered from the forward rates. This is the advantage to consider the forward rates rather than the spot rates. Also, comparing (4.10) with (4.9), we have

$$r(t) = f(t,t). \tag{4.12}$$

Hence, the spot rates (and so the money-market account) are also obtained from the forward rates.

It should be noted that these relationships between the interest rates hold for any circumstances. That is, they hold true even in a stochastic environment. For more relationships, see Exercise 4.3. Exercise 4.4 considers such relationships in the discrete-time case.

Later, we will prove (see Example 5.1) that, for the deterministic interest-rate economy, we must have

$$r(T) = f(t,T), \quad t \leq T. \tag{4.13}$$

Hence, from (4.11), we obtain

$$v(t,T) = \exp\left\{-\int_t^T r(s)\mathrm{d}s\right\} = \frac{B(t)}{B(T)}, \quad t \leq T. \tag{4.14}$$

That is, the discount bond prices can be determined from the spot rates in the deterministic interest-rate economy. Recall, however, that this no longer holds if the interest rates are stochastic.

Example 4.2 The LIBOR (London Inter-Bank Offer Rate) rates are the interest rates most commonly used in the international financial markets. They are offered by commercial banks in London, so that the LIBOR rates should reflect credit risk. However, since practitioners utilize the LIBOR rates as default-free interest rates, the LIBOR rate $L(t, t+\delta)$ must satisfy the equation

$$1 + \delta L(t, t+\delta) = \frac{1}{v(t, t+\delta)}. \tag{4.15}$$

Here, $\delta > 0$ is the time length covered by the LIBOR interest rates. For example, $\delta = 0.25$ means 3 months, $\delta = 0.5$ means 6 months, etc. From (4.5) with $S(t)$ being replaced by $v(t, t+\delta)$ and T by $t+\delta$, we identify the LIBOR rate $L(t, t+\delta)$ to be the average interest rate covering the period $[t, t+\delta]$.

4.3 Forward and Futures Contracts

A *forward contract* is an agreement to buy or sell an asset at a certain future time, called the *maturity*, for a certain price, called the *delivery price*. The forward contract is usually traded on the OTC basis.

Suppose that two parties agree with a forward contract at time t, and one

FORWARD AND FUTURES CONTRACTS

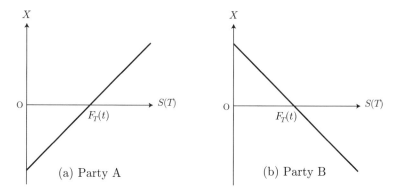

Figure 4.2 *The payoff functions of the forward contract*

of the parties, A say, wants to buy the underlying asset at maturity T for a delivery price $F_T(t)$. The other party, B say, agrees to sell the asset at the same date for the same price. At the time that the contract is entered into, the delivery price is chosen so that the value of the forward contract to both sides is zero. Hence, the forward contract costs nothing to enter. However, party A *must* buy the asset for the delivery price $F_T(t)$ at date T. Since party A can buy the asset at the market price $S(T)$, its gain (or loss) of the contract is given by

$$X = S(T) - F_T(t). \qquad (4.16)$$

Note that nobody knows the future price $S(T)$. Hence, the payoff X at date T is a random variable. If $S(T) = S$ is realized at the future date $T > t$, the payoff is given by

$$X(S) = S - F_T(t).$$

The function $X(S)$ with respect to S is called the *payoff function*. The payoff function for party B is

$$X(S) = F_T(t) - S.$$

Hence, the payoff functions of the forward contract are linear (see Figure 4.2).

Example 4.3 Suppose that a hamburger chain store company is subject to the risk of price fluctuation of meat. If the cost of meat becomes high, the company is ought to raise the hamburger price. However, then, the company will lose approximately 30% of existing consumers. If, on the other hand, the cost of meat becomes low, the company can lower the hamburger price to obtain new consumers; but the increase is estimated only about 5%. Because the downside risk is much more significant compared to the gain for this company, it decided to enter a forward contact to buy the meat at the current price x. Now, if the future price of meat is less than x, then the company loses money, because the company could buy the meat at a lower cost if it did not enter the contract. On the other hand, even when the price of meat becomes extremely high, the company can purchase the meat at the fixed cost x.

The *forward price* for a contract is defined as the delivery price that makes the contract have zero value. Hence, the forward and delivery prices coincide at the time that the contract is entered into. As time passes, the forward price changes according to the price evolution of the underlying asset (and the interest rates too) while the delivery price remains the same during the life of the contract. This is so, because the forward price $F_T(u)$ at time $u > t$ depends on the price $S(u)$, not on $S(t)$, while the delivery price $F_T(t)$ is determined at the start of the contract by $S(t)$. As we shall see in the next chapter, the time t forward price of an asset with no dividends is given by

$$F_T(t) = \frac{S(t)}{v(t,T)}, \quad t \leq T, \tag{4.17}$$

where $S(t)$ denotes the time t price of the underlying asset and $v(t,T)$ is the time t price of the default-free discount bond maturing at time T.

A *futures contract* is also an agreement between two parties to buy or sell an asset at maturity for a certain delivery price. Hence, the payoff of the futures contract at the maturity is the same as a forward contract (see Figure 4.2). Unlike forward contracts, however, futures contracts are usually traded on exchanges. Also, futures contracts are different from forward contracts by *marking to market* the account. That is, at the end of each trading day, the margin account is adjusted to reflect the investor's gain or loss. Further details of futures contracts will be found in Hull (2000).

The *futures price* for a contract is defined also as the delivery price that makes the contract have zero value at the time that the contract is entered into. Since a futures contract is similar to a forward contract, the futures price is often considered to be the same as the forward price. However, this is *not* true in general. A sufficient condition for the two prices to coincide with each other is that the interest rates are deterministic (see Example 13.6). The reader is referred to, e.g., Pliska (1997) for details.

4.4 Options

There are two types of options, call and put options. A *call option* gives the holder the right to buy the underlying asset by a certain date, called the *maturity*, for a certain price, called the *strike price*, while a *put option* gives the right to sell the underlying asset by a maturity for a strike price. *American* options can be exercised at any time up to the maturity, while *European* options can only be exercised at the maturity. Most of the options traded on exchanges are American. However, European options are easier to analyze than American options, and some of the properties of an American option are often deduced from those of its European counterpart.

The key feature that distinguishes options from forward and futures contracts is that options give the *right*. This means that the holder does not have to exercise options. To explain this, let $S(T)$ be the price of the underlying asset at the maturity T, and consider an investor who has bought a (European) call option with strike price K. If $S(T)$ is above the strike price K, the

OPTIONS

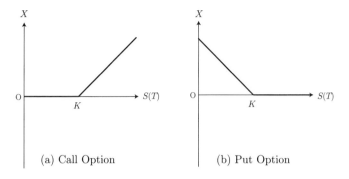

Figure 4.3 *The payoff functions of call and put options*

investor will buy the asset for the strike price K and sell it by the market price $S(T)$ immediately, whence his/her gain is $S(T) - K$. This is identical to the forward and futures contracts. If, on the other hand, $S(T)$ is below the strike price K, the investor will abandon the right and do nothing, whence his/her loss is 0. This makes options quite different from forward and futures contracts. The payoff function of the call option is therefore given by

$$X = \{S(T) - K\}_+, \tag{4.18}$$

where

$$\{x\}_+ = \max\{x, 0\} = \begin{cases} x, & x \geq 0, \\ 0, & x < 0. \end{cases} \tag{4.19}$$

Similarly, the payoff function of a put option with maturity T and strike price K is given by

$$X = \{K - S(T)\}_+. \tag{4.20}$$

These payoff functions are depicted in Figure 4.3. Observe the non-linearity of the payoff functions. As we shall see later, the non-linearity makes the valuation of options considerably difficult.

Example 4.4 Suppose that a pension fund is invested to an index of a financial market. The return of the index is relatively high, but subject to the risk of price fluctuation. At the target date T, if the index is below a promised level K, the fund may become shortage to repay the pension. Because of this downside risk, the company running the pension fund decided to purchase put options written on the index with strike price K. If the level $S(T)$ of the index at the target date T is less than K, then the company exercises the right to obtain the gain $K - S(T)$, which compensates the shortfall, since

$$S(T) + [K - S(T)] = K.$$

If, on the other hand, $S(T)$ is greater than K, the company abandons the right to do nothing; but in this case, the index may be sufficiently high for the purposes. The put option used to hedge downside risk in this way is called a *protective put*.

Note that the payoffs of options are always non-negative (see Figure 4.3). Hence, unlike forward and futures contracts, we cannot choose the strike price so that the value of an option is zero. This means that an investor must pay some amount of money to purchase an option. The cost is termed as the option *premium*. One of the goals in this book is to obtain the following *Black–Scholes formula* for a European call option:

$$c(S,t) = S\Phi(d) - Ke^{-r(T-t)}\Phi(d - \sigma\sqrt{T-t}), \qquad (4.21)$$

where

$$d = \frac{\log[S/K] + r(T-t)}{\sigma\sqrt{T-t}} + \frac{\sigma\sqrt{T-t}}{2}, \qquad (4.22)$$

$S(t) = S$ is the current price of the underlying security, K is the strike price, T is the maturity of the call option, r is the spot rate, σ is the volatility, and $\Phi(x)$ is the distribution function (3.14) of the standard normal distribution. The *volatility* is defined as the instantaneous standard deviation of the rate of return of the underlying security. Note that the spot rate as well as the volatility is constant in (4.21). The Black–Scholes formula will be discussed in detail later. See Exercise 4.7 for numerical calculations of the Black–Scholes formula. See also Exercises 4.8 and 4.9 for related problems.

4.5 Interest-Rate Derivatives

In this section, we introduce some of commonly traded interest-rate derivatives in the market. For example, any type of interest-rate options is traded in the market. Let $v(t,\tau)$ be the time t price of a bond (not necessarily a discount bond) with maturity τ. Then, the payoff of a European call option at the maturity T is given by

$$X = \{v(T,\tau) - K\}_+, \quad t \leq T < \tau,$$

where K denotes the strike price.

A forward contract of a discount bond is also considered. From (4.17), the forward price of the discount bond with maturity τ is given by

$$v_T(t,\tau) = \frac{v(t,\tau)}{v(t,T)}, \quad t \leq T < \tau. \qquad (4.23)$$

We shall discuss more about interest-rate derivatives in Chapter 13.

4.5.1 Interest-Rate Swaps and Swaptions

A *swap* is an agreement between two parties to exchange cashflows in the future according to a pre-arranged formula. The most common type of swap is the 'plain vanilla' interest-rate swap, where one party, A say, agrees to pay to the other party, B say, cashflows equal to interest at a pre-determined fixed rate on a notional principal for a number of years and, at the same time, party B agrees to pay to party A cashflows equal to interest at a floating rate on

INTEREST-RATE DERIVATIVES

the same notional principal for the same period of time. Hence, the interest-rate swap can be regarded as a portfolio of forward contracts. The reference floating rate in the international financial market is often the LIBOR rate. For example, if cashflows are exchanged semiannually, the 6-month LIBOR rate will be quoted. The fixed rate that is set to make the present values of the two cashflows equal is called the *swap rate*.

Example 4.5 Let the current time be $t = 0$ and consider an interest-rate swap that starts from $T_0 > 0$. For notational simplicity, we assume that the notional principal is unity and cashflows are exchanged at date $T_i = T_0 + i\delta$, $i = 1, 2, \ldots, n$. Let S be the swap rate. Then, in the fixed side, party A pays to party B the interest $S' = \delta S$ at dates T_i and so, the present value of the total payments is given by

$$V_{\text{FIX}} = S'v(0, T_1) + S'v(0, T_2) + \cdots + S'v(0, T_n) = \delta S \sum_{i=1}^{n} v(0, T_i), \quad (4.24)$$

where $v(t, T)$ denotes the time t price of the default-free discount bond with maturity T. On the other hand, in the floating side, suppose that party B pays to party A the interest $L'_i = \delta L(T_i, T_{i+1})$ at dates T_{i+1}, where $L(t, t+\delta)$ denotes the δ-LIBOR rate at time t. Since the future LIBOR rate $L(T_i, T_{i+1})$ is not observed now ($t = 0$), it is a random variable. However, as we shall show later, the present value for the floating side is given by

$$V_{\text{FL}} = v(0, T_0) - v(0, T_n). \quad (4.25)$$

Since the swap rate S is determined so that the present values (4.24) and (4.25) in both sides are equal, it follows that

$$S = \frac{v(0, T_0) - v(0, T_n)}{\delta \sum_{i=1}^{n} v(0, T_i)}. \quad (4.26)$$

Observe that the swap rate S does not depend on the LIBOR rates. The swap rates for various maturities are quoted in the swap market.

Let $S(t)$ be the time t swap rate of the plain vanilla interest-rate swap. The swap rate observed in the market changes, as interest rates change. Hence, the swap rate can be an underlying asset for an option. Such an option is called a *swaption*. Suppose that a swaption has maturity τ, and consider an interest-rate swap that starts at τ with notional principal 1 and payments being exchanged at $T_i = \tau + i\delta$, $i = 1, 2, \ldots, n$. Then, similar to (4.24), the value of the total payments for the fixed side at time τ is given by

$$V_{\text{FIX}}(\tau) = \delta S(\tau) \sum_{i=1}^{n} v(\tau, \tau + i\delta).$$

If an investor (for the fixed side) purchases a (European) swaption with strike rate K, then his/her payoff for the swaption at future time τ is given by

$$\delta \{S(\tau) - K\}_+ \sum_{i=1}^{n} v(\tau, \tau + i\delta), \quad (4.27)$$

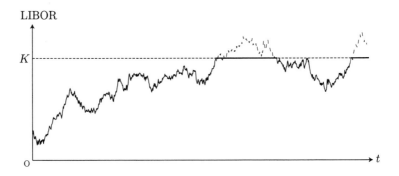

Figure 4.4 *Capping the LIBOR rate*

since the value of the total payments for the fixed side is fixed to be

$$\delta K \sum_{i=1}^{n} v(\tau, \tau + i\delta).$$

Note that (4.27) is the payoff of a call option. Similarly, in the floating side, the payoff is

$$\delta \{K - S(\tau)\}_+ \sum_{i=1}^{n} v(\tau, \tau + i\delta),$$

which is the payoff of a put option.

4.5.2 Caps and Floors

As in the case of swaps, let T_i be the time epoch that the ith payment is made. We denote the LIBOR rate that covers the interval $[T_i, T_{i+1}]$ by $L_i \equiv L(T_i, T_{i+1})$. An interest-rate derivative whose payoff is given by

$$\delta_i \{L_i - K\}_+, \quad \delta_i = T_{i+1} - T_i, \tag{4.28}$$

is called a *caplet*, where K is called the *cap rate*. A caplet is a call option written on the LIBOR rate. A *cap* is a sequence of caplets. If we possess a cap, then we can obtain $\{L_i - K\}_+$ from the cap. Hence, the effect is to 'cap' the LIBOR rate at the cap rate K from above, since

$$L_i - \{L_i - K\}_+ \leq K.$$

Figure 4.4 depicts a picture for capping the LIBOR rate by a cap.

A put option written on the LIBOR rate, i.e. the payoff is given by

$$\delta_i \{K - L_i\}_+, \quad \delta_i = T_{i+1} - T_i,$$

is called a *floorlet*, and a sequence of floorlets is a *floor*. If we possess a floor, then we can obtain $\{K - L_i\}_+$ from the floor, and the effect is to 'floor' the LIBOR rate at K from below.

4.6 Exercises

Exercise 4.1 Suppose that the current time is 0 and that the current forward rate of the default-free discount bonds is given by

$$f(0,t) = 0.01 + 0.005 \times t, \quad t \geq 0.$$

Consider a coupon-bearing bond with 5 yen coupons and a face value of 100 yen. Coupons will be paid semiannually and the maturity of the bond is 4 years. Supposing that the bond is issued today and the interest is the continuous compounding, obtain the present value of the coupon-bearing bond.

Exercise 4.2 In Example 4.1, suppose that the rate of return (yield) is calculated in the continuous compounding. Prove that the yield is a unique solution of the equation

$$q = Ce^{-y/2} + Ce^{-y} + \cdots + Ce^{-9y/2} + (1+C)e^{-5y}.$$

Explain how the yield y is obtained by the bisection method.

Exercise 4.3 Prove the following identities.

(1) $Y(t,T) = \dfrac{\int_t^T f(t,s)ds}{T-t}$ (2) $f(t,T,\tau) = \dfrac{\int_T^\tau f(t,s)ds}{\tau - T}$

Exercise 4.4 In the discrete-time case, prove the following.

(1) $B(t) = \displaystyle\prod_{i=1}^{t}(1+r(i-1))$ (2) $\dfrac{1}{v(t,T)} = \displaystyle\prod_{s=t}^{T-1}(1+f(t,s))$

Here, we use the convention $\prod_{i=1}^{0} a_i = 1$.

Exercise 4.5 Suppose that the current time is 0 and that the current term structure of the default-free discount bonds is given by

T	0.5	1.0	2.0	5.0	10.0
$v(0,T)$	0.9875	0.9720	0.9447	0.8975	0.7885

Obtain (a) the yields $Y(0,2)$ and $Y(0,5)$, (b) the forward yield $f(0,1,2)$, (c) the LIBOR rate $L(0,0.5)$ as the semiannual interest rate, and (d) the forward price of $v(0,10)$ maturing at 2 years later.

Exercise 4.6 Draw graphs of the payoff functions of the following positions: (1) Sell a call option with strike price K. (2) Sell a put option with strike price K. (3) Buy a call option with strike price K_1 and, at the same time, buy a put option with strike price K_2 and the same maturity (you may assume $K_1 < K_2$, but try the other case too).

Exercise 4.7 In the Black–Scholes formula (4.21), suppose that $S = K = 100$, $t = 0$, and $T = 0.4$. Draw the graphs of the option premiums with respect to the volatility σ, $0.01 \leq \sigma \leq 0.6$, for the cases that $r = 0.05, 0.1$, and 0.2.

Exercise 4.8 Let $\phi(x)$ denote the density function (3.13) of the standard normal distribution. Prove that
$$\phi(d) = \frac{K}{S}e^{-r(T-t)}\phi(d - \sigma\sqrt{T-t}),$$
where d is given by (4.22). Using this, prove the following for the Black–Scholes formula (4.21):
$$\begin{aligned} c_S &= \Phi(d) > 0, \\ c_T &= \frac{S\sigma}{2\sqrt{\tau}}\phi(d) + Kre^{-r\tau}\Phi(d - \sigma\sqrt{\tau}) > 0, \\ c_K &= -e^{-r\tau}\Phi(d - \sigma\sqrt{\tau}) < 0, \\ c_\sigma &= S\sqrt{\tau}\phi(d) > 0, \quad \text{and} \\ c_r &= K\tau e^{-r\tau}\Phi(d - \sigma\sqrt{\tau}) > 0. \end{aligned}$$
Here, $\tau = T - t$ and c_x denotes the partial derivative of c with respect to x.

Exercise 4.9 (Put–Call Parity) Consider call and put options with the same strike price K and the same maturity T written on the same underlying security $S(t)$. Show that
$$\{S(T) - K\}_+ + K = \{K - S(T)\}_+ + S(T).$$

Exercise 4.10 In the same setting as Exercise 4.5, obtain the swap rate with maturity $T = 10$. Assume that the swap starts now and the interests are exchanged semiannually. Interpolate the discount bond prices by a straight line if necessary.

CHAPTER 5

A Discrete-Time Model for Securities Market

This chapter overviews a general discrete-time model for securities market in order to introduce various concepts important in financial engineering. In particular, self-financing and replicating portfolios play key roles for the pricing of derivative securities. The reader should consult Pliska (1997) for complete discussions of the discrete-time model.

5.1 Price Processes

A family of random variables $\{X(t); t \in \mathcal{T}\}$ (or $\{X(t)\}$ for short) parameterized by time $t \in \mathcal{T}$ is called a *stochastic process*. A stochastic process is a mathematical tool to model a system that varies randomly in time. For example, we cannot predict the future behavior of security prices completely. If we could, nobody would make a gain, and there would be no incentives to trade financial securities. In fact, as we shall see, uncertainty is the most important factor for finance theory. The theory of stochastic processes has been used as the key tool to describe uncertainty.

Let $(\Omega, \mathcal{F}, P)$ be a probability space, and consider a stochastic process $\{X(t)\}$ defined on it. For each $t \in \mathcal{T}$, $X(t)$ is a random variable and the dependence structure between $X(t)$ is given in terms of the joint distribution

$$P\{X(t_1) \leq x_1, X(t_2) \leq x_2, \ldots, X(t_n) \leq x_n\},$$

where $t_1 < t_2 < \cdots < t_n$, $x_i \in \mathbf{R}$, $i = 1, 2, \ldots, n$, and $n = 1, 2, \ldots$. On the other hand, for each $\omega \in \Omega$, $X(t, \omega)$ is a realization of $X(t)$ and a function of time t. The real-valued function is called a *sample path* or a *realization* of the process $\{X(t)\}$.

Let the set of time epochs consist of $\mathcal{T} = \{0, 1, \ldots, T\}$, where $T < \infty$ and where the time $t = 0$ denotes the current time. Consider a financial market in which there are available $n + 1$ securities, the one (numbered 0) being a risk-free security (e.g. the default-free money-market account) and the others (numbered $1, 2, \ldots, n$) being risky securities (e.g. stocks), where $n \geq 1$.

Throughout this book, the financial market is assumed to be *frictionless*, i.e. there are no transactions costs or taxes, all securities are perfectly divisible, short sales of all securities are allowed without restriction, and the borrowing and lending rates of the risk-free security are the same. Moreover, all investors in the market are *price-takers* in the sense that their actions do not affect the probability distribution of prices of the available securities.

Let $S_i(t)$ denote the time t price of security i, $i = 0, 1, \ldots, n$, where the initial price $S_i(0)$ is known by all investors in the market. While the future value $S_0(t)$ of the risk-free security is a positive random variable, the future risky security prices $S_i(t)$, $i = 1, 2, \ldots, n$, are non-negative random variables. We denote the security prices in vector form* by

$$\mathbf{S}(t) = (S_0(t), S_1(t), \ldots, S_n(t))^\top, \quad t = 0, 1, \ldots, T.$$

The multivariate stochastic process $\{\mathbf{S}(t); t = 0, 1, \ldots, T\}$ (or $\{\mathbf{S}(t)\}$ for short) defined on $(\Omega, \mathcal{F}, P)$ is called a *price process*. In this chapter, we do not specify the stochastic process $\{\mathbf{S}(t)\}$, i.e. the dependence structure for the random variables, explicitly. Important special cases of price processes in financial engineering will be explained in the following chapters.

For a moment, we denote the information about security prices available in the market at time t by $\mathcal{F}_t$. For example, $\mathcal{F}_t$ is the smallest σ-field containing $\{\mathbf{S}(u); u = 0, 1, \ldots, t\}$. However, the information can be any as far as the time t security prices can be known based on the information and

$$\mathcal{F}_0 \subset \mathcal{F}_1 \subset \cdots \subset \mathcal{F}_T \subset \mathcal{F}. \tag{5.1}$$

Here, two σ-fields $\mathcal{G}$ and $\mathcal{H}$ are ordered as $\mathcal{G} \subset \mathcal{H}$ if and only if $A \in \mathcal{G}$ implies $A \in \mathcal{H}$. The sequence of information $\{\mathcal{F}_t; t = 0, 1, \ldots, T\}$ (or $\{\mathcal{F}_t\}$ for short) satisfying (5.1) is called a *filtration*.

Definition 5.1 A random variable X is said to be $\mathcal{F}_t$-*measurable* (or measurable with respect to $\mathcal{F}_t$) if $\{x_1 < X \leq x_2\} \in \mathcal{F}_t$ for any $x_1 < x_2$.

If the random variable X is $\mathcal{F}_t$-measurable, you can determine whether or not the event $\{x_1 < X \leq x_2\}$ occurs simply by examining $\mathcal{F}_t$ for any $x_1 < x_2$. That is, roughly speaking, we know the value of X given the information $\mathcal{F}_t$. The time t security prices $S_i(t)$ must be $\mathcal{F}_t$-measurable in our setting. For time $u > t$, the event $\{x_1 < S_i(u) \leq x_2\}$ cannot be determined by examining $\mathcal{F}_t$; we know its probability only.

Let $\theta_i(t)$, $t = 1, 2, \ldots, T$, denote the number of security i carried from time $t - 1$ to t, and let $\boldsymbol{\theta}(t) = (\theta_0(t), \theta_1(t), \ldots, \theta_n(t))^\top$. The vector $\boldsymbol{\theta}(t)$ is called a *portfolio* at time t, and the process $\{\boldsymbol{\theta}(t); t = 1, 2, \ldots, T\}$ is called a portfolio process. Note that the choice of $\boldsymbol{\theta}(t)$ may change according to the actual price process. Since we do not know the future price process now, the portfolios $\boldsymbol{\theta}(t)$, $t = 2, 3, \ldots, T$, are multivariate random variables. Of course, when determining the portfolio, we cannot use the future information about the price process. That is, the information we can use to decide $\boldsymbol{\theta}(t)$ must be $\mathcal{F}_{t-1}$ only. In other words, the time t portfolio $\boldsymbol{\theta}(t)$ is $\mathcal{F}_{t-1}$-measurable.

Definition 5.2 A stochastic process $\{X(t); t = 0, 1, \ldots, T\}$ is said to be *adapted* to the filtration $\{\mathcal{F}_t\}$, if each $X(t)$ is measurable with respect to $\mathcal{F}_t$. The process $\{X(t)\}$ is called *predictable* if $X(t)$ is measurable with respect to $\mathcal{F}_{t-1}$ for all $t = 1, 2, \ldots, T$.

---

* Throughout this book, every vector is a column vector and the transpose ⊤ is used to denote a row vector.

PRICE PROCESSES

In the discrete-time securities market, the price process $\{\mathbf{S}(t)\}$ is adapted to the filtration $\{\mathcal{F}_t\}$, while the portfolio process $\{\boldsymbol{\theta}(t)\}$ is predictable with respect to $\{\mathcal{F}_t\}$.

Let $d_i(t)$, $t = 1, 2, \ldots, T$, denote the dividend that security i pays out at time t. The cumulative dividend paid by security i until time t is denoted by $D_i(t) = \sum_{s=1}^{t} d_i(s)$. Of course, the future dividends are random variables, and so the dividend processes $\{d_i(t)\}$ are stochastic processes adapted to the filtration $\{\mathcal{F}_t\}$.

The *value process* $\{V(t); t = 0, 1, \ldots, T\}$ is a stochastic process defined by

$$V(t) = \sum_{i=0}^{n} \theta_i(t)\{S_i(t) + d_i(t)\}, \quad t = 1, 2, \ldots, T, \tag{5.2}$$

with $V(0) = \sum_{i=0}^{n} \theta_i(1) S_i(0)$. Hence, $V(0)$ is the initial portfolio value and $V(t)$, $t \geq 1$, is the time t value of the portfolio before any transactions are made at that time. The dividends are added to the portfolio value, since the gains can be used to construct a new portfolio at the next time.† Note that the value process $\{V(t)\}$ is adapted to the filtration $\{\mathcal{F}_t\}$, since the price and dividend processes are adapted to $\{\mathcal{F}_t\}$ and the portfolio process is predictable.

Definition 5.3 (Self-Financing Portfolio) A portfolio process $\{\boldsymbol{\theta}(t); t = 1, 2, \ldots, T\}$ is said to be *self-financing* if

$$V(t) = \sum_{i=0}^{n} \theta_i(t+1) S_i(t), \quad t = 0, 1, \ldots, T-1. \tag{5.3}$$

In Equation (5.3), the left-hand side is the portfolio value at time t that represents the value of the portfolio *before* any transactions are made, while the right-hand side is the time t value of the portfolio just *after* any transactions take place at that time. The self-financing portfolio means that the values of the portfolio just before and after any transactions are equal at any time. Intuitively, if no money is added to or withdrawn from the portfolio, then any change in the portfolio's value must be due to a gain or loss in the investment.

In some cases, it is more convenient to consider the *rate of return* of a security rather than the price itself. To be more specific, let $S_i(t)$ be the time t price of security i, and define

$$R_i(t) = \frac{\Delta S_i(t) + d_i(t+1)}{S_i(t)}, \quad t = 0, 1, \ldots, T-1, \tag{5.4}$$

where we denote

$$\Delta S_i(t) \equiv S_i(t+1) - S_i(t)$$

for the change in the value between time t and time $t+1$. The process $\{R_i(t); t = 0, 1, \ldots, T-1\}$ is called the *return process* of security i. Note that, given the rate of return $R_i(t)$, we have

$$S_i(t+1) = (1 + R_i(t)) S_i(t) - d_i(t+1). \tag{5.5}$$

† Throughout this book, we will assume that the security prices are ex-dividend.

Hence, if we know the initial price $S_i(0)$, the dividend process $\{d_i(t)\}$, and the return process $\{R_i(t)\}$, then we can obtain the price process $\{S_i(t)\}$ recursively from (5.5).

Consider the change in the portfolio value between time t and time $t+1$, i.e. $\Delta V(t) \equiv V(t+1) - V(t)$. If the portfolio is self-financing, we must have from (5.2) and (5.3) that

$$\Delta V(t) = \sum_{i=0}^{n} \theta_i(t+1)\{\Delta S_i(t) + d_i(t+1)\}, \quad t = 0, 1, \ldots, T-1. \quad (5.6)$$

Conversely, assuming (5.6) implies that the portfolio process $\{\boldsymbol{\theta}(t)\}$ is self-financing. Hence, we can use (5.6) as an alternative definition of self-financing portfolios.

From (5.4), we have

$$\Delta S_i(t) + d_i(t+1) = S_i(t)R_i(t).$$

Thus, letting

$$w_i(t) \equiv \frac{\theta_i(t+1)S_i(t)}{V(t)}, \quad i = 0, 1, \ldots, n,$$

the rate of return of the portfolio at time t is given by

$$R(t) \equiv \frac{\Delta V(t)}{V(t)} = \sum_{i=0}^{n} w_i(t)R_i(t), \quad t = 0, 1, \ldots, T-1. \quad (5.7)$$

Here, $\sum_{i=0}^{n} w_i(t) = 1$, since $V(t) = \sum_{i=0}^{n} \theta_i(t+1)S_i(t)$. Hence, the rate of return of the portfolio is a linear combination of the rates of return of the securities under consideration. This is one of the reasons why we often consider the return processes rather than the price processes. Note, however, that the weights $w_i(t)$ are $\mathcal{F}_t$-measurable random variables.

5.2 The Portfolio Value and Stochastic Integral

We continue to consider the same setting as in the previous section. Suppose that the portfolio process $\{\boldsymbol{\theta}(t)\}$ is self-financing, and let

$$G_i(t) \equiv S_i(t) + D_i(t), \quad t = 0, 1, \ldots, T,$$

for each $i = 0, 1, \ldots, n$. The quantity $G_i(t)$ represents the *gain* obtained from security i until time t. Since

$$\Delta G_i(t) = \Delta S_i(t) + d_i(t+1),$$

it follows from (5.6) that

$$\Delta V(t) = \sum_{i=0}^{n} \theta_i(t+1)\Delta G_i(t), \quad t = 0, 1, \ldots, T-1, \quad (5.8)$$

whence

$$V(t) = V(0) + \sum_{i=0}^{n} \sum_{u=0}^{t-1} \theta_i(u+1)\Delta G_i(u), \quad t = 1, 2, \ldots, T. \quad (5.9)$$

Here, the sum

$$I_i(t) \equiv \sum_{u=0}^{t-1} \theta_i(u+1)\Delta G_i(u), \quad t>0, \tag{5.10}$$

is called the (discrete) *stochastic integral* of $\{\theta_i(t)\}$ with respect to the stochastic process $\{G_i(t)\}$. Equation (5.9) is the key formula, showing that the future portfolio value is a sum of the initial portfolio value and the stochastic integrals with respect to the gain processes. In particular, when the securities pay no dividends, the portfolio value is reduced to a sum of the initial portfolio value and the stochastic integrals with respect to the price processes. That is, we have

$$V(t) = V(0) + \sum_{i=0}^{n} \sum_{u=0}^{t-1} \theta_i(u+1)\Delta S_i(u), \quad t=1,2,\ldots,T.$$

Conversely, suppose that the future portfolio value $V(t)$ is represented by (5.9) in terms of a portfolio process $\{\boldsymbol{\theta}(t)\}$. Then, it can be shown that the portfolio is self-financing. The proof is left in Exercise 5.3. We thus have the following.

Theorem 5.1 *The future portfolio value $V(t)$ is represented by (5.9) in terms of a portfolio process $\{\boldsymbol{\theta}(t); t=1,2,\ldots,T\}$ if and only if the portfolio process is self-financing.*

In some cases, it is more convenient to consider a denominated price process than the price process itself. To be more specific, let $\{Z(t); t=0,1,\ldots,T\}$ be a positive process, and consider the denominated prices $S_i^*(t) \equiv S_i(t)/Z(t)$. Such a denominating positive process is called a *numeraire*. The numeraire should be carefully chosen according to the problem under consideration. Here, we choose the risk-free security $S_0(t)$ as the numeraire. Then, $S_0^*(t) = 1$ for all $t=0,1,\ldots,T$. In the following, it is assumed that the risk-free security $S_0(t)$ pays no dividends, i.e. $d_0(t) = 0$ for all $t=1,2,\ldots,T$.

Let $d_i^*(t) = d_i(t)/S_0(t)$, and consider the denominated value process $\{V^*(t); t=0,1,\ldots,T\}$, where $V^*(t) = V(t)/S_0(t)$. From (5.2), the denominated value process is given by

$$V^*(t) = \sum_{i=0}^{n} \theta_i(t)\{S_i^*(t) + d_i^*(t)\}, \quad t=1,2,\ldots,T, \tag{5.11}$$

with $V^*(0) = \sum_{i=0}^{n} \theta_i(1)S_i^*(0)$. Note from (5.3) that the portfolio process $\{\boldsymbol{\theta}(t)\}$ is self-financing if and only if

$$V^*(t) = \sum_{i=0}^{n} \theta_i(t+1)S_i^*(t), \quad t=1,2,\ldots,T, \tag{5.12}$$

in the denominated setting.

Now, suppose that the portfolio process $\{\boldsymbol{\theta}(t)\}$ is self-financing. Then, since

$\Delta S_0^*(t) = 0$, we have from (5.11) and (5.12) that

$$V^*(t+1) - V^*(t) = \sum_{i=1}^{n} \theta_i(t+1)\{\Delta S_i^*(t) + d_i^*(t+1)\},$$

where $\Delta S_i^*(t) = S_i^*(t+1) - S_i^*(t)$. Hence, letting

$$\Delta G_i^*(t) \equiv \Delta S_i^*(t) + d_i^*(t+1), \quad t = 0, 1, \ldots, T-1, \qquad (5.13)$$

it follows that

$$V^*(t) = V^*(0) + \sum_{i=1}^{n} \sum_{u=0}^{t-1} \theta_i(u+1)\Delta G_i^*(u), \quad t = 1, 2, \ldots, T. \qquad (5.14)$$

It should be noted here that the portfolio process $\{\boldsymbol{\theta}(t)\}$ in (5.14) is the same as that in (5.9). Also, in (5.13), $\Delta G_i^*(t) \neq G_i^*(t+1) - G_i^*(t)$ unless the risky securities pay no dividends; in which case, we have

$$V^*(t) = V^*(0) + \sum_{i=1}^{n} \sum_{u=0}^{t-1} \theta_i(u+1)\Delta S_i^*(u), \quad t = 1, 2, \ldots, T.$$

Make sure that, in (5.14), the number of securities is just n, not $n+1$ as in (5.9).

5.3 No-Arbitrage and Replicating Portfolios

A *contingent claim* is a random variable X representing a payoff at some future time T. You can think of a contingent claim as part of a contract that a buyer and a seller make at time $t = 0$. The seller promises to pay the buyer the amount $X(\omega)$ at time T if the state $\omega \in \Omega$ occurs in the economy. Hence, when viewed at time $t = 0$, the payoff X is a random variable. A typical example of contingent claims is European-type derivatives. The problem of interest is to determine the value of the payoff X at time $t = 0$. In other words, what is the fair price that the buyer should pay the seller at time $t = 0$? As we shall see below, the fair price of a contingent claim can be determined by no-arbitrage pricing arguments.

In the following two definitions, $V(T)$ denotes the portfolio value given by (5.9).

Definition 5.4 (Replicating Portfolio) A contingent claim X is said to be *attainable* if there exists some self-financing trading strategy $\{\boldsymbol{\theta}(t); t = 1, 2, \ldots, T\}$, called a *replicating portfolio*, such that $V(T) = X$. That is,

$$X = V(0) + \sum_{i=0}^{n} \sum_{t=0}^{T-1} \theta_i(t+1)\Delta G_i(t) \qquad (5.15)$$

for some self-financing portfolio process $\{\boldsymbol{\theta}(t)\}$. In this case, the portfolio process is said to *generate* the contingent claim X.

Let p be the price of X at time $t = 0$, and suppose that $p > V(0)$. Then,

NO-ARBITRAGE AND REPLICATING PORTFOLIOS

an astute individual would sell the contingent claim for p at time $t = 0$, follows the trading strategy $\{\boldsymbol{\theta}(t)\}$ at a cost $V(0)$, and obtain a positive profit $p - V(0)$. The individual has made a risk-free profit, since at time T the value $V(T)$ of the portfolio, if followed by the trading strategy, is exactly the same as the obligation X of the contingent claim in every state of the economy. Similarly, if $p < V(0)$, then an astute individual takes the opposite position so as to make a risk-free profit $V(0) - p > 0$.

Suppose $p = V(0)$. Then, we cannot use the trading strategy to make a risk-free profit. However, as shown in Pliska (1997), we still need a condition to guarantee that $V(0)$ is the correct price of X. For this purpose, the following condition suffices, the one of the key notions for the pricing of derivative securities.

Definition 5.5 (Arbitrage Opportunity) An *arbitrage opportunity* is the existence of some self-financing trading strategy $\{\boldsymbol{\theta}(t); t = 1, 2, \ldots, T\}$ such that (a) $V(0) = 0$, and (b) $V(T) \geq 0$ and $V(T) > 0$ with positive probability.

Note that an arbitrage opportunity is a risk-free means of making profit. You start with nothing and, without any chance of going into debt, there is a chance of ending up with a positive amount of money. If such an opportunity were to exist, then everybody would jump in with this trading strategy, affecting the prices of the securities. This economic model would not be in equilibrium. Hence, for the securities market model to be sensible from the economic standpoint, there cannot exist any arbitrage opportunities.

Example 5.1 In this example, we prove (4.13), i.e.

$$r(T) = f(t, T), \quad t \leq T,$$

for the deterministic interest-rate economy. To this end, suppose that the current time is t and assume $r(T) < f(t, T)$. Then, you sell one unit of the discount bond with maturity T and, at the same time, purchase the discount bond maturing at $T+1$ with the amount of $(1+f(t,T))$ units. From the result of Exercise 4.4(2), we have

$$1 + f(t, T) = \frac{v(t, T)}{v(t, T+1)},$$

whence this trading strategy has the zero initial cost. On the other hand, at date T, you pay 1 dollar and, at the same time, the value of your second bond becomes $v(T, T+1) = 1/(1 + r(T))$. The profit at time T is then given by

$$(1 + f(t, T))v(T, T+1) - 1 = \frac{f(t, T) - r(T)}{1 + r(T)} > 0,$$

which shows that there is an arbitrage opportunity. If $r(T) > f(t, T)$, you will have the opposite position to make profit. See Exercise 5.5 for another example of an arbitrage opportunity.

We are now ready to state one of the most important results in financial engineering.

Theorem 5.2 (No-Arbitrage Pricing) *For a given contingent claim X, suppose that there exists a replicating trading strategy $\{\boldsymbol{\theta}(t); t = 1, 2, \ldots, T\}$ as in (5.15). If there are no arbitrage opportunities in the market, then $V(0)$ is the correct price of the contingent claim X.*

According to Theorem 5.2, we need to find the initial cost $V(0)$ of the replicating portfolio $\{\boldsymbol{\theta}(t)\}$ given by (5.15) in order to price the contingent claim X under the assumption of no arbitrage opportunities. In general, such a portfolio is difficult to find. However, as in the following examples, there are important classes of contingent claims for which we can find replicating portfolios with ease.

Theorem 5.2 tells us more. Consider two self-financing portfolios, P1 and P2 say, that replicate each other. That is, suppose that the value of P1 at the maturity is the same as that of P2. Then, in order to prevent arbitrage opportunities, the initial costs of the two portfolios must be the same. We will use this idea in the following examples.

Example 5.2 (Forward Price) Based on the no-arbitrage argument, this example obtains the forward price (4.17). Suppose that the asset is a stock with *no dividends*, and let $S(t)$ be the time t price of the stock. Consider a forward contract with maturity T and delivery price K for the stock, and let f be the time t value of the forward contract, where $t < T$. Moreover, let $v(t, T)$ be the time t price of the default-free discount bond maturing at time T. Here, we consider the following two portfolios:

P1: Buy one forward contract and K units of the discount bond, and

P2: Buy one underlying security.

From (4.16), the gain or loss of the forward contract at maturity T is given by $S(T) - K$, while the payoff from the discount bond is K. It follows that the value of portfolio P1 at the maturity is

$$S(T) - K + K = S(T),$$

which is the same as the value of P2 at the maturity whatever happens. Therefore, by the no-arbitrage condition, the present values of the two portfolios are the same, and we obtain

$$f + Kv(t, T) = S(t).$$

Since the forward price $F_T(t)$ is the delivery price for which $f = 0$, it follows that

$$F_T(t) = \frac{S(t)}{v(t, T)}, \quad t \leq T,$$

whence we obtain (4.17). Note that we have made no assumptions on the prices $S(t)$ and $v(t, T)$. As far as the default-free discount bond $v(t, T)$ is available in the market, the above formula holds for the forward price of any asset $S(t)$ that pays no dividends.

Example 5.3 (Put–Call Parity) Recall that the payoff functions of options are non-linear, which makes the valuation of options difficult (see Exercise 5.8). However, there always exists a simple relationship between call

NO-ARBITRAGE AND REPLICATING PORTFOLIOS

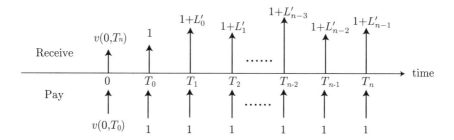

Figure 5.1 *The cashflows paid by the floating side*

and put options with the same strike price and the same maturity written on the same underlying security. This relationship can be obtained by a simple no-arbitrage argument as above.

Consider European call and put options written on security $S(t)$ with the same strike price K and the same maturity T, and let c and p be the premiums of the call and put options, respectively. We consider the following two portfolios:

P1: Buy one call and K units of the discount bond maturing at T, and

P2: Buy one put and one underlying security.

At the maturity T, the value of portfolio P1 becomes
$$\{S(T) - K\}_+ + K,$$
whereas the value of P2 is given by
$$\{K - S(T)\}_+ + S(T).$$
Here, we denote $\{x\}_+ = \max\{x, 0\}$. Note that the two values are the same from the result of Exercise 4.9. Hence, by the no-arbitrage condition, the present values of the two portfolios must be the same, and we obtain
$$c + Kv(t, T) = p + S(t), \quad t < T. \tag{5.16}$$
This relationship is called the *put–call parity* for European options.

Example 5.4 This example obtains the present value (4.25) of the cashflows for the floating side of the interest-rate swap considered in Example 4.5. To this end, we construct a trading strategy that replicates the cashflows paid by party B as follows. Consulting Figure 5.1, we consider the following trading strategy:

(1) At time 0, purchase the default-free discount bond with maturity T_0, and at the same time, sell the discount bond with maturity T_n.

(2) We receive 1 dollar at time T_0 from the discount bond. Invest this money into the LIBOR market with the interest rate $L(T_0, T_1)$.

(3) At time T_1, we receive $(1 + L'_0)$ dollars, where $L'_0 = \delta L(T_0, T_1)$ denotes the interest. Pay L'_0 as the interest, and invest 1 dollar into the LIBOR market with the interest rate $L(T_1, T_2)$.

⋮

(4) At time T_{n-1}, pay the interest $L'_{n-2} = \delta L(T_{n-2}, T_{n-1})$ and invest 1 dollar into the LIBOR market with the interest rate $L(T_{n-1}, T_n)$.

(5) At the maturity T_n, pay the interest $L'_{n-1} = \delta L(T_{n-1}, T_n)$. Also, pay 1 dollar to the bond holder.

Clearly, all the cashflows equal to the floating interest rates are paid by this trading strategy. The initial cost of the trading strategy consists of the two discount bonds, which is given by (4.25), as desired. Note that no assumptions have been made about the discount bond prices $v(t,T)$ and the LIBOR rates $L(t, t+\delta)$. See Exercise 5.9 for the case of equity swaps.

5.4 Martingales and the Asset Pricing Theorem

We have seen that the fair price of a contingent claim is the initial cost of a self-financing portfolio that generates the claim under the no-arbitrage condition. However, we have not mentioned how to construct such a portfolio in general and, in particular, we do not know how to calculate the initial cost. The initial price $V(0)$ in (5.15) can be calculated with the aid of the martingale concept.

Before proceeding, we need to extend the notion of conditional expectation $E[X|Y]$ defined by (2.28). Recall that $\mathcal{F}_t$ represents the information available at time t. A random variable X is said to be $\mathcal{F}_t$-measurable if $\{x_1 < X \leq x_2\} \in \mathcal{F}_t$ for any $x_1 < x_2$. X is called *integrable* if $E[||X||] < \infty$. A stochastic process $\{X(t)\}$ is called integrable if $E[||X(t)||] < \infty$ for all t.

Definition 5.6 (Conditional Expectation) Let X be an integrable random variable. The *conditional expectation* of X given $\mathcal{F}_t$ is an $\mathcal{F}_t$-measurable random variable Z that satisfies

$$E[YZ] = E[YX] \quad \text{for any } \mathcal{F}_t\text{-measurable } Y. \tag{5.17}$$

The conditional expectation Z is denoted by $E[X|\mathcal{F}_t]$ or $E_t[X]$ for short.

Recall that the conditional expectation $E[X|Y]$ is a random variable associated with the expectation of X given the information about the random variable Y. Roughly speaking, $E_t[X]$ is a random variable associated with the expectation of X given the information about all $\mathcal{F}_t$-measurable random variables. But, no matter how you understand the conditional expectation, all we need to remember in financial engineering is the following. The next result is parallel to Proposition 2.7. The proof is left in Exercise 5.10.

Proposition 5.1 *Let X and Y be integrable random variables.*
(1) *If X is independent of $\mathcal{F}_t$, then $E_t[X] = E[X]$.*
(2) *For any $\mathcal{F}_t$-measurable random variable Z, we have*

$$E_t[XZ] = ZE_t[X].$$

(3) *For real numbers a and b, we have*

$$E_t[aX + bY] = aE_t[X] + bE_t[Y].$$

In Proposition 5.1(1), X is said to be independent of $\mathcal{F}_t$, if X is independent of all $\mathcal{F}_t$-measurable random variables. That is, the probability distribution of X is not affected by giving the information $\mathcal{F}_t$. Also, in Property (2), since random variable Z is $\mathcal{F}_t$-measurable, it has no randomness given the information $\mathcal{F}_t$ and so, Z can be treated as if it were a constant. Property (3) is called the *linearity* of conditional expectation, as before.

Proposition 5.2 *Let X and Y be any random variables.*

(1) *If $f_1(x) \le f_2(x)$ for all x, then $E_t[f_1(X)] \le E_t[f_2(X)]$, for which the conditional expectations exist.*

(2) *For any convex function, we have $f(E_t[X]) \le E_t[f(X)]$, for which the conditional expectations exist.*

Proposition 5.2 is parallel to Proposition 2.8 and the proof is omitted. Property (1) is called the *monotonicity* of conditional expectation, while Property (2) is *Jensen's inequality* for conditional expectations.

The next result is parallel to Proposition 2.9 and the formula is often called the *chain rule* of conditional expectations. Note that E_0 means the unconditional expectation E, since $\mathcal{F}_0 = \{\emptyset, \Omega\}$ includes no information at all.

Proposition 5.3 *Let $h(x)$ be any real-valued function. For $s < t$, we have*

$$E_s[E_t[h(X)]] = E_s[h(X)], \tag{5.18}$$

for which the conditional expectations exist.

We are now in a position to introduce the martingale concept.

Definition 5.7 (Martingale) An integrable stochastic process $\{X(t); t = 0, 1, \cdots, T\}$ defined on the probability space $(\Omega, \mathcal{F}, P)$ with filtration $\{\mathcal{F}_t\}$ is called a *martingale* if

$$E_t[X(t+1)] = X(t), \quad t = 0, 1, \ldots, T-1. \tag{5.19}$$

Recall that $E_t[X(t+1)]$ is the expectation of $X(t+1)$ given the information $\mathcal{F}_t$ up to time t. Hence, Property (5.19) implies that, given the information $\mathcal{F}_t$, the mean of the future value $X(t+1)$ is the same as the current value $X(t)$. This is the notion of the so-called *fair game* in mathematics. See Exercises 5.11 and 5.12 for basic properties of martingales.

If the process $\{X(t)\}$ is a martingale, each $X(t)$ must be $\mathcal{F}_t$-measurable by the definition of conditional expectations. It follows from (5.19) that

$$E_t[\Delta X(t)] = 0, \quad t = 0, 1, \ldots, T-1, \tag{5.20}$$

where $\Delta X(t) = X(t+1) - X(t)$. That is, an integrable process $\{X(t)\}$ is a martingale if and only if the conditional expectation of the change in the value is zero.

Example 5.5 (Doob's Martingale) For an integrable random variable X and a given filtration $\{\mathcal{F}_t\}$, define

$$X(t) = E_t[X], \quad t = 0, 1, \ldots, T. \tag{5.21}$$

It is easily seen from Jensen's inequality (Proposition 5.2(2)) that
$$|X(t)| = |E_t[X]| \leq E_t[|X|].$$
Hence, from the chain rule (5.18), we conclude that the process $\{X(t)\}$ is integrable. Property (5.19) follows from (5.18) again, since
$$E_t[X(t+1)] = E_t[E_{t+1}[X]] = E_t[X] = X(t).$$
Therefore, the process $\{X(t)\}$ is a martingale.

Importance of martingales in financial engineering is mostly due to the following result. Consider the discrete stochastic integral
$$I(t) = \sum_{n=0}^{t-1} \theta(n+1)\Delta X(n), \quad t = 1, 2, \ldots, T, \tag{5.22}$$
with $I(0) = 0$. If the process $\{X(t)\}$ is a martingale and if $\{I(t)\}$ is integrable for the predictable process $\{\theta(t)\}$, then the stochastic integral $\{I(t)\}$ is also a martingale.[‡] To see this, we have
$$\Delta I(t) = \theta(t+1)\Delta X(t),$$
where $\Delta I(t) = I(t+1) - I(t)$. It follows that
$$E_t[\Delta I(t)] = \theta(t+1)E_t[\Delta X(t)] = 0,$$
since $\theta(t+1)$ is $\mathcal{F}_t$-measurable and the process $\{X(t)\}$ is a martingale. The next proposition summarizes.

Proposition 5.4 *Suppose that the process $\{X(t)\}$ is a martingale and the process $\{\theta(t)\}$ is predictable. If $\{I(t)\}$ defined by (5.22) is integrable, then the stochastic integral $\{I(t)\}$ is a martingale.*

The initial cost $V(0)$ in (5.15) can now be calculated as follows. Suppose that $S_0(0) = 1$, and consider the denominated processes $\{G_i^*(t); t = 0, 1, \cdots, T\}$ with $S_0(t)$ as the numeraire. Then, from (5.14) and (5.15), we obtain
$$\frac{X}{S_0(T)} = V(0) + \sum_{i=1}^{n} \sum_{t=0}^{T-1} \theta_i(t+1)\Delta G_i^*(t). \tag{5.23}$$
Suppose now that there exists a probability measure P^* under which the processes $\{G_i^*(t)\}$ are martingales. That is, under P^*, we have
$$E_t^*[S_i^*(t+1) + d_i^*(t+1)] = S_i^*(t), \quad t = 0, 1, \ldots, T-1, \tag{5.24}$$
for all $i = 1, 2, \ldots, n$. Here, E_t^* denotes the conditional expectation operator under P^*, not under the original probability measure P, given $\mathcal{F}_t$. Then, from Proposition 5.4 and (5.23), we obtain
$$V(0) = E^*\left[\frac{X}{S_0(T)}\right], \tag{5.25}$$
where $E^* = E_0^*$, under the regularity condition.

[‡] Some authors call this result the *martingale transformation*.

MARTINGALES AND THE ASSET PRICING THEOREM

Definition 5.8 (Risk-Neutral Probability Measure) Given a probability space $(\Omega, \mathcal{F}, P)$ with filtration $\{\mathcal{F}_t; t = 0, 1, \ldots, T\}$, a probability measure P^* is said to be *risk-neutral* if

(RN1) P^* is *equivalent* to P, i.e. $P(A) > 0$ if and only if $P^*(A) > 0$ for all $A \in \mathcal{F}$, and

(RN2) Equation (5.24) holds for all i and t with $S_0(t) = B(t)$, the money-market account.

If, in particular, the securities pay no dividends, then from (5.24) the denominated price processes $\{S_i^*(t)\}$, where $S_i^*(t) = S_i(t)/B(t)$, are martingales under P^*. By this reason, the risk-neutral probability measure is often called a *martingale measure*.

The existence of such a probability measure and the uniqueness of the price are guaranteed by the following crucial result, called the *fundamental theorem of asset pricing*. Note that the theorem fails to hold when the time horizon is infinite. See Pliska (1997) for details.

Theorem 5.3 (Asset Pricing Theorem) *There are no arbitrage opportunities if and only if there exists a risk-neutral probability measure. If this is the case, the price of an attainable contingent claim X is given by (5.25) with $S_0(t) = B(t)$ for every replicating trading strategy.*

According to Theorem 5.3, the no-arbitrage pricing of contingent claim X amounts to the following two steps: Let $S_0(t) = B(t)$.

(AP1) Find a risk-neutral probability measure P^*, and

(AP2) Calculate the expectation (5.25) under P^*.

This pricing scheme is called the *risk-neutral method*. The problem of how to find a risk-neutral probability measure for a particular case will be discussed in the following chapters.

Example 5.6 The default-free discount bond is a security that promises to pay a single cashflow of face value F at maturity T. Since the face value is repaid no matter what happens, once a risk-neutral probability measure P^* is found out, the price of the discount bond is given, from (5.25) with $S_0(t) = B(t)$, as

$$v(0, T) = E^* \left[\frac{F}{B(T)} \right], \quad t \leq T,$$

under the assumptions of Theorem 5.3. If, in particular, the face value is taken as $F = 1$ and the current time is t, we have

$$v(t, T) = E_t^* \left[\frac{B(t)}{B(T)} \right], \quad t \leq T, \tag{5.26}$$

since $B(t)$ is $\mathcal{F}_t$-measurable.

Finally, we state a stronger version of the fundamental theorem of asset pricing without proof.

Definition 5.9 (Complete Market) A securities market is said to be *complete* if every contingent claim is attainable; otherwise, the market is said to be *incomplete*.

Theorem 5.4 *A securities market is complete if and only if there exists a unique risk-neutral probability measure.*

5.5 American Options

While European options (or contingent claims) can only exercise the right at the maturity, the payoff of an American option can occur at any time on or before the maturity. This section considers the pricing of American options.

Let $Y(t)$ denote the time t payoff of an American option with maturity T. For example, we have $Y(t) = \{S_i(t) - K\}_+$, where $\{x\}_+ = \max\{x, 0\}$, if a call option with strike price K is written on the security $S_i(t)$. Similarly, the payoff may be defined as

$$Y(t) = \{K - I(t)\}_+, \quad I(t) \equiv \frac{1}{n}\sum_{i=1}^{n} S_i(t),$$

if a put option is written on the index $I(t)$ of the risky securities $S_i(t)$. An American option can only be exercised once. If it is never exercised, then no payoff occurs. Throughout this section, we denote the time t value of the American option by $Z(t)$.

Note that, since the exercise decision can be postponed until the maturity T, the value $Z(t)$ is at least as large as the value $V(t)$ of the European option that has payoff $X = Y(T)$ at the maturity. In addition, the possibility of being able to obtain a desirable payoff at an earlier time tends to make American options more valuable than the European counterparts. However, there are important situations that the two values coincide with each other in theory. For example, if $V(t) \geq Y(t)$ for all t, then $Z(t) = V(t)$ for all t, and it is optimal to wait until maturity. The proof is left in Exercise 5.13. See Example 5.7 below for such an example.

In order to consider the possibility of early exercises, it is necessary to introduce the notion of stopping times. See Exercise 5.14 for examples of stopping times.

Definition 5.10 (Stopping Time) For a given probability space $(\Omega, \mathcal{F}, P)$ with filtration $\{\mathcal{F}_t; t = 0, 1, \ldots, T\}$, a *stopping time* is a random variable τ taking values in the set $\{0, 1, \ldots, T, \infty\}$ such that the event $\{\tau = t\}$ belongs to $\mathcal{F}_t$ for each $t \leq T$.

If τ is a stopping time with respect to the filtration $\{\mathcal{F}_t\}$, we can evaluate whether or not the event $\{\tau = t\}$ occurs simply by examining $\mathcal{F}_t$, the information available at time t. We denote the set of stopping times that take values in the set $\{t, t+1, \ldots, T\}$ by $\mathcal{T}_t$. The next result is often useful for dealing with American options. We state it without proof.

Proposition 5.5 (Optional Sampling Theorem) *Consider a stochastic*

AMERICAN OPTIONS

process $\{X(t)\}$ defined on a probability space $(\Omega, \mathcal{F}, P)$ with filtration $\{\mathcal{F}_t\}$. If the process is a martingale, then for any stopping time $\tau \in \mathcal{T}_0$ we have

$$E[X(\tau)] = E[X(t)], \quad t = 0, 1, \ldots, T.$$

Recall that the price of a European option X is given by (5.25) if X is attainable. A similar situation exists for American options. An American option Y is said to be *attainable* if, for each stopping time $\tau \leq T$, there exists a self-financing portfolio that replicates the claim $Y(\tau)$. In the rest of this section, we take $S_0(t) = B(t)$, the money-market account, as the numeraire and assume that there exists a risk-neutral probability measure P^*. The denominated prices are denoted by, e.g. $S_i^*(t) = S_i(t)/B(t)$, as before.

Example 5.7 (American Call) Consider a stock that pays no dividends. This example demonstrates that, for an American call option written on the stock, it is optimal to wait until maturity. To see this, note that the time t price of the European call option with strike price K and maturity T is given, from (5.25) with $S_0(t) = B(t)$, by

$$V(t) = B(t) E_t^* \left[\frac{\{S(T) - K\}_+}{B(T)} \right], \quad t \leq T,$$

where $S(t)$ denotes the time t stock price. Since the function $\{x\}_+ = \max\{x, 0\}$ is convex in x, Jensen's inequality (Proposition 5.2(2)) shows that

$$B(t) E_t^* \left[\frac{\{S(T) - K\}_+}{B(T)} \right] \geq \{B(t) E_t^*[S^*(T)] - K v(t, T)\}_+,$$

where we have used (5.26). But, since the denominated process $\{S^*(t)\}$ is a martingale under P^*, we have $E_t^*[S^*(T)] = S^*(t)$. Since $v(t, T) \leq 1$, it follows that

$$E_t^* \left[\frac{\{S(T) - K\}_+}{B(T)} \right] \geq \frac{\{S(t) - K\}_+}{B(t)}, \quad t \leq T, \qquad (5.27)$$

which reveals that the process $\{\{S(t) - K\}_+/B(t); t = 0, 1, \ldots, T\}$ is a submartingale (see Exercise 5.15). Now, applying something similar to the above optional sampling theorem, we conclude that

$$E_t^* \left[\frac{\{S(T) - K\}_+}{B(T)} \right] \geq E_t^* \left[\frac{\{S(\tau) - K\}_+}{B(\tau)} \right]$$

for any stopping time $\tau \leq T$. Hence, no matter how you choose τ, the value of the American option cannot exceed the European counterpart. Alternatively, (5.27) implies that we are in a situation that $V(t) \geq Y(t)$ for all t. Hence, it is optimal to wait until maturity for American calls, as desired.

Given a risk-neutral probability measure P^*, define the adapted process $\{Z(t)\}$ by

$$Z(t) = \max_{\tau \in \mathcal{T}_t} E_t^*[Y^*(\tau)], \quad t = 0, 1, \ldots, T, \qquad (5.28)$$

where $Y^*(t) = Y(t)/B(t)$. Then, as is shown below, the stopping time

$$\tau(t) \equiv \min\{s \geq t : Z(s) = Y^*(s)\} \qquad (5.29)$$

maximizes the right-hand side of (5.28). Note that the right-hand side is the maximum (denominated) payoff obtained by an optimal exercise strategy. Hence, the fair price of the American option is given by $Z(0)$. The reason is based on the no-arbitrage argument similar to the one given for the European case. The proof is left to the reader.

Theorem 5.5 *Suppose that there exists a risk-neutral probability measure P^*, that the adapted process $\{Z(t); t = 0, 1, \ldots, T\}$ defined by (5.28) exists, and that the American option is attainable. Then, the value of the American option is equal to $Z(0)$ and the optimal exercise strategy $\tau(0)$ is given by (5.29).*

Finally, we describe a method for computing the value process $\{Z(t)\}$ defined by (5.28). The computation will be done backwards in time, based on the dynamic programming approach. See, e.g., Puterman (1994) for details of the dynamic programming theory.

As the first step, let $t = T$. Then, $Z(T) = Y^*(T)$ and $\tau(T) = T$. Next, for $t = T - 1$, we have

$$\begin{aligned} Z(T-1) &= \max\left\{Y^*(T-1), E_{T-1}^*[Z(T)]\right\} \\ &= \max\left\{Y^*(T-1), E_{T-1}^*[Y^*(T)]\right\}, \end{aligned}$$

which coincides with (5.28). The optimal exercise strategy (5.29) is also verified, since $\tau(T-1) = T-1$ if and only if $Z(T-1) = Y^*(T-1)$. In general, we have

$$Z(t-1) = \max\left\{Y^*(t-1), E_{t-1}^*[Z(t)]\right\}, \quad (5.30)$$

where the calculation is performed backwards, i.e. $t = T, T-1, \ldots, 2, 1$. It is not difficult to prove that $Z(t-1)$ given by (5.30) satisfies the equation (5.28), provided that the exercise strategy (5.29) is employed. The proof is left in Exercise 5.16. Hence, the value process $\{Z(t)\}$ can be computed according to (5.30) backwards in time with the initial condition $Z(T) = Y^*(T)$.

5.6 Change of Measure

We saw in (3.20) that we can define a new probability measure using the moment generating function of a random variable. In this section, we describe a general framework of this procedure, called the *change of measure*.

Consider a probability space $(\Omega, \mathcal{F}, P)$, and suppose that a random variable X defined on it is positive and integrable. Let $\eta = X/E[X]$, and define

$$\widetilde{P}(A) = E[\eta 1_A], \quad A \in \mathcal{F}, \quad (5.31)$$

where 1_A denotes the indicator function of event A. It is readily seen that $\widetilde{P}$ is a probability measure. Also, (5.31) can be rewritten as

$$\int_A \mathrm{d}\widetilde{P}(\omega) = \int_A \eta(\omega)\mathrm{d}P(\omega), \quad A \in \mathcal{F}. \quad (5.32)$$

Hence, $\widetilde{P}$ is equivalent to P, since $\eta > 0$. Conversely, it can be shown that, for given equivalent probability measures P and $\widetilde{P}$, there always exists a positive

CHANGE OF MEASURE

random variable η such that $E[\eta] = 1$ and (5.31) holds true. This result is known as the Radon–Nikodym theorem. Also, because of the expression in (5.32), the random variable η is usually denoted by $\mathrm{d}\widetilde{P}/\mathrm{d}P$ and called the *Radon–Nikodym derivative*.

Now, consider a contingent claim X with maturity T generated by the replicating portfolio $\{\boldsymbol{\theta}(t)\}$ given in (5.15). In order to evaluate the initial cost $V(0)$, we take $S_0(t) = v(t,T)$, the default-free discount bond maturing at time T, instead of the money-market account $B(t)$. The denominated processes are denoted by

$$S_i^T(t) = \frac{S_i(t)}{v(t,T)}, \quad t = 0, 1, \ldots, T,$$

in order to emphasize that the numeraire is now the discount bond $v(t,T)$. Note that $S_i^T(t)$ is the forward price given by (4.17) if $S_i(t)$ pays no dividends. Using the exact same arguments as for (5.23), we can obtain

$$X = \frac{V(0)}{v(0,T)} + \sum_{i=1}^{n} \sum_{t=0}^{T-1} \theta_i(t+1) \Delta G_i^T(t), \qquad (5.33)$$

since $v(T,T) = 1$, where $\Delta G_i^T(t) = \Delta S_i^T(t) + d_i^T(t+1)$, etc. The proof is left to the reader.

Suppose that there exists a probability measure P^T under which the processes $\{G_i^T(t)\}$ are martingales. That is, under P^T, we have

$$E_t^T \left[S_i^T(t+1) + d_i^T(t+1) \right] = S_i^T(t), \quad t = 0, 1, \ldots, T-1. \qquad (5.34)$$

Here, E_t^T denotes the conditional expectation operator under P^T given $\mathcal{F}_t$. Then, from Proposition 5.4 and (5.33), we obtain

$$V(0) = v(0,T) E^T[X], \qquad (5.35)$$

where $E^T = E_0^T$, under the regularity condition.

Definition 5.11 (Forward-Neutral Probability Measure) Given a probability space $(\Omega, \mathcal{F}, P)$ with filtration $\{\mathcal{F}_t; t = 0, 1, \ldots, T\}$, a probability measure P^T is said to be *forward-neutral* if

(FN1) P^T is equivalent to P (hence, so is to P^*), and

(FN2) Equation (5.34) holds for all i and t with $S_0(t) = v(t,T)$, the default-free discount bond.

If, in particular, the securities pay no dividends, then from (5.34) the denominated price processes $\{S_i^T(t)\}$ are martingales under P^T. Since $S_i^T(t) = S_i(t)/v(t,T)$ is the forward price of $S_i(t)$, the probability measure P^T is called forward-neutral.

According to the asset pricing theorem (Theorem 5.3), the following method to price contingent claims is permissible: Suppose that there is available the default-free discount bond and that the risk-neutral pricing method is valid in the market. Then,

(FAP1) Find a forward-neutral probability measure P^T, and

(FAP2) Calculate the expectation (5.35) under P^T.

This pricing scheme is called the *forward-neutral method*. The problem of how to find a forward-neutral probability measure P^T for a particular case will be discussed later.

Note that, when the interest rates are deterministic, the risk-neutral method (5.25) with $S_0(t) = B(t)$ and the forward-neutral method (5.35) with $S_0(t) = v(t,T)$ are the same, since from (4.14) we have

$$v(t,T) = \frac{B(t)}{B(T)}, \quad t \leq T.$$

On the other hand, in a stochastic interest-rate economy, while the risk-neutral method requires the joint distribution of $(X, B(T))$ to evaluate the expectation (5.25), we need the *marginal* distribution of X only in the forward-neutral method (5.35). This is the advantage of the forward-neutral method to the risk-neutral method in a stochastic interest-rate economy.

Suppose that the contingent claim X is attainable, and there are no arbitrage opportunities in the market. Then, its time 0 price is given by (5.25) with $S_0(t) = B(t)$ for every replicating trading strategy. It follows from (5.25) and (5.35) that

$$v(0,T)E^T[X] = E^*\left[\frac{X}{B(T)}\right].$$

Taking $X = 1_A$ for any $A \in \mathcal{F}$, we then obtain

$$P^T(A) = E^*\left[\frac{1_A}{v(0,T)B(T)}\right], \quad A \in \mathcal{F}.$$

Comparing this with (5.31), we conclude that the Radon–Nikodym derivative is given by

$$\frac{dP^T}{dP^*} = \frac{1}{v(0,T)B(T)} = \frac{v(T,T)/v(0,T)}{B(T)/B(0)}. \tag{5.36}$$

See Geman, El Karoui, and Rochet (1995) for more general results in the change of measure.

5.7 Exercises

Exercise 5.1 Suppose that $\Omega = \{\omega_1, \omega_2\}$, $n = 2$, and $T = 1$. As the security price processes, we assume that $S_0(0) = 1$, $S_1(0) = S_2(0) = 5$ and

$$S_0(1,\omega_1) = 1.05, \quad S_1(1,\omega_1) = 5.5, \quad S_2(1,\omega_1) = 4.5,$$
$$S_0(1,\omega_2) = 1.02, \quad S_1(1,\omega_2) = 5.0, \quad S_2(1,\omega_2) = 5.4.$$

The securities pay no dividends. Denoting any self-financing portfolio by $(\theta_0, \theta_1, \theta_2)^\top$, calculate the value process $\{V(t); t = 0, 1\}$ for each ω_i, $i = 1, 2$. Verify that (5.9) holds for each state.

Exercise 5.2 In the same setting as Exercise 5.1, consider the denominated prices with numeraire $S_0(t)$. Obtain the denominated value process $\{V^*(t); t = 0, 1\}$ for each state. Also, verify (5.14).

EXERCISES

Exercise 5.3 Prove the 'only if' part in Theorem 5.1.

Exercise 5.4 In Definition 5.5, consider the denominated prices with numeraire $S_0(t)$ that pays no dividends. Prove that the self-financing portfolio $\{\boldsymbol{\theta}(t)\}$ is an arbitrage opportunity if and only if (a) $V^*(0) = 0$, and (b) $V^*(T) \geq 0$ and $V^*(T) > 0$ with positive probability.

Exercise 5.5 Suppose that $n = T = 1$ and $\Omega = \{\omega_1, \omega_2\}$ with $P(\omega_i) > 0$. For the security price processes, we assume that $S_0(0) = 1$, $S_0(1, \omega_i) = 1.1$ for $i = 1, 2$, $S_1(0) = 5$, $S_1(1, \omega_1) = 5.4$ and $S_1(1, \omega_2) = 5.2$. Prove that there exists an arbitrage opportunity in this securities market.

Exercise 5.6 Prove that there are no arbitrage opportunities in the securities market considered in Exercise 5.1. Obtain the option premium for a call option written on the security $S_1(1)$ with strike price $K = 5.2$ and maturity $T = 1$.

Exercise 5.7 (Leverage) In the securities market considered in Exercise 5.1, suppose that $P(\omega_1) = P(\omega_2) = 1/2$. Calculate the mean rate of return for security $S_1(1)$. Also, compute the mean rate of return for the call option considered in Exercise 5.6. Explain why the mean rate of return for the option is much higher than that for the underlying security.

Exercise 5.8 (Merton's Bounds) Consider a European call option written on the security considered in Example 5.3. Let $C(t)$ be the time t price of the call option with maturity T and strike price K. Using a similar argument given in Example 5.2, prove that

$$\max\{S(t) - Kv(t,T), 0\} \leq C(t) \leq S(t).$$

Exercise 5.9 (Equity Swap) In the framework of Example 4.5, we want to use the rate of return of an equity instead of the LIBOR rate. Such a swap is called an *equity swap*. Let $I(t)$ be the time t price of an equity. In the 'plain vanilla' equity swap, two parties agree to exchange the cashflows based on the fixed interest rate S and those based on the return

$$\frac{I(T_{i+1}) - I(T_i)}{I(T_i)} = \frac{I(T_{i+1})}{I(T_i)} - 1, \quad i = 0, 1, \ldots, n-1.$$

Suppose that the equity swap starts at $t = 0$. Prove that the swap rate S is given by (4.26), i.e.

$$S = \frac{1 - v(0, T_n)}{\delta \sum_{i=1}^{n} v(0, T_i)}.$$

It should be noted that the equity swap rate S does not depend on the equity return. See Chance and Rich (1998) and Kijima and Muromachi (2001) for details of equity swaps.

Exercise 5.10 In Proposition 5.1(2), let $U = E_t[XZ]$. Then, from Proposition 5.3 and (5.17), we have

$$E[YU] = E[YXZ] \quad \text{for any } \mathcal{F}_t\text{-measurable } Y.$$

Using this and the fact that YZ is also $\mathcal{F}_t$-measurable, prove the property (2). Also, prove the other properties in Propositions 5.1 and 5.3 by using similar arguments.

Exercise 5.11 Suppose that the process $\{X(t)\}$ is a martingale with respect to the filtration $\{\mathcal{F}_t\}$. Prove that $E_t[X(t+n)] = X(t)$ for any $n \geq 1$ and that $E[X(t)] = E[X(0)]$ for any $t \geq 1$.

Exercise 5.12 (Doob's Decomposition) For an integrable process $\{X(t)\}$ adapted to the filtration $\{\mathcal{F}_t\}$, define
$$D(t) = X(t) - E_{t-1}[X(t)], \quad t = 1, 2, \ldots, T,$$
with $D(0) = 0$. Show that the process $\{M(t)\}$ defined by $M(t) = \sum_{k=0}^{t} D(k)$ is a martingale and the process $\{A(t)\}$ defined by $A(t) = X(t) - M(t)$ is predictable.

Exercise 5.13 Let $Y(t)$ be the time t payoff of an American option and let $Z(t)$ be its time t value. Also, let $V(t)$ denote the time t value of the European option that has payoff $X = Y(T)$ at the maturity T. Show that, if $V(t) \geq Y(t)$ for all t, then $Z(t) = V(t)$ for all t, and it is optimal to wait until maturity.

Exercise 5.14 For a process $\{X(t)\}$ defined on a probability space $(\Omega, \mathcal{F}, P)$ with filtration $\{\mathcal{F}_t; t = 0, 1, \ldots\}$, let $\tau = \min\{t : X(t) \in A\}$ for some $A \subset \mathbf{R}$. Prove that τ is a stopping time with respect to $\{\mathcal{F}_t\}$. Also, prove that if τ_1 and τ_2 are stopping times then so are both $\max\{\tau_1, \tau_2\}$ and $\min\{\tau_1, \tau_2\}$.

Exercise 5.15 (Submartingale) A process $\{X(t)\}$ adapted to the filtration $\{\mathcal{F}_t; t = 0, 1, \ldots, T\}$ is called a *submartingale* if the following two conditions are satisfied:

(SM1) For each $t = 0, 1, \ldots, T$, we have $E[\max\{X(t), 0\}] < \infty$, and

(SM2) $E_t[X(t+1)] \geq X(t)$ for $t = 0, 1, \ldots, T-1$.

Prove that, if $\{Y(t)\}$ is a martingale such that $E\left[e^{Y(t)}\right] < \infty$ for each t, then the process $\{X(t)\}$ defined by $X(t) = e^{Y(t)}$ is a submartingale. How about the process defined by $Z(t) = Y^2(t)$?

Exercise 5.16 Prove that $Z(t-1)$ given by (5.30) satisfies (5.28) when the exercise strategy (5.29) is used.

CHAPTER 6

Random Walks

Consider a particle moving around the integers. At time 0, the particle is at the origin and moves one step right or left with some probability. Such a mathematical model is called a *random walk* and used to describe random phenomena in diverse areas of physics, economics, and finance. In particular, random walks are the basis of the binomial model for the securities market, the main theme of the next chapter. This chapter provides a concise summary of the theory of random walks useful in finance. Note, however, that the theory of random walks is very rich, and the results provided here are indeed only 'the tip of the iceberg'. See, e.g., Feller (1957) for details.

Let $(\Omega, \mathcal{F}, P)$ be a probability space on which we work throughout this chapter. All the random variables considered here are defined on the probability space. However, until actually required, we do not specify it explicitly.

6.1 The Mathematical Definition

For IID random variables $X_1, X_2, \ldots$, define the *partial sum*

$$W_n \equiv X_1 + X_2 + \cdots + X_n, \quad n = 1, 2, \ldots, \tag{6.1}$$

with $W_0 = 0$. It is readily seen that

$$W_{n+1} = W_n + X_{n+1}, \quad n = 0, 1, 2, \ldots. \tag{6.2}$$

Hence, the random variables $W_1, W_2, \ldots$ are not independent, but $W_{n+1} - W_n$ is independent of W_n. More generally, $W_n - W_m$, $m < n$, is independent of W_m. The difference $W_n - W_m$ is called an *increment* and, hence, the process $\{W_n, n = 0, 1, \ldots\}$ is said to have *independent increments*.

An important special class of the partial-sum processes is a random walk. In order to give the definition, we need to extend the Bernoulli distributions given by (3.1). Suppose that the underlying IID random variables $X_1, X_2, \ldots$ follow the distribution

$$P\{X_n = x\} = 1 - P\{X_n = y\} = p, \quad 0 < p < 1, \tag{6.3}$$

where we assume $y \leq 0 < x$. The Bernoulli distribution (3.1) is a special case of (6.3) with $x = 1$ and $y = 0$.

Definition 6.1 (Random Walk) The partial-sum process $\{W_n\}$ with underlying IID random variables X_n defined by (6.3) is called a *random walk*.[*]

---

[*] In some probability textbooks, the process $\{W_n\}$ is called a random walk only if $x = 1$ and $y = -1$.

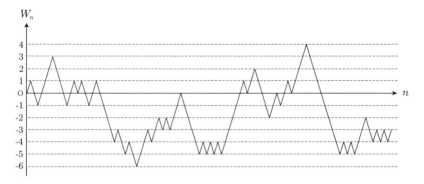

Figure 6.1 *A sample path of the random walk with state space $\mathcal{Z}$*

In particular, the random walk is said to be *symmetric* if $x = 1$, $y = -1$, and $p = 1/2$.

For the random walk $\{W_n\}$, it is readily seen that each W_n follows the binomial distribution $B(n, p)$, i.e.

$$P\{W_n = kx + (n-k)y\} = b_k(n, p), \quad k = 0, 1, \ldots, n, \tag{6.4}$$

where $b_k(n, p)$ denotes the binomial probability (3.2). Also, the mean and the variance of W_n are given, respectively, as

$$E[W_n] = n(px + qy)$$

and

$$V[W_n] = npq(x - y)^2,$$

where $q = 1 - p$. See Exercise 6.1 for the moment generating function (MGF) of W_n.

In the following, unless stated otherwise, we consider a random walk $\{W_n\}$ with $x = 1$ and $y = -1$ for notational simplicity. The sample space of the random variable W_n is therefore given by

$$\{-n, -n+1, \ldots, -1, 0, 1, \ldots, n-1, n\},$$

provided that there is no restriction on its movement. The union of the sample spaces is called the *state space* of the process. Hence, if the random walk has no restriction on its movement, its state space is given by the set of all integers

$$\mathcal{Z} \equiv \{0, \pm 1, \pm 2, \ldots\}.$$

A typical sample path of the random walk is depicted in Figure 6.1.

6.2 Transition Probabilities

Associated with the random walk $\{W_n\}$ is the conditional probability

$$u_{ij}(m, n) = P\{W_n = j | W_m = i\}, \quad m < n, \tag{6.5}$$

TRANSITION PROBABILITIES

where $i, j \in \mathcal{Z}$. This probability is called the *transition probability* from state i at time m to state j at time n. In this section, we derive the transition probabilities of the random walk $\{W_n\}$ with $x = 1$ and $y = -1$. The general case can be treated similarly.

Since X_{n+1} is independent of W_n and since X_n are IID, the *one-step* transition probability is given by

$$u_{ij}(n, n+1) = \begin{cases} p, & j = i+1, \\ q, & j = i-1, \\ 0, & \text{otherwise}, \end{cases} \tag{6.6}$$

where $q \equiv 1 - p$. The transition probability is *time-homogeneous* (or homogeneous for short) in the sense that it does not depend on time n. It is also spatially homogeneous, since it depends only on the difference $j - i$. Hence, we assume without loss of generality that $W_0 = 0$. If $W_0 = i_0 \neq 0$, we consider a random walk $\{\hat{W}_n\}$ defined by $\hat{W}_n = W_n - i_0$, and the desired results are obtained by adding i_0 to $\{\hat{W}_n\}$.

Suppose first that the random walk $\{W_n\}$ has no restriction on its movement. Since the random walk is temporally as well as spatially homogeneous, it is enough to consider the transition probabilities

$$u_j(n) \equiv P\{W_n = j | W_0 = 0\}, \quad n = 0, 1, 2, \ldots.$$

In particular, if the random walk is symmetric, i.e. $p = 1/2$, then we denote it by $v_j(n)$.

Now, since X_n takes values either 1 or -1, the random walk never returns to the initial state 0 in a time step of odd number. In general, it is readily seen that $u_j(n) = 0$ if $n + j$ is odd. Moreover, even though $n + j$ is an even number, the random walk can never reach such states j as $j < -n$ or $j > n$. It follows that

$$u_j(n) = 0, \quad n + j \text{ is odd or } |j| > n.$$

For state j that the random walk can reach at time n, the transition probability is given as follows. The next result is obtained from (6.4) by setting $x = 1$ and $y = -1$. See Exercises 6.2 and 6.3 for related results.

Proposition 6.1 *Suppose that the random walk $\{W_n\}$ has no restriction on its movement. Then, the transition probability $u_j(n)$ is given by*

$$u_j(n) = b_{(n+j)/2}(n, p), \quad n = 0, 1, 2, \ldots,$$

where $j = 0, \pm 1, \ldots, \pm n$, and where $b_k(n, p)$ denotes the binomial probability (3.2) with the understanding that $b_k(n, p) = 0$ unless $k = 0, 1, \ldots, n$.

In particular, for a symmetric random walk, the transition probability is given by

$$v_j(n) = {}_nC_{(n+j)/2} \, 2^{-n}, \quad n = 0, 1, 2, \ldots,$$

where $j = 0, \pm 1, \ldots, \pm n$, and where the binomial coefficient ${}_nC_k$ is understood to be 0 unless $k = 0, 1, \ldots, n$.

The transition probability $u_j(n)$ satisfies the following recursive relation.

Note from (6.2) that

$$\begin{aligned}u_j(n+1) &= P\{X_{n+1}=1,\,W_n=j-1|W_0=0\}\\&+P\{X_{n+1}=-1,\,W_n=j+1|W_0=0\}.\end{aligned}$$

Since X_{n+1} is independent of W_n, it follows that

$$u_j(n+1) = p\,u_{j-1}(n) + q\,u_{j+1}(n), \quad n=0,1,2,\ldots, \qquad (6.7)$$

where the initial condition is given by

$$u_0(0)=1;\quad u_j(0)=0,\ j\neq 0. \qquad (6.8)$$

The transition probability $u_j(n)$ is calculated recursively using (6.7) and (6.8). It is easy to check that $u_j(n)$ given in Proposition 6.1 satisfies the recursion (6.7) and the initial condition (6.8).

In general, if $W_0=i$, then the transition probability is given by

$$u_{ij}(n) = u_{j-i}(n), \quad n=0,1,2,\ldots, \qquad (6.9)$$

due to the spatial homogeneity, where $u_{ij}(n)=u_{ij}(0,n)$. This can be proved formally as follows. From (6.1), we have

$$W_n - W_0 = \sum_{i=1}^n X_i, \quad n=0,1,\ldots.$$

Hence, defining $\hat{W}_n = W_n - W_0$, the random walk $\{\hat{W}_n\}$ starts with $\hat{W}_0=0$, and its transition probability is given by $u_j(n)$.

The recursive relation (6.7) can also be derived by conditioning on W_1. That is, since $W_1 = W_0 + X_1$, we obtain

$$\begin{aligned}u_{ij}(n+1) &= P\{W_{n+1}=j,\,X_1=1|W_0=i\}\\&+P\{W_{n+1}=j,\,X_1=-1|W_0=i\}\\&= p\,P\{W_{n+1}=j|W_1=i+1\}\\&+q\,P\{W_{n+1}=j|W_1=i-1\}.\end{aligned}$$

But, since the random walk is time-homogeneous, it follows that

$$P\{W_{n+1}=j|W_1=i+1\} = P\{W_n=j|W_0=i+1\},$$

whence

$$u_{ij}(n+1) = p\,u_{i+1,j}(n) + q\,u_{i-1,j}(n), \quad n=0,1,\ldots. \qquad (6.10)$$

The initial condition is given by

$$u_{ij}(0) = \delta_{ij}, \quad i,j\in\mathcal{Z},$$

where δ_{ij} denotes Kronecker's delta meaning that $\delta_{ij}=1$ if $i=j$ and $\delta_{ij}=0$ otherwise. Due to the spatial homogeneity (6.9), we obtain (6.7) from (6.10) with $i=0$.

Recall that the recursive relation (6.7) has been obtained by conditioning on W_{n-1}, while (6.10) is derived by conditioning on W_1. Because of this reason, the recursion (6.7) is called the *forward* equation and (6.10) the *backward*

THE REFLECTION PRINCIPLE

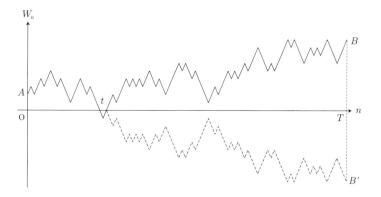

Figure 6.2 *A sample path of the random walk and its mirror image*

equation. In the case of random walks, the forward and backward equations coincide with each other.

6.3 The Reflection Principle

Throughout this section, we assume that the random walk $\{W_n\}$ is symmetric, i.e. $p = 1/2$ with $x = 1$ and $y = -1$.

Figure 6.2 depicts typical sample paths of the random walk. In particular, the sample path $\overline{AB}$ shows that the random walk starts from state A, reaches the x-axis at time t, and is at state B at time $T > t$. Associated with this sample path is the one denoted by $\overline{AB'}$, where the random walk starts from state A, reaches the x-axis at time t, but now is at state B' which is symmetric to B about the x-axis. That is, the sample path $\overline{AB'}$ after time t is a *mirror image* of the sample path $\overline{AB}$. Note that the mirror image is unique and can be constructed whenever the random walk reaches the x-axis. Also, the probability that the sample path $\overline{AB}$ occurs is equal to the probability that the sample path $\overline{AB'}$ occurs due to the symmetry. This observation is the simplest form of the so-called *reflection principle*.

Proposition 6.2 (Reflection Principle) *In a symmetric random walk with initial state $i > 0$, the probability that it reaches the x-axis at some time before T and is at state $j \geq 0$ at time T is the same as the probability that it crosses the x-axis to reach state $-j \leq 0$ at time T.*

Suppose that the state 0 is *absorbing*. That is, the random walk stays at state 0 forever once it reaches that state. Since the random walk is *skip-free* to both directions, this implies that the random walk never reaches negative states if it starts from a positive state. The transition probabilities $u_{ij}(n)$ obtained in the previous section are modified by the absorbing boundary. In order to distinguish the absorbing case, we denote the transition probabilities of this case by $a_{ij}(n)$ in what follows. Note that the state space is restricted to the positive integers in the absorbing case.

For a symmetric random walk $\{W_n\}$ with absorbing state 0, we define
$$M_n \equiv \min_{0 \leq k \leq n} W_k, \quad n = 0, 1, 2, \ldots. \tag{6.11}$$
Since the state 0 is absorbing, it follows that
$$a_{ij}(n) = P\{W_n = j, M_n > 0 | W_0 = i\}, \quad n = 0, 1, 2, \ldots, \tag{6.12}$$
where $i, j > 0$. That is, the restriction due to the absorbing state is replaced by the condition $\{M_n > 0\}$. The next result can be proved by applying the reflection principle.

Proposition 6.3 *For a symmetric random walk $\{W_n\}$ with absorbing state 0, the transition probabilities are given by*
$$a_{ij}(n) = v_{i-j}(n) - v_{i+j}(n), \quad n = 0, 1, 2, \ldots,$$
where $i, j > 0$.

Proof. Suppose that the symmetric random walk $\{W_n\}$ has no restriction on the movement. Then, the event $\{W_n = j\}$ is decomposed as
$$\{W_n = j\} = \{W_n = j, M_n > 0\} + \{W_n = j, M_n \leq 0\},$$
where $+$ means a union of disjoint events. Hence, from (6.12), we obtain
$$a_{ij}(n) = P\{W_n = j | W_0 = i\} - P\{W_n = j, M_n \leq 0 | W_0 = i\}.$$
Here, $P\{W_n = j, M_n \leq 0 | W_0 = i\}$ is the probability that the symmetric random walk starting from state $i > 0$ reaches the x-axis and then returns to state $j > 0$ at time n. Hence, from the reflection principle (Proposition 6.2), this probability is equal to the probability that it crosses the x-axis to reach state $-j$ at time n. That is,
$$P\{W_n = j, M_n \leq 0 | W_0 = i\} = P\{W_n = -j | W_0 = i\} = v_{-j-i}(n).$$
The proposition follows at once by the symmetry of the transition probabilities $v_j(n)$ (see Exercise 6.2). □

Let T_j denote the first time that the random walk $\{W_n\}$ reaches state j. Such a random variable is called the *first passage time* and an example of stopping times. Since the random walk is skip-free, the event $\{T_0 = n\}$ conditional on $W_0 = i > 0$ occurs if and only if it is in state 1 at time $n - 1$ without visiting state 0 and the next transition is a negative direction. That is, we have
$$P\{T_0 = n | W_0 = i\} = P\{X_n = -1, W_{n-1} = 1, M_{n-1} > 0 | W_0 = i\}, \tag{6.13}$$
where $i > 0$ and where M_n denotes the minimum (6.11) of the symmetric random walk $\{W_n\}$. Furthermore, from the independence of X_n and W_{n-1}, the right-hand side of (6.13) is equal to $q \, a_{i1}(n - 1)$. The distribution of the first passage time is therefore given, from Proposition 6.3, as
$$P\{T_0 = n | W_0 = i\} = \frac{1}{2}\{v_{i-1}(n-1) - v_{i+1}(n-1)\}, \quad n = 0, 1, 2, \ldots, \tag{6.14}$$

where $i > 0$. In particular, when $i = 1$, since

$$v_0(2r) - v_2(2r) = \{{}_{2r}C_r - {}_{2r}C_{r+1}\}2^{-2r} = \frac{(2r)!}{r!(r+1)!}2^{-2r}, \quad r = 1, 2, \ldots,$$

we have

$$P\{T_0 = 2n + 1 | W_0 = 1\} = \frac{(2n)!}{n!(n+1)!}2^{-2n-1} \quad (6.15)$$

and $P\{T_0 = 2n | W_0 = 1\} = 0$ for $n = 0, 1, 2, \ldots$.

From (6.15), we obtain

$$\begin{aligned}
P\{T_0 \leq 2n + 1 | W_0 = 1\} &= \sum_{r=0}^{n} P\{T_0 = 2r + 1 | W_0 = 1\} \\
&= \frac{1}{2} + \sum_{r=1}^{n} \frac{(2r)!}{r!(r+1)!} 2^{-2r-1} \\
&= 1 - {}_{2n+1}C_{n+1}\, 2^{-2n-1},
\end{aligned}$$

where we have used the well-known identity

$$\sum_{r=1}^{n} \frac{(2r)!}{r!(r+1)!} 2^{-2r} = 1 - \frac{(2n+1)!}{n!(n+1)!} 2^{-2n}.$$

It follows that, for a symmetric random walk, we have

$$P\{T_0 \leq 2n + 1 | W_0 = 1\} = 1 - v_1(2n+1), \quad n = 0, 1, 2, \ldots.$$

Since $v_1(n) \to 0$ as $n \to \infty$ (see Exercise 6.3), it follows that

$$P\{T_0 < \infty | W_0 = 1\} = 1.$$

That is, letting r_i denote the probability that the symmetric random walk starting from state i eventually reaches state 0, this result reveals that $r_1 = 1$. Because of the spatial homogeneity, we thus have $r_i = 1$ for all $i > 0$. See Exercise 6.4 for more general results.

Another application of the reflection principle is the following. See Exercise 6.5 for a related problem.

Proposition 6.4 *Suppose that a symmetric random walk $\{W_n\}$ has no restriction on its movement, and let $M_n = \min_{0 \leq k \leq n} W_k$. Then, the joint distribution of (W_n, M_n) is given by*

$$P\{W_n = i, M_n = j | W_0 = 0\} = v_{2j-i}(n) - v_{2j-i-2}(n), \quad n = 0, 1, 2, \ldots,$$

where $i \geq j$ and $j \leq 0$. Moreover,

$$P\{M_n = j | W_0 = 0\} = v_j(n) + v_{j-1}(n), \quad n = 0, 1, 2, \ldots,$$

where $j \leq 0$ and where either $v_j(n)$ or $v_{j-1}(n)$ is 0 according that $j + n$ is odd or even.

Proof. Consulting Figure 6.3, an application of the reflection principle yields

$$P\{M_n \leq j, W_n = i | W_0 = 0\} = P\{W_n = 2j - i | W_0 = 0\} = v_{2j-i}(n).$$

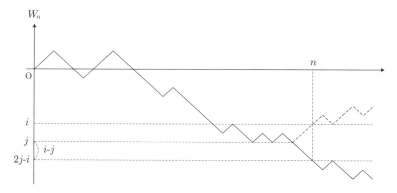

Figure 6.3 *Another example of the reflection principle*

The first result of the proposition follows since

$$P\{M_n = j, W_n = i | W_0 = 0\}$$
$$= P\{M_n \leq j, W_n = i | W_0 = 0\} - P\{M_n \leq j-1, W_n = i | W_0 = 0\}.$$

Noting that $\lim_{j \to \pm\infty} v_j(n) = 0$, the marginal distribution of M_n is given by

$$P\{M_n = j | W_0 = 0\} = \sum_{i=j}^{\infty} \{v_{2j-i}(n) - v_{2j-i-2}(n)\} = v_j(n) + v_{j-1}(n),$$

whence the second result, completing the proof. □

6.4 The Change of Measure Revisited

Consider a random walk $\{W_n\}$ with no restriction on the movement. If the upward transition probability is p, we call it a p-random walk. Let

$$m(\theta) \equiv E\left[e^{\theta X_n}\right] = p e^{\theta} + q e^{-\theta}, \quad n = 1, 2, \ldots,$$

which exists for any θ (see Exercise 6.1). Here, $q = 1 - p$ denotes the downward transition probability. Suppose that $W_0 = i$, and let

$$Y_n \equiv m^{-n}(\theta) e^{\theta(W_n - i)}, \quad n = 1, 2, \ldots. \tag{6.16}$$

Since $W_n - i = \sum_{i=1}^{n} X_i$, we obtain

$$Y_n = \prod_{i=1}^{n} m^{-1}(\theta) e^{\theta X_i}, \quad n = 1, 2, \ldots. \tag{6.17}$$

Note that, since X_n are independent, the random variables $e^{\theta X_n}$ are also independent. It follows that

$$E[Y_n] = \prod_{i=1}^{n} m^{-1}(\theta) E\left[e^{\theta X_n}\right] = 1, \quad n = 1, 2, \ldots. \tag{6.18}$$

In fact, the process $\{Y_n\}$ is a martingale. The proof is left in Exercise 6.8. See Exercise 6.7 for a related problem.

Let $T < \infty$ be fixed, and define

$$P^*(A) = E[1_A Y_T], \quad A \in \mathcal{F}, \tag{6.19}$$

where 1_A denotes the indicator function. Since $E[Y_T] = 1$, it is readily seen that P^* is a probability measure. Hence, (6.19) defines another 'change of measure' formula; cf. (5.31).

The p-random walk $\{W_n\}$ is transferred through (6.19) to a symmetric random walk, and vice versa, as follows.

Proposition 6.5 *For some $\theta \in \mathbf{R}$, let*

$$p^* = \frac{p\,e^\theta}{p\,e^\theta + q\,e^{-\theta}}.$$

Then, the p-random walk $\{W_n\}$ is transferred to a p^*-random walk under P^* defined by (6.19).

Proof. For any $n \leq T$, the probability $P^*\{X_n = 1\}$ is obtained by (6.19) as

$$P^*\{X_n = 1\} = E^*\left[1_{\{X_n=1\}}\right] = E\left[1_{\{X_n=1\}} \prod_{i=1}^{T} m^{-1}(\theta)\,e^{\theta X_i}\right].$$

Since $X_1, \ldots, X_T$ are independent under P, it follows from (6.18) that

$$P^*\{X_n = 1\} = E\left[m^{-1}(\theta) 1_{\{X_n=1\}} e^{\theta X_n}\right] \prod_{i \neq n} E\left[m^{-1}(\theta)\,e^{\theta X_i}\right]$$

$$= E\left[m^{-1}(\theta) 1_{\{X_n=1\}} e^{\theta X_n}\right].$$

Therefore, we obtain

$$P^*\{X_n = 1\} = m^{-1}(\theta) E\left[1_{\{X_n=1\}} e^{\theta X_n}\right] = \frac{p\,e^\theta}{p\,e^\theta + q\,e^{-\theta}} = p^*.$$

Similarly, we obtain $P^*\{X_n = -1\} = q^* \equiv 1 - p^*$. Moreover, it can be readily seen that $X_1, \ldots, X_T$ are also independent under P^*. Hence, $\{W_n\}$ is a p^*-random walk under P^*. □

Corollary 6.1 *For $0 < p < 1$, let $q = 1 - p$.*

(1) *Let $\theta = \log \sqrt{p/q}$. Then, a symmetric random walk is transferred to a p-random walk under P^* defined by (6.19).*

(2) *Let $\theta = \log \sqrt{q/p}$. Then, a p-random walk is transferred to a symmetric random walk under P^* defined by (6.19).*

Proof. For Part (1), we have

$$p^* = \frac{\frac{1}{2} e^\theta}{\frac{1}{2} e^\theta + \frac{1}{2} e^{-\theta}} = \frac{\sqrt{p/q}}{\sqrt{p/q} + \sqrt{q/p}} = \frac{p}{p+q} = p.$$

The result follows from Proposition 6.5. Part (2) can be proved similarly. □

Recall that the reflection principle (Proposition 6.2) applies only to symmetric random walks. The distributions of the first passage times as well as the minimum of a random walk obtained in the previous section are therefore restricted only to the symmetric case. In the following, we will demonstrate that those results can be transferred to the asymmetric case by the change of measure. Let $\{W_n\}$ be a symmetric random walk under P, and define θ as in Corollary 6.1(1). Then, under the new probability measure P^* defined by (6.19), the random walk $\{W_n\}$ is a p-random walk.

Our first application of the change of measure formula (6.19) is to obtain the transition probabilities $a_{ij}(n)$ of the p-random walk when state 0 is absorbing. To this end, let $n \leq T$. Using a similar argument that leads to (6.12), the transition probabilities are given by

$$a_{ij}(n) = P^*\{W_n = j,\ M_n > 0 | W_0 = i\}, \quad n = 0, 1, 2, \ldots,$$

where $i,\ j > 0$. It follows from (6.19) that

$$a_{ij}(n) = E\left[1_{\{W_n=j,\ M_n>0\}} Y_T | W_0 = i\right].$$

Since $W_T = W_n + \sum_{i=n+1}^T X_i$ and since $X_{n+1}, \ldots, X_T$ are independent of W_n under P, we obtain

$$\begin{aligned} a_{ij}(n) &= E\left[1_{\{W_n=j,\ M_n>0\}} Y_n | W_0 = i\right] E\left[\prod_{i=n+1}^T m^{-1}(\theta) e^{\theta X_i}\right] \\ &= m^{-n}(\theta) E\left[1_{\{W_n=j,\ M_n>0\}} e^{\theta(W_n - i)} | W_0 = i\right] \\ &= m^{-n}(\theta) e^{\theta(j-i)} P\{W_n = j,\ M_n > 0 | W_0 = i\}. \quad (6.20) \end{aligned}$$

But, since the random walk $\{W_n\}$ is symmetric under P, we have from Proposition 6.3 that

$$P\{W_n = j,\ M_n > 0 | W_0 = i\} = v_{j-i}(n) - v_{j+i}(n).$$

Here, we have used the symmetry $v_j(n) = v_{-j}(n)$ in the symmetric random walk (see Exercise 6.2). Also, since $\theta = \log \sqrt{p/q}$, we obtain

$$e^{\theta j} = \left(\frac{p}{q}\right)^{j/2}$$

and

$$m(\theta) = \frac{e^\theta + e^{-\theta}}{2} = \frac{1}{2}\left(\sqrt{\frac{p}{q}} + \sqrt{\frac{q}{p}}\right) = (4pq)^{-1/2}.$$

It follows from (6.20) that

$$a_{ij}(n) = 2^n\, p^{(n+j-i)/2} q^{(n-j+i)/2} \{v_{j-i}(n) - v_{j+i}(n)\}.$$

Therefore, we obtain from Proposition 6.1 that

$$a_{ij}(n) = u_{j-i}(n) - \left(\frac{q}{p}\right)^i u_{j+i}(n), \quad n = 0, 1, 2, \ldots, \quad (6.21)$$

where $i, j > 0$ and where $u_j(n)$ denotes the transition probabilities of the p-random walk when there is no restriction on the movement.

Second, by a similar argument, we can obtain the distribution of the minimum $M_n = \min_{1 \le k \le n} W_k$ of a p-random walk $\{W_n\}$. Let $n \le T$, and use the same change of measure as above. Then,

$$\begin{aligned} P^*\{M_n = j, W_n = i | W_0 = 0\} &= E\left[1_{\{M_n=j, W_n=i\}} Y_T | W_0 = 0\right] \\ &= m^{-n}(\theta) E\left[1_{\{M_n=j, W_n=i\}} e^{\theta W_n} | W_0 = 0\right] \\ &= m^{-n}(\theta) e^{\theta i} P\{M_n = j, W_n = i | W_0 = 0\}. \end{aligned}$$

But, since the random walk $\{W_n\}$ is symmetric under P, we obtain from Proposition 6.4 that

$$P\{M_n = j, W_n = i | W_0 = 0\} = v_{i-2j}(n) - v_{i+2-2j}(n),$$

where $i \ge j$ and $j \le 0$. It follows from Proposition 6.1 that

$$P^*\{M_n = j, W_n = i | W_0 = 0\} = \left(\frac{p}{q}\right)^j u_{i-2j}(n) - \left(\frac{p}{q}\right)^{j-1} u_{i+2-2j}(n), \tag{6.22}$$

where $i \ge j$ and $j \le 0$. Therefore, letting

$$\bar{U}_j(n) \equiv P^*\{W_n > j | W_0 = 0\} = \sum_{i=j+1}^{\infty} u_i(n),$$

we obtain

$$P^*\{M_n = j | W_0 = 0\} = \left(\frac{p}{q}\right)^j \bar{U}_{-j-1}(n) - \left(\frac{p}{q}\right)^{j-1} \bar{U}_{-j+1}(n), \quad j \le 0.$$

Simple algebra then leads to

$$P^*\{M_n > j | W_0 = 0\} = \bar{U}_j(n) - \left(\frac{p}{q}\right)^j \bar{U}_{-j}(n), \quad j < 0. \tag{6.23}$$

The proof is left in Exercise 6.11. See also Exercise 6.10 for another application of the change of measure formula.

6.5 The Binomial Securities Market Model

Let $S(t)$ be the time t price of a risky security (stock), and suppose $S(0) = S$. It is assumed throughout this section that the stock pays no dividends. The binomial model has been formally introduced in Example 3.1. In this section, we define the binomial security price model in terms of a random walk.

Let $\{X_i\}$ be a family of IID Bernoulli random variables with $x = 1$ and $y = 0$ in (6.3), and let $\{W_n\}$ be the associated random walk. From the result in Example 3.1, we have

$$S(t) = u^{W_t} d^{t-W_t} S, \quad t = 0, 1, 2, \ldots, T, \tag{6.24}$$

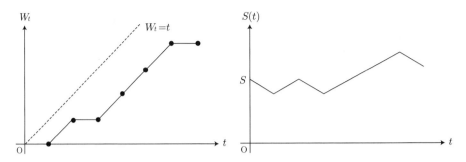

Figure 6.4 *The random walk $\{W_t\}$ and the price process $\{S(t)\}$*

where $T < \infty$. The quantity u is called the *up-factor* of the binomial model, whereas d is called the *down-factor*. Figure 6.4 depicts the evolution of the random walk $\{W_t\}$ and that of the price process $\{S(t)\}$. See Exercise 6.12 for an alternative construction of the price process using another random walk.

Since $W_{t+1} = W_t + X_{t+1}$ by definition, we obtain from (6.24) that

$$S(t+1) = u^{X_{t+1}} d^{1-X_{t+1}} S(t), \quad t = 0, 1, \ldots, T-1.$$

It follows that

$$\frac{\Delta S(t)}{S(t)} = u^{X_{t+1}} d^{1-X_{t+1}} - 1, \tag{6.25}$$

where $\Delta S(t) = S(t+1) - S(t)$. Also, it is readily seen that

$$P\{S(t) = u^k d^{t-k} S\} = b_k(t, p), \quad k = 0, 1, 2, \ldots, t, \tag{6.26}$$

where $b_k(n, p)$ denotes the binomial probability (3.2).

In addition to the risky security $S(t)$, we assume that there is also available a risk-free security whose time t price is denoted by

$$B(t) = R^t, \quad t = 0, 1, 2, \ldots, T, \tag{6.27}$$

where $R > 1$ is a constant. The quantity $B(t)$, the money-market account, represents the time t value of the risk-free bank account when one dollar is deposited at time 0. Hence, $r \equiv R - 1 > 0$ denotes the *interest rate*. Note that, in the notation of Chapter 5, we have $n = 1$, $S_0(t) = B(t)$, and $S_1(t) = S(t)$.

Since the securities pay no dividends, the value process $\{V(t)\}$ defined in (5.2) is given by

$$V(t) = b(t)B(t) + \theta(t)S(t) \quad t = 1, 2, \ldots, T, \tag{6.28}$$

where $V(0) = b(1) + \theta(1)S$. If the portfolio process $\{(b(t), \theta(t)); t = 1, 2, \ldots, T\}$ is self-financing, we have from (5.6) that

$$\Delta V(t) = b(t+1)\Delta B(t) + \theta(t+1)\Delta S(t) \quad t = 0, 1, \ldots, T-1.$$

Taking the money-market account as the numeraire, the denominated value process satisfies

$$\Delta V^*(t) = \theta(t+1)\Delta S^*(t) \quad t = 0, 1, \ldots, T-1, \tag{6.29}$$

THE BINOMIAL SECURITIES MARKET MODEL

where $S^*(t) = S(t)/B(t)$; see (5.12).

In the rest of this section, we calculate the expected value of the stock price $S(T)$. The pricing of a derivative security in the binomial model is deferred to the next chapter.

Let $h(x)$ be any real-valued function. The function $h(x)$ can be the payoff function of a derivative security written on the stock. That is, an investor will receive $h(S(T))$ at the maturity T of the claim. The discounted, expected value of $h(S(T))$ is calculated, from (2.19) and (6.26), as

$$E\left[\frac{h(S(T))}{B(T)}\right] = R^{-T}\sum_{k=0}^{T} h\left(u^k d^{T-k} S\right) b_k(T, p).$$

Note that the right-hand side is discounted by $B(T) = R^T$ in order to obtain the *present value* of the payoff.

In particular, if $h(x) = x$, we have

$$E\left[S^*(T)\right] = SR^{-T}\sum_{k=0}^{T} {}_T C_k (up)^k (dq)^{T-k} = S\left(\frac{up+dq}{R}\right)^T, \quad (6.30)$$

where $q = 1 - p$ and $S^*(T) = S(T)/B(T)$.

On the other hand, if $h(x) = \{x - K\}_+$, i.e. the claim is a call option with strike price K, then the discounted, expected payoff is given by

$$E\left[\frac{\{S(T)-K\}_+}{B(T)}\right] = E\left[S^*(T)\right]\overline{B}_k(T, p') - KR^{-T}\overline{B}_k(T, p), \quad (6.31)$$

where $p' = up/(up + dq)$,

$$k = \min\left\{a : a > \log[K/Sd^T]/\log[u/d]\right\},$$

and $\overline{B}_k(n, p)$ denotes the survival probability of the binomial distribution with parameter (n, p); see Section 3.1.

It is convenient to employ the change of measure formula (6.19) to prove (6.31), which is highlighted in the next example.

Example 6.1 Let A be the event that the call option becomes in-the-money at the maturity T, i.e. $A = \{S(T) \geq K\}$. Using the indicator function 1_A of the event A, we then have

$$E\left[\{S(T) - K\}_+\right] = E\left[S(T)1_A\right] - KE\left[1_A\right].$$

Note that

$$E\left[1_A\right] = P\{S(T) \geq K\} = \overline{B}_k(T, p).$$

On the other hand, let $\theta = \log[u/d]$ so that

$$m(\theta) = E\left[e^{\theta X_n}\right] = p\frac{u}{d} + q.$$

It is easily seen that the change of measure formula (6.19) also holds for this situation (see Exercise 6.13). That is, we obtain

$$p' = \frac{pe^\theta}{m(\theta)} = \frac{up}{up+dq},$$

and the process $\{W_n\}$ is a p'-random walk under the probability measure

$$P'(A) = E[Y_T 1_A] = \frac{E[e^{\theta W_T} 1_A]}{m^T(\theta)}.$$

Since $S(T) = d^T e^{\theta W_T} S$ from (6.24), it follows that

$$E[S(T)1_A] = S d^T m^T(\theta) P'(A) = S(up+dq)^T P'\{S(T) \geq K\}.$$

The result follows at once from (6.30). See Exercise 6.14 for an alternative derivation of (6.31).

6.6 Exercises

Exercise 6.1 Obtain the moment generating function (MGF) of W_n in Definition 6.1. Using the MGF, calculate its mean and variance; cf. (3.6).

Exercise 6.2 Prove Proposition 6.1. Also, prove that

$$u_{-j}(n) = \left(\frac{q}{p}\right)^j u_j(n), \quad n = 0, 1, 2, \ldots,$$

where $j = 0, \pm 1, \cdots, \pm n$. In particular, we have $v_{-j}(n) = v_j(n)$ for a symmetric random walk.

Exercise 6.3 Using Stirling's formula

$$n! \approx \sqrt{2\pi}\, n^{n+0.5} e^{-n},$$

prove that, for $u_j(n)$ in Proposition 6.1, we have

$$u_0(2n) = \frac{(2n)!}{(n!)^2}(pq)^n \approx \frac{(4pq)^n}{\sqrt{\pi n}}, \quad 0 < p < 1,$$

where $q = 1 - p$. Hence, since $4pq \leq 1$, we conclude that $\lim_{n \to \infty} u_j(n) = 0$ for any $j \in \mathcal{Z}$.

Exercise 6.4 For a p-random walk, let r_i be the probability that the random walk starting from state $i > 0$ eventually reaches state 0. Prove that r_i satisfies the relation

$$r_i = q\, r_{i-1} + p\, r_{i+1}, \quad i = 1, 2, \ldots,$$

with $r_0 = 1$, where $q = 1 - p$. Show that this equation can be solved as

$$1 - r_i = \frac{1 - (q/p)^i}{1 - q/p}(1 - r_1), \quad i = 1, 2, \ldots,$$

provided that $p \neq q$. Furthermore, assuming that $\lim_{i \to \infty} r_i = 0$, prove that $r_i = 1$ if $q > p$ and $r_i = (q/p)^i$ if $q < p$.

Exercise 6.5 Suppose that a p-random walk $\{W_n\}$ has no restriction on its movement, and let $M_n^* = \max_{0 \leq k \leq n} W_k$. Obtain the joint distribution of (W_n, M_n^*) as well as the marginal distribution of M_n^*.

Exercise 6.6 Consider a p-random walk $\{W_n\}$ starting from state $i > 0$. State 0 is said to be *reflecting* if the random walk at state 1 moves up to state 2 with probability p and returns there with probability $q = 1 - p$. Prove that the transition probabilities of a *symmetric* random walk, $r_{ij}(n)$ say, with reflecting boundary at state 0 satisfy

$$r_{ij}(n) = v_{i-j}(n) + v_{i+j-1}(n), \quad n = 0, 1, 2, \ldots,$$

where $i, j > 0$ and $v_j(n)$ denote the transition probabilities of a symmetric random walk with no restriction on its movement.

Exercise 6.7 For a random walk $\{W_n\}$ with underlying random variables X_n, let $\mathcal{F}_n$ be the σ-field generated from $\{X_1, X_2, \ldots, X_n\}$. Prove that any symmetric random walk is a martingale with respect to $\{\mathcal{F}_n\}$. Also, for a p-random walk $\{W_n\}$, let

$$Z_n = (W_n - n\mu)^2 - \sigma^2 n, \quad n = 0, 1, 2, \ldots,$$

where $\mu = E[X_1]$ and $\sigma^2 = V[X_1]$. Prove that the process $\{Z_n\}$ is also a martingale with respect to $\{\mathcal{F}_n\}$.

Exercise 6.8 (Wald's Identity) Prove that the process $\{Y_n\}$ defined by (6.16) is a martingale. Hence, from the optional sampling theorem (Proposition 5.5), we have

$$E\left[m^{-\tau}(\theta) e^{\theta(W_\tau - i)}\right] = 1$$

for any finite stopping time τ.

Exercise 6.9 Using the change of measure formula (3.20) with $X = W_T - i$, derive the probability measure given by (6.19).

Exercise 6.10 Apply the change of measure arguments to (6.13) so as to obtain the first-passage-time distribution to state 0 for a p-random walk.

Exercise 6.11 Using the symmetry result obtained in Exercise 6.2, prove that (6.23) holds.

Exercise 6.12 In this construction, we let $\{X_i\}$ be a family of IID Bernoulli random variables with $x = 1$ and $y = -1$ in (6.3). Denoting the associated random walk by $\{W_t\}$, show that the binomial model (6.24) can be expressed as

$$S(t) = u^{(t+W_t)/2} d^{(t-W_t)/2} S, \quad t = 0, 1, 2, \ldots, T.$$

Exercise 6.13 Let $\{W_n\}$ be the p-random walk given in the binomial model (6.24), and define Y_T as in (6.16). For some θ, let

$$p^* = \frac{p e^\theta}{p e^\theta + q},$$

where $q = 1 - p$. Mimicking the proof of Proposition 6.5, show that the p-random walk $\{W_n\}$ is transferred to a p^*-random walk under P^* defined by the transformation (6.19).

Exercise 6.14 Let $A = \{S(T) \geq K\}$ as in Example 6.1. Using

$$E[\{S(T) - K\}_+] = \sum_{k=0}^{T} \left(Su^k d^{T-k} - K\right) 1_{\{Su^k d^{T-k} \geq K\}} b_k(T, p),$$

prove Equation (6.31) directly.

CHAPTER 7

The Binomial Model

In the previous chapter, we saw that a random walk provides a simple model for a security price process. Namely, let $S(t)$ be the time t price of a financial security, and suppose that $X_t \equiv S(t)/S(t-1)$ follows the Bernoulli distribution (6.3). Since $S(t) = X_t S(t-1)$, the model describes the situation that the security price goes up with probability p and goes down with probability $q = 1 - p$, independently of the past. In this chapter, we provide detailed discussions of the *binomial model*, an important class of discrete securities market models.

7.1 The Single-Period Model

Consider the binomial security price model (6.24). This section intends to obtain the fair price of a contingent claim written on the security (stock). For this purpose, we need to construct a self-financing portfolio that generates the claim (see Theorem 5.2). In order to demonstrate this idea explicitly, we first consider the single-period case, i.e. $T = 1$. A general multi-period model will be treated later.

Before proceeding, we will make sure that the assumption $d < R < u$ is needed to rule out arbitrage opportunities. Recall that an *arbitrage opportunity* is a chance that investors can make money without any risk when starting with nothing.

Suppose that $R < d$ in the single-period binomial model. Then, at time 0, you sell the money-market account for 1 dollar as many as you want, w say, and purchase w/S units of the stock, so the net cost to construct the portfolio is zero. At time 1, you must repay Rw dollars for the money-market account, but at the same time, the value of your stock is at least $dS \times w/S = dw$, which is greater than Rw. Hence, since you make a positive profit at time 1 for sure with zero cost, this is an arbitrage opportunity. Similarly, it is readily shown that the case $R > u$ makes an arbitrage opportunity too. Therefore, we must have $d < R < u$ as well as $R \geq 1$ in order to rule out arbitrage opportunities.

In the binomial model, we replicate a derivative security in terms of the stock and the money-market account. That is, let $B = b(1)$ and $w = \theta(1)$ in (6.28). The current value of the portfolio is equal to $V(0) = wS + B$. The time 1 value of the portfolio depends on the value of the stock. If $S(1) = uS$ then the portfolio value becomes

$$V(1) = wS(1) + RB = wuS + RB,$$

while it becomes $wdS + RB$ in the case that $S(1) = dS$.

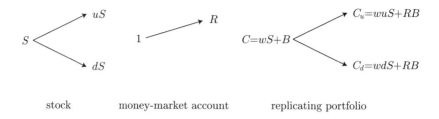

Figure 7.1 *The binomial model and the replicating portfolio*

The value of the derivative is also dependent on the value of the stock $S(1)$. Suppose that the value becomes C_u when $S(1) = uS$, while it becomes C_d in the case that $S(1) = dS$. If, in particular, the derivative is a call option maturing at time 1 with strike price K, then the value of the derivative is equal to its payoff, i.e. $C_u = \{uS - K\}_+$ and $C_d = \{dS - K\}_+$, where $\{x\}_+ = \max\{x, 0\}$.

The portfolio is said to *replicate* the derivative if, whatever the outcome of the underlying stock price is, the value of the portfolio coincides with that of the derivative security. Hence, $wuS + RB = C_u$ for the case that $S(1) = uS$ while $wdS + RB = C_d$ for the case that $S(1) = dS$ (see Figure 7.1). Note that we have two equations for two variables w and B to be determined. That is, the replicating portfolio can be constructed by solving the following equations simultaneously:

$$\begin{cases} C_u = wuS + RB, \\ C_d = wdS + RB. \end{cases} \quad (7.1)$$

It is readily verified that the simultaneous equation is solved as

$$w = \frac{C_u - C_d}{(u-d)S} \quad \text{and} \quad B = \frac{uC_d - dC_u}{(u-d)R},$$

whence, from Theorem 5.2, the fair price of the derivative, C say, is given by

$$C = \frac{C_u - C_d}{u - d} + \frac{uC_d - dC_u}{(u-d)R}. \quad (7.2)$$

See Exercise 7.1 for a numerical example.

In order to relate the pricing formula (7.2) to the risk-neutral valuation method, let

$$p^* = \frac{R - d}{u - d}, \quad q^* = \frac{u - R}{u - d}, \quad (7.3)$$

so that $p^* + q^* = 1$. Under the assumption $d < R < u$, we observe that $0 < p^* < 1$ and

$$E^*[S^*(1)] = \frac{uS}{R}p^* + \frac{dS}{R}q^* = S^*(0),$$

where E^* is the expectation operator associated with p^*. Hence, p^* defined by (7.3) is the unique risk-neutral probability (see Definition 5.8). Also, it is

THE SINGLE-PERIOD MODEL

easily verified that the formula (7.2) can be rewritten as

$$C = R^{-1}(p^* C_u + q^* C_d) = R^{-1} E^*[C(1)], \qquad (7.4)$$

where $C(1)$ is a random variable representing the time 1 value of the derivative such that $C(1) = C_u$ if $S(1) = uS$ and $C(1) = C_d$ if $S(1) = dS$. The valuation formula (7.4) is equivalent to the risk-neutral pricing formula (5.25).

At this point, it will be helpful to calculate the *mean excess return per unit of risk* for each security. The mean excess return for the stock is given by

$$E\left[\frac{\Delta S(0)}{S}\right] - r = up + dq - R,$$

where p denotes the upward probability, $q = 1 - p$ the downward probability, and $r = R - 1$. The variance of the return for the stock is

$$V\left[\frac{\Delta S(0)}{S}\right] = pq(u-d)^2.$$

Since risk is defined by the standard deviation of the return, the mean excess return per unit of risk for the stock, λ_S say, is obtained as

$$\lambda_S = \frac{up + dq - R}{\sqrt{pq}(u-d)}. \qquad (7.5)$$

It should be noted that these calculations are carried out under the physical probability p, not under the risk-neutral probability p^*.

We next calculate the mean excess return per unit of risk for the derivative. Let C be the current price of the derivative. Then, as for (7.5), the mean excess return per unit of risk for the derivative, λ_C say, is obtained as

$$\lambda_C = \frac{C_u p + C_d q - RC}{\sqrt{pq}(C_u - C_d)}. \qquad (7.6)$$

It is not difficult to show that $\lambda_S = \lambda_C$. The proof is left in Exercise 7.2. In the finance literature, the mean excess return per unit of risk for the derivative is called the *market price of risk*. We thus have established the following important result.

Theorem 7.1 *Suppose that a contingent claim is attainable in the binomial model. Then, the mean excess return per unit of risk for the contingent claim is the same as that of the underlying stock.*

Theorem 7.1 suggests another pricing method for an attainable contingent claim $C(1)$. Namely, the price of the claim can be obtained by equating the two mean excess returns per unit of risk. That is,

$$\frac{E[S(1)] - RS}{\sqrt{V[S(1)]}} = \frac{E[C(1)] - RC}{\sqrt{V[C(1)]}}. \qquad (7.7)$$

It follows that

$$C = \frac{1}{R}\left[E[C(1)] - \frac{\sqrt{V[C(1)]}}{\sqrt{V[S(1)]}}(E[S(1)] - RS)\right]. \qquad (7.8)$$

Note that the price does not reflect investors' attitude towards risk, but depends only on the values observed in the market.

An interpretation of the result (7.8) is as follows. Suppose that an investor wants to purchase the contingent claim $C(1)$. What is the fair price of the claim? The discounted, expected payoff $E[C(1)]/R$ is certainly a candidate to the price, since the investor can expect that amount of money when he/she enters the contract. However, the investor loses money when the actual payoff is below the expected payoff. If the investor does not enter the contract, he/she possesses that money for sure, which can be used for other purposes. There exists an investor, called a *risk-averter*, who prefers money for sure to uncertain gambling if the expected outcome is the same as the sure money. The risk-averter requires a *risk premium*, π say, for the uncertain outcome. That is, the price of the claim $C(1)$ must be given by

$$C = \frac{1}{R}\{E[C(1)] - \pi\}.$$

Comparing this with (7.8), we conclude that

$$\pi = \lambda_C \sqrt{V[C(1)]}, \qquad (7.9)$$

where λ_C denotes the market price of risk and $\lambda_C = \lambda_S$ in the attainable case. Hence, the risk premium π is equal to the market price of risk times the volatility of the contingent claim.

7.2 The Multi-Period Model

What is the most important in the previous discussions is that the same argument holds for any *triangle* structure depicted in Figure 7.1. That is, given the two values of a contingent claim at period $n+1$, the value of the claim at period n is obtained by solving a simultaneous equation similar to (7.1). Notice that the payoff of a contingent claim is realized only when it is exercised and the (European) claim is exercised only at the maturity. Hence, the value of a contingent claim is known only at the maturity, and the premium of the claim is obtained by solving the simultaneous equations backwards starting from the maturity.

To be more specific, consider the binomial model (6.24) with the underlying Bernoulli random variables X_n. It is convenient to describe the model by using the lattice structure depicted in Figure 7.2. It should be noted that the upward probability written at each node is the risk-neutral probability p^*, not the physical probability p. The following method to value contingent claims is known as the *binomial pricing method*, originally developed by Cox, Ross, and Rubinstein (1979).

In the binomial lattice model (see Figure 7.2), the ith node from the bottom at time t is labeled by (t,i), $i = 0, 1, \ldots, t$. Hence, the node (t,i) reveals that the stock price has gone up i times until time t. Let $c(t,i)$ denote the value of a contingent claim at node (t,i). Since, in general, we have from (7.4) that

$$C(t) = R^{-1} E_t^*[C(t+1)], \quad t = 0, 1, \ldots, T-1, \qquad (7.10)$$

THE MULTI-PERIOD MODEL

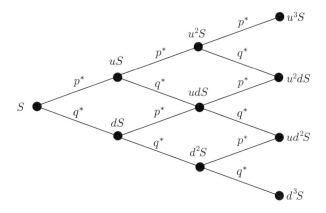

Figure 7.2 *The binomial lattice (3 periods)*

it is readily seen that the following recursive relation holds:

$$c(t,i) = R^{-1}[p^*c(t+1,i+1) + q^*c(t+1,i)], \quad t = 0,1,\ldots,T-1, \quad (7.11)$$

where $q^* = 1 - p^*$ and $i = 0,1,\ldots,t$. The recursion can be solved backwards starting from

$$c(T,i) = h\left(u^i d^{T-i} S\right), \quad i = 0,1,\ldots,T, \quad (7.12)$$

where $h(x)$ is the payoff function, since its payoff is realized only at the maturity T.

It should be noted that, according to the risk-neutral valuation method (5.25), the current value of the contingent claim in this setting is given by

$$C = R^{-T} E^*[h(S(T))]. \quad (7.13)$$

This can be formally proved by (7.10) and the chain rule of conditional expectations (see Proposition 5.3).

In particular, consider a European call option, i.e. $h(x) = \{x - K\}_+$. Note that the equation (6.30) holds with the physical probability p being replaced by the risk-neutral probability p^*. But, from (7.3), we have $up^* + dq^* = R$, so that $E^*[S^*(T)] = S$ from (6.30). Hence, from (6.31), we have proved the following famous result. The next result has been obtained by Cox, Ross, and Rubinstein (1979), which is the discrete counterpart of the Black–Scholes formula (4.21).

Theorem 7.2 *The premium of a European call option with strike price K and maturity T is given by*

$$C = S\overline{B}_k(T,p') - KR^{-T}\overline{B}_k(T,p^*), \quad p^* = \frac{R-d}{u-d},$$

where S is the current price of the underlying stock, $p' = up^*/R$,

$$k = \min\{a : a > \log[K/Sd^T]/\log[u/d]\},$$

and $\overline{B}_k(n,p)$ denotes the survival probability of the binomial distribution with parameter (n,p).

As in the single-period case, we can easily construct the replicating portfolio at each node. The simultaneous equation corresponding to (7.1) at node (t,i) is given by

$$\begin{cases} c(t+1,i+1) &= wuS(t,i) + RB, \\ c(t+1,i) &= wdS(t,i) + RB, \end{cases} \quad (7.14)$$

where $S(t,i) = u^i d^{t-i} S$ denotes the stock price at node (t,i); see (6.24). It is easily seen that

$$\theta(t,i) \equiv wS(t,i) = \frac{c(t+1,i+1) - c(t+1,i)}{u-d}$$

and

$$c(t,i) - \theta(t,i) = \frac{uc(t+1,i) - dc(t+1,i+1)}{(u-d)R}.$$

We note that, if we hold the replicating portfolio at each node, the short position for the contingent claim has no risk. Because of this reason, the replicating portfolio is often called the *hedging portfolio*. See Exercise 7.4 for a numerical example.

The key recursion (7.11) holds even when the quantities R and p^* are different over nodes. That is, the binomial lattice method is valid even for the case that the parameters R, u, and d are dependent on state $S(t)$ as well as time t, provided that the triangle structure remains available. This is so, since the formula (7.10) always holds and, in the binomial model, the risk-neutral probability is given by (7.3). That is, at node (t,i), the risk-neutral probability $p^*(t,i)$ is obtained as

$$p^*(t,i) = \frac{R(t,i) - d(t,i)}{u(t,i) - d(t,i)}, \quad (7.15)$$

where $u(t,i)$ and $d(t,i)$ are the up-factor and down-factor, respectively, of the binomial model and $R(t,i) - 1$ is the interest rate at node (t,i). The value of a contingent claim written on the stock can be solved backwards using (7.11) and the initial condition $C(T,i) = h(S(T,i))$.

Example 7.1 In this example, we consider a binomial model in which the up-factor and down-factor are dependent on the current price. Namely, suppose that the up-factor is given by $u(S) = uS^\alpha$, when $S(t) = S$, for some $\alpha \geq 0$ and the down-factor is given by $d(S) = dS^{-\alpha}$. The binomial lattice for this case is depicted in Figure 7.3.

Suppose that $S(0) = S > 0$. The security prices $S(t,i)$ at nodes (t,i) are given, respectively, by $S(1,1) = uS^{1+\alpha}$, $S(1,0) = dS^{1-\alpha}$, and

$$S(2,3) = u^{2+\alpha} S^{(1+\alpha)^2}, \quad S(2,2) = u^{1-\alpha} d S^{1-\alpha^2},$$
$$S(2,1) = u d^{1-\alpha} S^{1-\alpha^2}, \quad S(2,0) = d^{2-\alpha} S^{(1-\alpha)^2}.$$

The proof is left to the reader. Now, suppose for simplicity that the interest rate is constant, $r = R - 1$ say. Then, the risk-neutral probability at node

THE MULTI-PERIOD MODEL

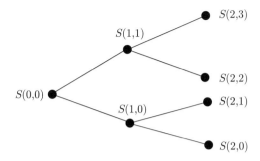

Figure 7.3 *The binomial lattice when the up-factor and down-factor are dependent on the price (2 periods)*

Table 7.1 *The number 2^t as a function of t*

t	2^t	t	2^t
6	64	24	16,777,216
8	256	26	67,108,864
10	1,024	28	268,435,456
12	4,096	30	1,073,741,824
14	16,384	32	4,294,967,296
16	65,536	34	17,179,869,184
18	262,144	36	68,719,476,736
20	1,048,576	38	274,877,906,944
22	4,194,304	40	1,099,511,627,776

$(1,1)$, for example, is given by (7.15) as

$$p^*(1,1) = \frac{R - d(uS^{1+\alpha})^{-\alpha}}{u(uS^{1+\alpha})^\alpha - d(uS^{1+\alpha})^{-\alpha}}.$$

See Exercise 7.5 for the pricing of options in this model.

The model given in Example 7.1 looks very flexible to describe a real situation. However, there are 2^t nodes at time t, because the upward state from state $(t-1, i-1)$ differs from the downward state from state $(t-1, i)$. Note that the number 2^t grows very rapidly as t increases (see Table 7.1). Hence, from practical standpoints, it is important to construct a recombining lattice model to value derivative securities. See Exercise 7.6 for another model that is not recombining.

Definition 7.1 The binomial lattice is said to be *recombining* if the upward state from state $(t, i-1)$ is equal to the downward state from state (t, i) for all t and i. That is,

$$u(t, i-1)S(t, i-1) = d(t, i)S(t, i)$$

for all t and i, where $S(t, i)$ denotes the stock price at node (t, i).

7.3 The Binomial Model for American Options

The binomial lattice can be used to value American options as well. In this section, we consider an American-type derivative with payoff function $h(S)$.

As in the previous section, we denote the stock price at node (t,i) by $S(t,i)$. In particular, if the stock price process follows the binomial model (6.24), we have
$$S(t,i) = u^i d^{t-i} S, \quad i = 0, 1, \ldots, t,$$
where $t = 0, 1, \ldots, T$.

At each node, an investor who has purchased the American derivative has the right to exercise. If exercised at node (t,i), the value of the derivative is given by $h(S(t,i))$. On the other hand, if the investor postpones the right, the value of the derivative at node (t,i) is the same as (7.11), i.e.
$$A(t,i) \equiv R^{-1}[p^* C(t+1, i+1) + q^* C(t+1, i)].$$

Here, $C(t,i)$ denotes the value of the American-type derivative at node (t,i), and p^* is the risk-neutral probability. It follows from (5.30) that
$$C(t,i) = \max\{h(S(t,i)), A(t,i)\}, \quad i = 0, 1, \ldots, t, \qquad (7.16)$$
for all $t = 0, 1, \ldots, T-1$. This is so, because an astute individual chooses an early exercise only when the payoff $h(S(t,i))$ is equal to the present value of the derivative.

The value of the American derivative at the maturity is the same as the European counterpart, i.e. $h(S(T))$. Hence, starting from the maturity, the value of the American derivative can be evaluated backwards using (7.16). We illustrate this procedure in the following example. Note that a hedging portfolio (i.e. a replicating portfolio) can be obtained by the same way as in the previous section. Also, the quantities R and p^* can vary over nodes in the binomial lattice.

We have already seen that the value of an American call option written on the stock that pays no dividends is the same as the European counterpart. This result is proved, if we can show that
$$A(t,i) \geq \{S(t,i) - K\}_+ \qquad (7.17)$$
at each node (t,i). The proof is based on induction. When $t = T$, (7.17) is obviously true. Assuming that the result is true for some $t+1$, we have
$$\begin{aligned} RA(t,i) &\geq p^*\{S(t+1,i+1) - K\}_+ + q^*\{S(t+1,i) - K\}_+ \\ &\geq \{p^* S(t+1,i+1) + q^* S(t+1,i) - K\}_+, \end{aligned}$$
where the second inequality follows since the function $\{x - K\}_+$ is convex in x. But, since the process $\{S^*(t)\}$ is a martingale under p^*, we obtain
$$p^* S(t+1, i+1) + q^* S(t+1, i) = RS(t,i),$$
whence
$$A(t,i) \geq \{S(t,i) - K/R\}_+ \geq \{S(t,i) - K\}_+$$
and (7.17) holds for all (t,i).

American put options may be exercised early, as the next example shows.

Example 7.2 In the binomial model (6.24), suppose that $u = 1.2$, $d = u^{-1}$, $R = 1.1$, and $S = 5$, and consider an American put option written on the stock with strike price $K = 5$ and maturity $T = 3$. Referring to the binomial lattice depicted in Figure 7.2, we have

$$C(3,0) = 2.1066, \quad C(3,1) = 0.8333, \quad C(3,2) = 0, \quad C(3,3) = 0$$

at the maturity. Next, at time $t = 2$, we have

$$A(2,0) = 1.0616, \quad A(2,1) = 0.2066, \quad A(2,2) = 0$$

so that, at node $(2,0)$, an astute individual chooses an early exercise. The option value at time $t = 2$ can be calculated by (7.16) as

$$C(2,0) = 1.5277, \quad C(2,1) = 0.2066, \quad C(2,2) = 0.$$

Repeating similar algebra, we obtain the premium of the American put option as $C(0,0) = 0.2404$. The calculation is left to the reader.

7.4 The Trinomial Model

In this section, we investigate what if there are three possibilities in the movement of the underlying stock price. More specifically, in the single-period model, suppose that $S(1) = uS$ with probability p, $S(1) = mS$ with probability q, or $S(1) = dS$ with probability $1 - p - q$, where $d < m < u$ and $0 < p + q < 1$. Of course, in order to rule out arbitrage opportunities, we need to assume that $d < R < u$ and $R \geq 1$, where $r = R - 1$ is the default-free interest rate.

Consider a contingent claim written on the stock $S(1)$. We try to replicate the claim by the stock and the money-market account. That is, let $B = b(1)$ and $w = \theta(1)$ in (6.28). The current value of the portfolio is equal to $V(0) = wS + B$. As in the binomial case, the time 1 value of the portfolio is equal to $V(1) = wuS + RB$ if $S(1) = uS$, $V(1) = wmS + RB$ if $S(1) = mS$, or $V(1) = wdS + RB$ if $S(1) = dS$. Suppose that the value of the claim at time 1 is given by C_u when $S(1) = uS$, C_m when $S(1) = mS$, or C_d when $S(1) = dS$. This trinomial model is depicted in Figure 7.4 (cf. Figure 7.2 for the binomial case).

Suppose that the portfolio replicates the contingent claim. Then, we must have the following simultaneous equation:

$$\begin{cases} C_u = wuS + RB, \\ C_m = wmS + RB, \\ C_d = wdS + RB. \end{cases} \tag{7.18}$$

Note that we have three equations for two variables w and B to be determined. It can be shown that the system of equations has a solution if and only if

$$\frac{C_u - C_d}{u - d} = \frac{C_u - C_m}{u - m} = \frac{C_m - C_d}{m - d}. \tag{7.19}$$

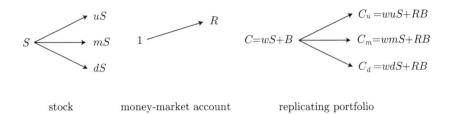

Figure 7.4 *The trinomial model and the replicating portfolio*

That is, the contingent claim is attainable if and only if the condition (7.19) holds. If this is the case, the simultaneous equation is solved as

$$w = \frac{C_u - C_m}{(u - m)S} \quad \text{and} \quad B = \frac{uC_m - mC_u}{(u - m)R},$$

whence, from Theorem 5.2, the fair price of the claim, C say, is obtained as

$$C = \frac{C_u - C_m}{(u - m)} + \frac{uC_m - mC_u}{(u - m)R}$$

under the no-arbitrage condition. Note that the value is the same as (7.2) under the condition (7.19). The proof is left in Exercise 7.11.

Suppose that there are no arbitrage opportunities in the market. Then, from Theorem 5.3, there exists a risk-neutral probability measure P^*, which we denote by $(p^*, q^*, 1 - p^* - q^*)$. That is, under the risk-neutral probability measure, we have $S(1) = uS$ with probability p^*, $S(1) = mS$ with probability q^*, and $S(1) = dS$ with probability $1 - p^* - q^*$. The requirement that the underlying stock price is a martingale under P^* is stated as

$$(u - d)p^* + (m - d)q^* = R - d. \tag{7.20}$$

Similarly, the requirement that the value of the contingent claim is a martingale under P^* is stated as

$$(C_u - C_d)p^* + (C_m - C_d)q^* = CR - C_d.$$

It is readily checked that, under the attainable condition (7.19), the latter equation is the same as (7.20). Moreover, from (7.4), we have

$$\begin{aligned}
C &= R^{-1}\left[p^* C_u + q^* C_m + (1 - p^* - q^*)C_d\right] \\
&= \frac{C_u - C_m}{(u - m)} + \frac{uC_m - mC_u}{(u - m)R} + \left[\frac{m - d}{u - m}(C_u - C_m) - (C_m - C_d)\right]q^*.
\end{aligned}$$

But, under the condition (7.19), the last term in the above equation vanishes, which confirms the attainability of the claim. Since there are many probability distributions $(p^*, q^*, 1 - p^* - q^*)$ that satisfy (7.20), this market is incomplete (see Theorem 5.4).

7.5 The Binomial Model for Interest-Rate Claims

Let $v(t, T)$ denote the time t price of the default-free discount bond maturing at time T, $t \leq T$. The distinguishing feature of the discount bonds from other securities is the fact that $v(T, T) = 1$. That is, the discount bond promises to pay a fixed amount of money, called the face value, to holders at the maturity. The problems of interest are how to determine the current price of the bond and how to describe the stochastic behavior of the price process $v(t, T)$ with respect to $t = 0, 1, \ldots, T$. In this section, we consider these problems in the binomial setting. See Jarrow (1996) for more detailed discussions about discrete interest-rate models.

7.5.1 The Spot-Rate Model

Let $r(t)$ be the spot interest rate at time t, $t = 1, 2, \ldots$. So far, we have treated the interest rates as positive constants; however, in this subsection, we assume that $\{r(t)\}$ is a positive stochastic process. Let $R(t) = 1 + r(t) > 1$.

As we have already seen in (5.26), the time t price of the default-free discount bond with maturity T is given by

$$v(t, T) = E_t^* \left[\frac{B(t)}{B(T)} \right], \quad t \leq T,$$

under the no-arbitrage condition, where $B(t)$ denotes the time t price of the money-market account and E_t^* is the conditional expectation operator under a risk-neutral probability measure P^* given $\mathcal{F}_t$, the information available at time t. In this subsection, we consider a single-period binomial model in which there are two discount bonds with different maturities and obtain a replicating portfolio associated with the discount bonds.

For notational simplicity, we denote the time t price of the default-free discount bond with maturity T_i, $i = 1, 2$, where $T_1 \neq T_2$, by $C^i(t)$. Note that the discount bonds are considered to be derivative securities written on the spot rates. However, the spot rate is not a traded security. So, we need to replicate the money-market account $B(t)$ by these discount bonds.

To that end, suppose that $B(1) = R(1) = u$ with probability p and $B(1) = d$ with probability $q = 1 - p$, where $1 < d < u$. Let $C^i(0) = C^i$ and suppose that $C^i(1) = C_u^i$ if $R(1) = u$ or $C^i(1) = C_d^i$ if $R(1) = d$. In Equation (6.28), let $w_1 = b(1)$ and $w_2 = \theta(1)$, where $B(t) = C^1(t)$ and $S(t) = C^2(t)$. The current value of the replicating portfolio is equal to $V(0) = w_1 C^1 + w_2 C^2$. Also, as before, the time 1 value of the portfolio is equal to $V(1) = w_1 C_u^1 + w_2 C_u^2$ if $R(1) = u$ or $V(1) = w_1 C_d^1 + w_2 C_d^2$ if $R(1) = d$. This binomial model is depicted in Figure 7.5.

Suppose that the portfolio replicates the money-market account $B(1)$. Then, we must have the following simultaneous equation:

$$\begin{cases} u &= w_1 C_u^1 + w_2 C_u^2, \\ d &= w_1 C_d^1 + w_2 C_d^2. \end{cases} \quad (7.21)$$

Note that we have two equations for two variables w_1 and w_2 to be determined.

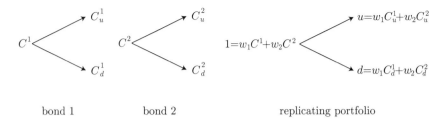

Figure 7.5 *The binomial model for discount bonds*

It is readily seen that the system of equations has a solution
$$w_1 = \frac{dC_u^2 - uC_d^2}{C_d^1 C_u^2 - C_u^1 C_d^2}, \quad w_2 = \frac{uC_d^1 - dC_u^1}{C_d^1 C_u^2 - C_u^1 C_d^2}.$$

Since $V(0) = B(0) = 1$ in order to prevent arbitrage opportunities, we must have
$$\frac{dC_u^2 - uC_d^2}{C_d^1 C_u^2 - C_u^1 C_d^2} C^1 + \frac{uC_d^1 - dC_u^1}{C_d^1 C_u^2 - C_u^1 C_d^2} C^2 = 1.$$

It follows after some algebra that
$$\frac{uC^1 - C_u^1}{uC_d^1 - dC_u^1} = \frac{uC^2 - C_u^2}{uC_d^2 - dC_u^2}. \tag{7.22}$$

The proof is left in Exercise 7.12.

Equation (7.22) tells us that the fraction $(uC - C_u)/(uC_d - dC_u)$ is the same for the two discount bonds with different maturities. Therefore, denoting this fraction by
$$\frac{uC - C_u}{uC_d - dC_u} = \frac{\lambda}{d}, \tag{7.23}$$

we must have
$$C(0) = (1 - \lambda)\frac{C_u}{u} + \lambda \frac{C_d}{d}. \tag{7.24}$$

Note that the λ defined by (7.23) plays a similar role to the market price of risk (see Theorem 7.1). However, in contrast to the previous case, the λ cannot be determined by the parameters of the model alone. This is so, because the spot rate $r(t) = R(t) - 1$ is not a traded security. We cannot use the spot rate to replicate the discount bonds.

Suppose that $0 < \lambda < 1$. Then, Equation (7.24) can be written as
$$C(0) = E^* \left[\frac{C(1)}{B(1)} \right], \tag{7.25}$$

where E^* denotes the expectation with respect to λ. That is, the λ is a risk-neutral probability since the expected, discounted value of $C(1)$ is the same as the current price $C(0)$. Also, repeated applications of (7.25) yield
$$v(0, T) = E^* \left[\frac{1}{B(T)} \right], \quad T \geq 0,$$

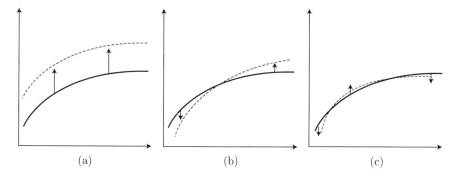

Figure 7.6 *Typical movements of the yield curve:* (a) *parallel shift,* (b) *twist, and* (c) *curvature*

since $C(T) = 1$ at the maturity. Hence, we recovered the key pricing formula (5.26) using the spot-rate model.

7.5.2 The Term Structure Model

Suppose that the current time is t, and the prices $v(t,T)$ of default-free discount bonds with various maturities T are observed in the market. The observed prices $v(t,T)$ are a function of $T = t, t+1, \ldots$, and this function is called the *term structure* of discount bonds.

At the next period, the bond prices change to $v(t+1, T)$, $T = t+1, t+2, \ldots$, with keeping some dependencies between them. Typical movements of the yield curve (the term structure) observed in the real market are depicted in Figure 7.6.* In order to describe such complicated movements of the term structure, the spot-rate model explained above is not flexible enough. Therefore, researchers have started to model the movement of the term structure directly. In this subsection, we explain such an idea using a two-period binomial model.

Consider a two-period economy labeled by $\{0, 1, 2\}$, in which there are traded three default-free discount bonds $v(0, t)$, $t = 0, 1, 2$. While the bond $v(0, 0) = 1$ has matured, the bond $v(0, 1)$ will mature at the next period. The price of the bond $v(0, 2)$ will change to $v(1, 2)$ at the next period. However, the change is random, and we assume that there are two possibilities. That is, $v(1, 2) = v_u(1, 2)$ if the underlying economic situation is in state u, or $v(1, 2) = v_d(1, 2)$ if the underlying situation is in state d. Here, we assume that

$$v_u(1,2) < v_d(1,2).$$

Note that the term structure at time 0 is

$$[1,\ v(0,1),\ v(0,2)]$$

---

* Recall that the yield curve and the discount bond prices are connected by (4.7).

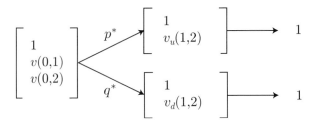

Figure 7.7 *The binomial model for the term structure (2 periods)*

while the time 1 term structure is
$$[1, v_u(1,2)] \quad \text{or} \quad [1, v_d(1,2)].$$
Figure 7.7 describes the movement of the term structure in the two-period binomial model.

Let (p^*, q^*) be a risk-neutral probability for the states u and d, respectively, where $q^* = 1 - p^*$. We shall determine the risk-neutral probability. Assuming it exists, the martingale property (5.24) is equivalent, in this case, to
$$v(0,2) = R_1^{-1}[p^* v_u(1,2) + q^* v_d(1,2)],$$
where $R_1 = 1 + r_1 = 1/v(0,1)$. It follows that
$$p^* = \frac{v_d(1,2) - v(0,2)/v(0,1)}{v_d(1,2) - v_u(1,2)}. \tag{7.26}$$
Recall that there are no arbitrage opportunities if and only if there is a risk-neutral probability. Therefore, as far as $0 < p^* < 1$ in (7.26) or, equivalently,
$$v_u(1,2) < \frac{v(0,2)}{v(0,1)} < v_d(1,2), \tag{7.27}$$
there are no arbitrage opportunities in the two-period binomial market. The interpretation of (7.27) should be clear. The three-period case is left in Exercise 7.13. See Ho and Lee (1986) for details.

7.6 Exercises

Exercise 7.1 Suppose that $u = 1.2$, $d = u^{-1}$, $R = 1.1$, and $S = 5$ in the single-period binomial model. Calculate the premium of a call option with strike price $K = 5$ and maturity $T = 1$. Also, calculate the premium of a put option with the same strike price and the same maturity so as to confirm the put-call parity (5.16).

Exercise 7.2 Confirm that the market price of risk λ_C given in (7.6) is equal to λ_S given in (7.5).

Exercise 7.3 In the same setting as Exercise 7.1, assume that the physical probability is any p, where $0 < p < 1$. Calculate the call option premium using (7.8), and compare the result with the premium obtained in Exercise 7.1.

Exercise 7.4 In the same setting as Exercise 7.1 except $T = 3$, construct the binomial lattice and calculate the option value $C(t, i)$ at each node (t, i). Also, calculate the hedging portfolio at each node.

Exercise 7.5 In Example 7.1, suppose that $u = 1.2$, $d = u^{-1}$, $R = 1.1$, $\alpha = 0.1$, and $S = 5$. Calculate the premiums of European call and put options with strike price $K = 5$ and maturity $T = 2$. Does the put-call parity hold even in this case?

Exercise 7.6 In the binomial model, suppose now that the stock pays dividends proportional to the stock price at the end of each period. That is, the ex-dividend price is given by $(1 - \delta)S$, when the cum-dividend price is S, for some δ, $0 < \delta < 1$. Construct the binomial lattice for the stock price process, and calculate the premiums of the call and put options in the setting of Exercise 7.5 with $\delta = 0.1$. Is the binomial lattice recombining? Also, does the put-call parity hold even in this case? *Hint*: The risk-neutral probability is determined through (5.24).

Exercise 7.7 In the binomial model, suppose that $R = (u + d)/2$ and $d = u^{-1}$, so that the risk-neutral probability is $p^* = 1/2$ (the resulting random walk is symmetric). For the stock price process $\{S(t)\}$, consider a European call option whose payoff is given by $\{M(T) - K\}_+$, where $M(t)$ denotes the maximum of the stock price until time t. Obtain the premium of this call option. *Hint*: Define $S(t)$ as given in Exercise 6.12.

Exercise 7.8 Consider the general binomial model (6.24) with $d = u^{-1}$. Using the change of measure formula, obtain the premium of the call option considered in Exercise 7.7.

Exercise 7.9 In the same setting as Example 7.2, consider an American call option. Calculate the option premium of the call option and confirm that it is optimal not to exercise.

Exercise 7.10 In the setting of Exercise 7.6, consider an American call option written on the stock that pays dividends. Calculate the premium of the call option. Is it optimal not to exercise for the call option?

Exercise 7.11 Confirm that the system of equations (7.18) has a solution if and only if the condition (7.19) holds. Prove that, under this condition, the fair price of the claim is given by (7.2).

Exercise 7.12 Confirm that Equation (7.22) holds. Also, derive a condition under which the λ defined by (7.23) becomes a risk-neutral probability.

Exercise 7.13 Extend the arguments given for the two-period economy in subsection 7.5.2 to the three-period case. In particular, determine the risk-neutral probability at each node.

CHAPTER 8

A Discrete-Time Model for Defaultable Securities

In the pricing of a corporate debt, we need to construct a stochastic model to describe the default time epoch of the debt. The hazard rate model is a promising tool for that purpose. This chapter considers a discrete-time model for the pricing of defaultable securities. See Bielecki and Rutkowski (2002) and Duffie and Singleton (2002) for details of credit risk valuation.

8.1 The Hazard Rate

Consider a discount bond issued by a corporate firm at time 0, and let X be a random variable representing the lifetime of the bond. It is assumed that X takes values on non-negative integers. The probability distribution of X is denoted by
$$p_n = P\{X = n\}, \quad n = 0, 1, 2, \ldots.$$
In the literature of credit risk valuation, the distribution function
$$F_n = P\{X \leq n\}, \quad n = 0, 1, 2, \ldots,$$
is called the *cumulative default probability*, whereas its complementary probability function
$$S_n = P\{X \geq n\}, \quad n = 0, 1, 2, \ldots,$$
is called the *survival probability*.

Note that
$$F_n + S_{n+1} = 1, \quad n = 0, 1, 2, \ldots,$$
and these probabilities are obtained from the probability distribution $\{p_n\}$ of X. That is,
$$F_n = \sum_{k=0}^{n} p_k \quad \text{and} \quad S_n = \sum_{k=n}^{\infty} p_k, \quad n = 0, 1, 2, \ldots.$$
Conversely, given the survival probability S_n, the probability distribution is recovered as
$$p_n = S_n - S_{n+1}, \quad n = 0, 1, 2, \ldots.$$
It is also recovered from the cumulative default probability as well.

Now, suppose that the current time is $t > 0$, and the bond is currently alive (see Figure 8.1). That is, we are given the information that the lifetime of the bond is not less than t. This information changes the probability distribution

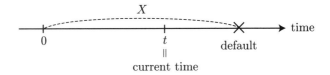

Figure 8.1 *The lifetime of a corporate bond*

of the lifetime. In fact, by the law of conditional probability, we have

$$P\{X = t+n | X \geq t\} = \frac{P\{X = t+n\}}{P\{X \geq t\}} = \frac{p_{t+n}}{\sum_{k=t}^{\infty} p_k}, \quad n = 0, 1, 2, \ldots. \quad (8.1)$$

Recall that $P\{X = t+n | X \geq t\} = P\{X = n\}$, i.e. X has the *memoryless* property (3.11), if and only if X is geometrically distributed. Hence, in general, the lifetime distribution changes according to the information that the bond is currently alive.

Definition 8.1 (Hazard Rate) The *hazard rate* is the probability that the default occurs at the next period given no default before that time. Mathematically, the hazard rate at time t for discrete random variable X is defined by

$$h_t \equiv P\{X = t | X \geq t\} = \frac{p_t}{S_t}, \quad t = 0, 1, 2, \ldots, \quad (8.2)$$

provided that $S_t > 0$. The function h_t with respect to t is called the *hazard function*. The random variable X is called *increasing hazard rate* (IHR for short) if the hazard function h_t is non-decreasing in t, while it is called *decreasing hazard rate* (DHR for short) if h_t is non-increasing in t.

Note that the hazard function is constant, i.e. a constant hazard rate (CHR for short), if and only if X is geometrically distributed. The proof is left in Exercise 8.1. Also, the hazard rate h_t is the special case of the conditional distribution (8.1) with $n = 0$. The hazard rate can be calculated from the cumulative default probability F_n or the survival probability S_n.

Conversely, any of these can be deduced from the hazard rates. To see this, we obtain from (8.2) that

$$\frac{S_{t+1}}{S_t} = 1 - h_t, \quad t = 0, 1, 2, \ldots,$$

since $p_n = S_n - S_{n+1}$. It follows that

$$S_{t+1} = \prod_{n=0}^{t} (1 - h_n), \quad t = 0, 1, 2, \ldots, \quad (8.3)$$

with $S_0 = 1$. The probability distribution is then obtained as

$$p_t = h_t \prod_{n=0}^{t-1} (1 - h_n), \quad t = 0, 1, 2, \ldots,$$

where $\prod_{n=0}^{-1}(1-h_n) = 1$. The conditional probability (8.1) can also be obtained from the hazard rates (see Exercise 8.3).

Example 8.1 Let X be a non-negative, discrete random variable with hazard rates h_t, and suppose that $P\{X < \infty\} = 1$. The random variable X can be generated as follows. Let Y be an exponentially distributed random variable with parameter 1, i.e.

$$P\{Y > x\} = e^{-x}, \quad x \geq 0.$$

Let $\Lambda_t = -\log \prod_{n=0}^{t}(1-h_n)$, and define

$$N = \min\{t : \Lambda_t \geq Y\}. \tag{8.4}$$

The random variable N is well defined, since Λ_t is non-decreasing in t and diverges as $t \to \infty$. Also, it is easily seen that the event $\{N > t\}$ is equivalent to the event $\{Y > \Lambda_t\}$. It follows that

$$P\{N > t\} = P\left\{Y > -\log \prod_{n=0}^{t}(1-h_n)\right\} = \prod_{n=0}^{t}(1-h_n) = S_{t+1},$$

whence N and X are equal in law. Summarizing, the lifetime X can be generated by the hazard rates h_t and the exponentially distributed random variable Y through (8.4).

8.2 A Discrete Hazard Model

Suppose that we are given the hazard function h_n of a lifetime X, where $0 < h_n < 1$, and want to simulate the default time epoch. For this purpose, we define independent (not necessarily identical) Bernoulli random variables Y_n according to

$$P\{Y_n = 1\} = 1 - P\{Y_n = 0\} = 1 - h_n, \quad n = 0, 1, 2, \ldots. \tag{8.5}$$

Now, as in Example 3.2, let

$$\tau = \min\{n : Y_n = 0\}. \tag{8.6}$$

The random variable τ is finite with probability one, since the probability of the event $\{Y_n = 1 \text{ for all } n = 0, 1, 2, \ldots\}$ is zero. Then, from (8.6) and the independence of Y_n's, we obtain

$$P\{\tau > t\} = P\{Y_n = 1 \text{ for all } n \leq t\} = \prod_{n=0}^{t}(1-h_n), \tag{8.7}$$

whence τ and X are equal in law. In particular, if the Bernoulli random variables Y_n are identically distributed, i.e. $h_n = h$ say, then the lifetime τ is geometrically distributed with parameter h, as expected.

The construction of default time epoch through (8.6) is more flexible than the usual one (e.g. the one given in Example 8.1), since it is possible to make the hazard function h_n a stochastic process. Note that, even in the stochastic case, the lifetime random variable X can be defined through (8.6). To explain this, consider a sample path of the stochastic hazard process $h_n = h_n(\omega)$, $n = 0, 1, 2, \ldots$. The values of Y_n's can be generated according to the Bernoulli trials (8.5) independently, and the lifetime X is determined by (8.6) accordingly.

In order to calculate the survival probability of X, we have from (8.7) that, given any realization of h_n,

$$P\{X > t | h_n, n = 0, 1, 2 \ldots\} = \prod_{n=0}^{t}(1 - h_n).$$

It follows from the law of total probability (2.31) that

$$S_{t+1} = E[P\{X > t | h_n, n = 0, 1, 2, \ldots\}] = E\left[\prod_{n=0}^{t}(1 - h_n)\right], \quad t = 0, 1, 2, \ldots, \tag{8.8}$$

with $S_0 = 1$. It should be noted that we do not assume that the hazard rates h_n are independent. What we have assumed here is that Y_n are independent, given any realization of h_n. This is the notion of *conditional independence*. Since h_n can be dependent on each other, the random variables Y_n may not be independent. The stochastic process $\{N_t; t = 0, 1, \ldots\}$ defined by $N_t = 1_{\{X \le t\}}$, where X is constructed as above, is often called the (discrete) *Cox process*. Note that the event $\{N_t = 0\}$ indicates no default before time t. In the literature of credit risk valuation, it is common to use Cox processes to generate default time epochs. See, e.g., Lando (1998), Kijima (2000), and Kijima and Muromachi (2000).

The Cox process can be seen as follows. Suppose that $\{h_t; t = 0, 1, \ldots\}$ is a stochastic process and a corporate bond is alive at time $t - 1$. At the next time epoch t, we first observe the realization of h_t and determine the value of Y_t according to the Bernoulli trial (8.5) independently. If the trial results in $Y_t = 0$, then put $X = t$ according to (8.6); otherwise go one step ahead. This procedure is repeated until the event $\{Y_n = 0\}$ eventually happens.

In order to confirm that this procedure indeed generates the lifetime X in law, it suffices to calculate the survival probability. Since $\{X \ge t\}$ if and only if $\{Y_n = 1$ for all $n < t\}$ and since the events $\{Y = 0\}$ are determined independently, we obtain (8.8). The detailed proof is left to the reader.

Example 8.2 Suppose that the stochastic hazard rates h_t take the same value h for all t, but the value h is a random variable. That is, we are uncertain about the value of the hazard rates, but we know that the hazard rates are unchanged over time. In this situation, under the assumption of conditional independence, given the value of $h_0 = h$, the lifetime is geometrically distributed and the conditional survival probability is given by

$$P\{X > t | h_0 = h\} = (1 - h)^t, \quad t = 0, 1, \ldots.$$

It follows that the survival probability is obtained as

$$S_t = E\left[(1 - h)^{t-1}\right], \quad t = 1, 2, \ldots.$$

If, in particular, h is equal to either λ or μ with probability p or $1 - p$,

respectively, we have
$$S_t = p(1-\lambda)^{t-1} + (1-p)(1-\mu)^{t-1}, \quad t = 1, 2, \ldots.$$

See Exercise 8.4 for another model.

8.3 Pricing of Defaultable Securities

In this section, we consider a discrete-time model for the pricing of corporate discount bonds in the no-arbitrage pricing framework. The following discussions are parallel to those in the binomial model considered in Chapter 7. A continuous-time model for the pricing of corporate bonds will be discussed in Chapter 13.

Throughout this section, we denote the time t price of the defaultable discount bond maturing at time T by $v^c(t,T)$. The time t price of the default-free discount bond maturing at time T is denoted by $v(t,T)$, as before. This section considers a two-period model only in order to avoid notational complexity. The default-free discount bond follows the binomial term structure model presented in the previous chapter. The defaultable discount bond is assumed to follow a similar binomial model, i.e. the price either goes up or goes down, in the case of no default. In addition to these possibilities, there is another possibility of default. It is assumed that default can occur at either $t = 1$ or $t = 2$.

Suppose that the current time is t and the prices $v(t,T)$ as well as $v^c(t,T)$ for various maturities T are observed in the market. The observed prices are functions of $T = t, t+1, \ldots$, and such a function is called the *term structure* of discount bonds.

Consider a two-period economy labeled by $\{0, 1, 2\}$, in which there are traded three defaultable discount bonds $v^c(0,t)$, $t = 0, 1, 2$, where the bond $v^c(0,0) = 1$ has matured. The bond $v^c(0,1)$ will either mature or default at the next period, while the price of the bond $v^c(0,2)$ will change to $v^c(1,2)$ in the case of no default. However, the change is random, and we assume that there are two possibilities in the case of no default. That is, $v^c(1,2) = v_u^c(1,2)$ if the underlying economic situation is in state (u, n), or $v^c(1,2) = v_d^c(1,2)$ if the state is (d, n). Here, 'n' indicates the case of no default. It is assumed that
$$v_u^c(1,2) < v_d^c(1,2).$$

Note that the term structure of the defaultable discount bonds at time 0 is
$$[1, v^c(0,1), v^c(0,2)],$$
while the time 1 term structure is given either by
$$[1, v_u^c(1,2)] \quad \text{or} \quad [1, v_d^c(1,2)]$$
in the case of no default.

On the other hand, when default occurs, we assume that claim holders will receive some constant $\delta > 0$ for sure at the maturity. In practice, the fraction δ is smaller than the face value $F = 1$ and the quantity δ is called the *recovery*

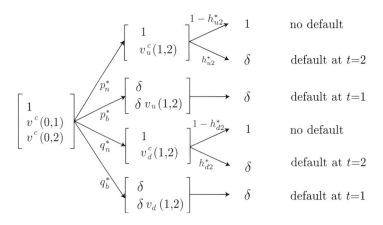

Figure 8.2 *Evolution of the term structure of the defaultable discount bonds*

rate. If default occurs at time $t = 1$, while the bond $v^c(0,1)$ pays the recovery rate δ to claim holders at the maturity $t = 1$, the bond $v^c(0,2)$ pays the recovery rate at the maturity $t = 2$. This recovery formulation is called the *recovery of treasure* (RT).

Let $C_x(1,2)$ denote the time $t = 1$ price of the defaulted bond with maturity $T = 2$ at node (x, b), where 'b' indicates the case of default. It is readily seen that, under the no-arbitrage pricing framework, we must have

$$C_x(1,2) = \delta v_x(1,2), \quad x = u, d,$$

where $v_x(t, T)$ denotes the time t price of the default-free discount bond maturing at time T in the state x. Hence, in the case of default at time $t = 1$, the time $t = 1$ term structure is given by

$$[\delta, \delta v_u(1,2)] \quad \text{or} \quad [\delta, \delta v_d(1,2)].$$

If default occurs at time $t = 2$ for the discount bond $v^c(1,2)$, it pays the recovery rate δ to claim holders at the maturity $t = 2$. Figure 8.2 depicts the evolution of the term structure of the defaultable discount bonds.

Recall that there are no arbitrage opportunities in the economy if and only if there exists a risk-neutral probability measure (see Theorem 5.3). In the following, we adopt this assumption and intend to find out a risk-neutral probability measure P^*. Let h^*_{x2} be the default probability of the bond at node (x, n) and at time $t = 2$ under P^*. At the node (x, n), the martingale property (5.24) is equivalent, in this case, to

$$v^c_x(1,2) = v_x(1,2)[(1 - h^*_{x2}) + h^*_{x2}\delta], \quad x = u, d. \tag{8.9}$$

It follows that

$$h^*_{x2} = \frac{1 - v^c_x(1,2)/v_x(1,2)}{1 - \delta}, \quad x = u, d.$$

Hence, in order to rule out arbitrage opportunities, we must have $0 < h^*_{x2} < 1$

so that
$$\delta v_x(1,2) < v_x^c(1,2) < v_x(1,2), \quad x = u, d.$$
That is, the defaultable discount bond price must be somewhere between the discounted recovery rate and the default-free discount bond price.

Let p_n^*, q_n^*, p_b^*, and q_b^* be the risk-neutral probabilities associated with the states (u,n), (d,n), (u,b), and (d,b), respectively. Consulting Figure 8.2, we obtain from (5.24) that
$$v^c(0,1) = v(0,1)[p_n^* + q_n^* + (p_b^* + q_b^*)\delta] \tag{8.10}$$
and
$$v^c(0,2) = v(0,1)\left[p_n^* v_u^c(1,2) + q_n^* v_d^c(1,2) + p_b^* \delta v_u(1,2) + q_b^* \delta v_d(1,2)\right]. \tag{8.11}$$
Hence, the term structure of the defaultable discount bonds can be calculated, once the risk-neutral probabilities are found out.

Note that the risk-neutral probabilities are any positive numbers satisfying (8.10), (8.11), and
$$p_n^* + q_n^* + p_b^* + q_b^* = 1.$$
In order to make this point clearer, suppose that the default occurs independently of the state of the economy. That is, denoting the upward and downward probabilities of the economy in the risk-neutral world by p^* and $q^* = 1 - p^*$, respectively, we have $p_n^* = p^*(1 - h_1^*)$, $q_n^* = q^*(1 - h_1^*)$, $p_b^* = p^* h_1^*$, and $q_b^* = q^* h_1^*$, where h_1^* is the default probability of the bond at time $t = 1$ under P^*. It follows from (8.10) that
$$v^c(0,1) = v(0,1)\left[1 - h_1^* + h_1^* \delta\right]. \tag{8.12}$$
Also, from (8.9) with $h_{u2}^* = h_{d2}^* = h_2^*$ and (8.11), we obtain
$$v^c(0,2) = v(0,2)\left[h_1^* \delta + (1 - h_1^*)\frac{v_x^c(1,2)}{v_x(1,2)}\right], \quad x = u, d. \tag{8.13}$$
The proof is left in Exercise 8.5.

In order to rule out arbitrage opportunities, we must have $0 < h_1^* < 1$ so that, from (8.12),
$$\delta v(0,1) < v^c(0,1) < v(0,1).$$
That is, again, the defaultable discount bond price must be greater than the discounted recovery rate and less than the default-free discount bond price. Also, from (8.13), we obtain
$$\delta v(0,2) < v^c(0,2) < \frac{v_x^c(1,2)}{v_x(1,2)} v(0,2)$$
and, from (8.12) and (8.13),
$$\frac{1 - v^c(0,1)/v(0,1)}{1 - \delta} = \frac{v_x^c(1,2)/v_x(1,2) - v^c(0,2)/v(0,2)}{v_x^c(1,2)/v_x(1,2) - \delta}.$$
The proof is left to the reader. See Jarrow and Turnbull (1995) for more details.

Let τ denote the default time epoch of the defaultable discount bond. From (8.9) and (8.13), we obtain

$$\begin{aligned}v^c(0,2) &= v(0,2)\left[(1-h_1^*)(1-h_2^*) + \{h_1^* + (1-h_1^*)h_2^*\}\delta\right]\\ &= v(0,2)E^*\left[1_{\{\tau>2\}} + \delta 1_{\{\tau\leq 2\}}\right],\end{aligned} \quad (8.14)$$

where E^* denotes the expectation operator under P^*.

Suppose that there is a risk-neutral probability measure P^*, as given. Then, in general, it is known that the time t price of a defaultable contingent claim X with maturity T is given by

$$C(t) = B(t)E_t^*\left[\frac{X}{B(T)}\left(1_{\{\tau>T\}} + \delta 1_{\{\tau\leq T\}}\right)\right], \quad t \leq T, \quad (8.15)$$

where $\tau > t$. Here, $B(t)$ is the time t price of the default-free money-market account and δ denotes the recovery rate, which can be a random variable in the general setting.

In particular, if the contingent claim is a defaultable discount bond maturing at time T, i.e. $X = 1$, then we obtain from (8.15) that

$$v^c(t,T) = v(t,T) - E_t^*\left[\frac{B(t)}{B(T)}L1_{\{\tau\leq T\}}\right], \quad (8.16)$$

where $\tau > t$ and $L = 1 - \delta$ is the *loss fraction* due to default (see Exercise 8.6). Equation (8.16) says that the price of the defaultable discount bond is equal to the price of the default-free discount bond minus the risk-adjusted expectation of the discounted loss at default.

Moreover, if the default-free money-market account $B(t)$ and the default time epoch τ are independent and if the recovery rate δ is constant, then we obtain from (8.15) that

$$v^c(t,T) = v(t,T)E_t^*\left[1_{\{\tau>T\}} + \delta 1_{\{\tau\leq T\}}\right] = v(t,T)\left[\delta + (1-\delta)P_t^*\{\tau > T\}\right]. \quad (8.17)$$

Hence, in this setting, the pricing of defaultable discount bonds is reduced to obtaining the survival probability $P_t^*\{\tau > T\}$ of the bond under the risk-neutral probability measure P^*. See Jarrow and Turnbull (1995) for details.

We next consider another recovery formulation called the *recovery of market value* (RMV). That is, instead of recovering some fraction at maturity, we assume that claim holders will receive at default epoch a fraction of the market value just before default. Denoting the recovery fraction by δ, this means that the time t price of the contingent claim is given by

$$C(t) = B(t)E_t^*\left[\frac{X}{B(T)}1_{\{\tau>T\}} + \frac{\delta C(\tau)}{B(\tau)}1_{\{\tau\leq T\}}\right]. \quad (8.18)$$

The difference between (8.18) and (8.15) is only due to the recovery formulation at default. That is, claim holders will receive at maturity the fraction δ for sure in (8.15), while they will receive at default epoch the fraction δ of the market value $C(\tau)$ just before default in (8.18). The RMV formulation was originally considered by Duffie and Singleton (1999) for defaultable securities.

We note that (8.18) is an integral equation for $C(t)$, which can be solved as

$$C(t) = M(t) E_t^* \left[\frac{X}{M(T)} \right]; \quad M(t) = \prod_{n=1}^{t} \frac{1 + r_n}{1 - Lh_n}, \quad (8.19)$$

where $L = 1 - \delta$ and r_n denotes the time n default-free spot rate. The proof is left in Exercise 8.7. Another recovery formulation is discussed in Exercise 8.8 with the price comparison in the recovery formulation.

8.4 Correlated Defaults

Since a typical financial institution possesses a portfolio of corporate bonds, it is essential to extend the above hazard model to the multivariate case. Let X_i be the lifetime of the ith corporate bond and suppose that the institution possesses n different bonds in total. Then, we need to determine the joint distribution of the lifetimes X_i, i.e.

$$P\{X_1 = t_1, X_2 = t_2, \ldots, X_n = t_n\}, \quad t_i \geq 0,$$

in order to understand their probabilistic properties. Recall that all the information of the probabilistic properties is determined by the joint distribution. However, when n gets large, it becomes extremely difficult to estimate the joint distribution from the observed data only. Hence, some assumptions should be invoked to determine the joint distribution based on easily obtainable data. In the literature of survival analyses, many such attempts have been made. See, e.g., Singpurwalla (1995) and Kijima, Li, and Shaked (2001). In this section, we sketch the idea of Kijima (2000) and Kijima and Muromachi (2000) to determine the joint distribution based on the Cox processes.

In order to keep the presentation as simple as possible, we take $n = 2$. Let h_t^i denote the hazard rate of bond i at time t, $t = 0, 1, 2, \ldots$, and suppose that $\{(h_t^1, h_t^2)\}$ is a two-dimensional stochastic process. We shall use the same idea as before in order to determine the default epochs X_1 and X_2. That is, as in (8.5) and (8.6), given any realization of (h_t^1, h_t^2), $t = 0, 1, 2, \ldots$, let Y_t^i be independent Bernoulli random variables such that

$$P\{Y_t^i = 1\} = 1 - P\{Y_t^i = 0\} = 1 - h_t^i, \quad t = 0, 1, 2, \ldots,$$

where $0 \leq h_t^i \leq 1$, and define

$$X_i = \min\{n : Y_n^i = 0\}, \quad i = 1, 2.$$

Then, the process $\{(N_t^1, N_t^2)\}$ defined by $N_t^i = 1_{\{X_i \leq t\}}$, $i = 1, 2$, follows a two-dimensional Cox process with the underlying stochastic hazard rate process $\{(h_t^1, h_t^2)\}$. The actual default times are determined under the *conditional independence*. A similar argument to the univariate case then leads to the following:

$$P\{X_1 > t_1, X_2 > t_2\} = E\left[\prod_{n=0}^{t_1} (1 - h_n^1) \prod_{k=0}^{t_2} (1 - h_k^2) \right], \quad t_1, t_2 = 0, 1, 2, \ldots.$$
(8.20)

The proof is left to the reader.

If the processes $\{h_t^1\}$ and $\{h_t^2\}$ are independent, then we have from (8.8) and (8.20) that

$$P\{X_1 > t_1, X_2 > t_2\} = P\{X_1 > t_1\}P\{X_2 > t_2\},$$

so that X_1 and X_2 are independent. But, if $\{h_t^1\}$ and $\{h_t^2\}$ are not independent, then they are not independent in general.

Example 8.3 This example calculates the joint survival probability (8.20) for a particular case. Suppose as in Example 8.2 that each stochastic hazard rate h_t^i takes the same value for all t, but the value $h_t^i = \lambda_i$ is a continuous random variable on $(0,1)$. Then, under the assumption of conditional independence, given any value of $\lambda_i \in (0,1)$, each lifetime is geometrically distributed so that the conditional joint survival probability is given by

$$P\{X_1 > t_1, X_2 > t_2 | h_0^1 = \lambda_1, h_0^2 = \lambda_2\} = (1-\lambda_1)^{t_1}(1-\lambda_2)^{t_2}.$$

It follows from (8.20) that

$$P\{X_1 > t_1, X_2 > t_2\} = \int_0^1 \int_0^1 (1-\lambda_1)^{t_1}(1-\lambda_2)^{t_2} f(\lambda_1, \lambda_2) d\lambda_1 d\lambda_2,$$

where $f(x,y)$ is the joint density function of (λ_1, λ_2). Note that, unless λ_1 and λ_2 are independent, the default times X_1 and X_2 are not independent. See Exercise 8.9 for another bivariate model.

For two random variables X_1 and X_2 given in (8.20), suppose that (h_t^1, h_t^2) are deterministic. Then, the conditional independence implies that X_1 and X_2 are independent. Denote the *first-to-default time* by τ, i.e. $\tau = \min\{X_1, X_2\}$. If the hazard rates are deterministic, then the hazard function $h(t)$ of τ is also deterministic and given by

$$h(t) = h_t^1 + h_t^2 - h_t^1 h_t^2, \quad t = 0, 1, 2, \ldots. \tag{8.21}$$

The proof is left to the reader.

Next, suppose that h_t^i are stochastic, and they are *not* independent. However, as above, we assume that, given any realization of $\{(h_t^1, h_t^2)\}$, the random variables X_j are conditionally independent, whence the joint survival probability is given by (8.20). Since $\{\tau > t\} = \{X_1 > t, X_2 > t\}$, we obtain

$$P\{\tau > t\} = E\left[\prod_{n=0}^{t}(1-h_n^1)(1-h_n^2)\right]. \tag{8.22}$$

The hazard rate for τ is calculated according to (8.2). Note that, since the hazard rates h_t^i are not deterministic, the hazard rate for τ is not given by (8.21) in general.

Finally, we consider the *conditional* hazard rate defined by

$$\hat{h}(t) = P\{\tau = t | \tau \geq t; (h_n^1, h_n^2), n \leq t-1\}.$$

Suppose that the two bonds are currently alive, i.e. $\tau \geq 1$, and we are inter-

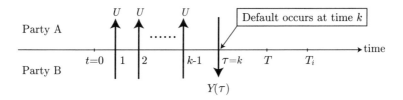

Figure 8.3 *Cashflows in a typical credit swap*

ested in the conditional hazard rate $\hat{h}(1)$ of τ. From (8.22), we obtain

$$\begin{aligned}\hat{h}(1) &= 1 - \frac{P\{\tau \geq 2|(h_0^1, h_0^2)\}}{P\{\tau \geq 1|(h_0^1, h_0^2)\}} \\ &= 1 - \frac{E\left[\prod_{n=0}^1 (1-h_n^1)(1-h_n^2)\big|(h_0^1, h_0^2)\right]}{(1-h_0^1)(1-h_0^2)} \\ &= E\left[h_1^1 + h_1^2 - h_1^1 h_1^2 | (h_0^1, h_0^2)\right],\end{aligned}$$

since $\{\tau = 1\} = \{\tau \geq 1\} - \{\tau \geq 2\}$. In general, we have

$$\hat{h}(t) = E\left[h_t^1 + h_t^2 - h_t^1 h_t^2 | (h_n^1, h_n^2), n \leq t-1\right], \quad t = 1, 2, \ldots. \qquad (8.23)$$

The proof is left in Exercise 8.10. Equation (8.23) should be compared with (8.21).

Example 8.4 (Credit Swap) This example considers a *credit swap* in which Party A pays Party B at the credit event, if it is before the maturity of the contract, the difference between the value of an asset just before default and its market value at default epoch. In compensation of the payment, Party B pays Party A an annuity at a rate, called the *credit swap premium*, until either the maturity of the contract or termination by the designated credit event, whichever happens first (see Figure 8.3). Such a credit swap (typically of the first-to-default feature when the credit event is default of several underlying assets) is frequently traded on the OTC (over-the-counter) basis. See Duffie (1999) for the survey of credit swap valuation.

To be more specific, suppose we have n underlying defaultable discount bonds. The time t price of the ith discount bond maturing at time T_i is denoted by $v_i(t, T_i)$, $i = 1, 2, \ldots, n$, if default has not occurred. Let X_i denote the default time of discount bond i, and let τ be the first-to-default time. We assume that the discount bonds are alive at time t, i.e. $X_i > t$, and that T_i are longer than the maturity T of the swap contract. The payment $Y(\tau)$ from Party A to Party B at the credit event, if before the maturity, is given by

$$Y(\tau) = v_i(\tau, T_i) - \phi_i(\tau) \quad \text{if } \tau = X_i \leq T, \qquad (8.24)$$

where $\phi_i(t)$ is the market value of discount bond i in the event of default at time t. Here, we implicitly assume that no simultaneous defaults will occur.

Suppose that there exists a risk-neutral probability measure P^* as given, and that the credit swap premium U is paid from Party B to Party A at time epochs $t = 1, 2, \cdots, T$, in the case of no default before the maturity T. If default occurs at time t, then the payment terminates at time $t-1$. Denoting the default-free money-market account by $B(t)$, the value of the annuity paid from Party B to Party A is given by

$$R_B = E^* \left[\sum_{t=1}^{T} \left(\sum_{k=1}^{t-1} \frac{U}{B(k)} \right) 1_{\{\tau=t\}} + \sum_{t=1}^{T} \frac{U}{B(t)} 1_{\{\tau>T\}} \right], \qquad (8.25)$$

where E^* denotes the expectation operator under P^*. On the other hand, the value of the payment at the credit event, if any, is

$$R_A = \sum_{i=1}^{n} E^* \left[\frac{Y(\tau)}{B(\tau)} 1_{\{\tau=X_i \leq T\}} \right]. \qquad (8.26)$$

Since we must have $R_A = R_B$ for the swap contract, the credit swap premium is determined from (8.25) and (8.26) so that

$$U = \frac{\sum_{i=1}^{n} E^* \left[Y(\tau) 1_{\{\tau=X_i \leq T\}} B^{-1}(\tau) \right]}{E^* \left[\sum_{t=1}^{T} \left(\sum_{k=1}^{t-1} B^{-1}(k) \right) 1_{\{\tau=t\}} + \sum_{t=1}^{T} B^{-1}(t) 1_{\{\tau>T\}} \right]}. \qquad (8.27)$$

Note that the swap premium U given by (8.27) can be evaluated if we have the joint survival probability of (X_1, X_2). See Kijima (2000) for details.

8.5 Exercises

Exercise 8.1 Let X be a non-negative, discrete random variable with $p_n = P\{X = n\}$, $n = 0, 1, \ldots$. Prove that the hazard function is constant, $h_t = h$ say, if and only if X is geometrically distributed with $p_n = h(1-h)^n$.

Exercise 8.2 (Weibull Distribution) Suppose that the hazard function for a non-negative, discrete random variable X is given by

$$h_t = 1 - q^{(t+1)^\alpha - t^\alpha}, \quad t = 0, 1, 2, \ldots,$$

for some $\alpha > 0$ and $0 < q < 1$. Obtain the survival probability. *Note*: h_t is increasing for $\alpha > 1$, decreasing for $0 < \alpha < 1$, and constant for $\alpha = 1$. See Nakagawa and Osaki (1975) for details.

Exercise 8.3 Prove that, for each $t = 0, 1, \ldots$, we have

$$P\{X = t+n | X \geq t\} = h_{t+n} \prod_{k=t}^{t+n-1} (1 - h_k), \quad n = 0, 1, 2, \ldots,$$

where $\prod_{k=t}^{t-1}(1 - h_k) = 1$.

Exercise 8.4 In contrast to Example 8.2, suppose that the hazard rates h_n are independent and equal to either λ or μ with probability p or $1-p$, respectively. Calculate the survival probability.

Exercise 8.5 Suppose that default occurs independently of the state of the economy under the risk-neutral probability measure P^*. Recalling that the

EXERCISES

upward probability p^* of the economy under P^* is given by (7.26), prove that (8.13) holds.

Exercise 8.6 Show that the discount bond price (8.11) can be rewritten as in (8.16).

Exercise 8.7 In the RMV recovery formulation, show that the price $C(t)$ given by (8.19) satisfies the integral equation (8.18). *Hint*: Use a Cox process argument. Note that, for $s > t$,

$$P_t^*\{\tau = s | h_n, n = t, t+1, \ldots, T\} = h_s \prod_{n=t+1}^{s-1} (1 - h_n).$$

Exercise 8.8 Suppose that claimholders will receive a fraction of contingent claim X, rather than the market value, at default epoch. Denoting the fraction by δ, confirm that the time t price of the contingent claim is given by

$$C(t) = B(t) E_t^* \left[\frac{X}{B(T)} 1_{\{\tau > T\}} + \frac{\delta X}{B(\tau)} 1_{\{\tau \leq T\}} \right].$$

This recovery formulation is called the *recovery of face value* (RFV) and originally considered by Madan and Unal (1998). Now, assuming that all the parameters except the recovery formulation are the same, prove that

$$v_{\text{RMV}}(t, T) \leq v_{\text{RT}}(t, T) \leq v_{\text{RFV}}(t, T),$$

where $v_a(t, T)$ denotes the time t price of the defaultable discount bond under recovery formulation a, $a = $ RMV, RT or RFV.

Exercise 8.9 Suppose that the hazard rates (h_n^1, h_n^2) are independent with respect to n and equal to either $(h_n^1, h_n^2) = (\lambda_1, \lambda_2)$ or (μ_1, μ_2) with probability p or $1-p$, respectively. Calculate the survival probability of the first-to-default time τ.

Exercise 8.10 Let τ denote the first-to-default time of (X_1, X_2). Prove that the conditional hazard rate of τ is given by (8.23).

CHAPTER 9

Markov Chains

A Markov process is a stochastic process characterized by the Markov property that the distribution of the future process depends only on the current state, not on the whole history. If the state space consists of countably many states, the Markov process is called a Markov chain and, if there are absorbing states, it is called an absorbing Markov chain. Absorbing Markov chains play a prominent role in finance, for example, by identifying default as an absorbing state. This chapter explains the importance of discrete-time Markov chains by showing various examples from finance. See, e.g., Çinlar (1975), Karlin and Taylor (1975), Anderson (1991), or Kijima (1997) for more information about Markov chains.

9.1 Markov and Strong Markov Properties

When drawing a die randomly, the resulting outcomes are statistically independent. However, in many practical situations, it is natural to assume that an outcome occurring in the future depends, to some extent, on its history. The point here is how far we should go back. An extreme case is to assume that the future depends on the whole history. However, analytical tractability will then be completely lost. A *Markov property* can be seen as an intermediate between them.

Throughout this chapter, we consider a stochastic process $\{X_n\}$ in discrete time defined on a finite state space $\mathcal{N}$. The Markov property asserts that the distribution of X_{n+1} depends only on the current state $X_n = i_n$, not on the whole history. Formally, the process $\{X_n\}$ is called a *Markov chain* if, for each n and every $i_0, \ldots, i_n$ and $j \in \mathcal{N}$,

$$P\{X_{n+1} = j | X_0 = i_0, X_1 = i_1, \ldots, X_n = i_n\} = P\{X_{n+1} = j | X_n = i_n\}. \tag{9.1}$$

Here, $P\{Y = y | X = x\}$ denotes the conditional probability defined by

$$P\{Y = y | X = x\} = \frac{P\{Y = y, X = x\}}{P\{X = x\}},$$

where $P\{X = x\} > 0$. The Markov property was proposed by Markov (1906) as part of his work on generalizing the classical limit theorems of probability. Recall that the independence assumption is equivalent to

$$P\{X_{n+1} = j | X_0 = i_0, \ldots, X_n = i_n\} = P\{X_{n+1} = j\}.$$

The difference between the independence assumption and the Markov property (9.1) does not perhaps seem significant at first sight. However, as we shall see below, the Markov property enables us to develop a rich system of concepts and theorems and to derive many useful results in applications.

Given the history $\{X_0 = i_0, \ldots, X_n = i_n\}$, the Markov property suggests that the current state $\{X_n = i_n\}$ is enough to determine all the distributions of the future (see Exercise 9.1). That is,

$$P\{X_{n+1} = i_{n+1}, \ldots, X_{n+m} = i_{n+m} | X_0 = i_0, \ldots, X_n = i_n\} \qquad (9.2)$$
$$= P\{X_{n+1} = i_{n+1}, \ldots, X_{n+m} = i_{n+m} | X_n = i_n\}, \quad m = 1, 2, \ldots.$$

Thus, once the current state X_n is known, prediction of future distributions cannot be improved by adding any knowledge of the past. Note, however, that this does not imply that the past lacks information about future behaviors (this is true for the independent case). The past *does* affect the future through the present state.

Another consequence of the Markov property (9.1) is that the past and the future are *conditionally independent* given the present state. That is,

$$P\{X_0 = i_0, \ldots, X_{n-1} = i_{n-1}, X_{n+1} = i_{n+1}, \ldots, X_{n+m} = i_{n+m} | X_n = i_n\}$$
$$= P\{X_0 = i_0, \ldots, X_{n-1} = i_{n-1} | X_n = i_n\} \qquad (9.3)$$
$$\times P\{X_{n+1} = i_{n+1}, \ldots, X_{n+m} = i_{n+m} | X_n = i_n\}.$$

Conversely, (9.3) characterizes the Markov property (9.1) and, hence, the conditional independence (9.3) is an equivalent notion to the Markov property. The proof is left in Exercise 9.2.

The Markov property may hold at some stopping times (see Definition 5.10) as well. The validity of the Markov property at stopping times is called the *strong Markov property*. The process satisfying the strong Markov property is characterized by

$$P\{X_{\tau+1} = j | X_0 = i_0, \ldots, X_{\tau-1} = i_{\tau-1}, X_\tau = i\} = P\{X_{\tau+1} = j | X_\tau = i\} \qquad (9.4)$$

for every stopping time τ. Here, the random variable X_τ is defined as $X_{\tau(\omega)}(\omega)$ for each $\omega \in \Omega$. It can be shown from (9.4) that the past and the future of the process are conditionally independent given the present, where the stopping time τ plays the role of 'present'. It is known that *any* discrete-time Markov chain has the strong Markov property. See Kijima (1997) for the proof. For a continuous-time Markov process, numerous sufficient conditions are known for the strong Markov property. See, e.g., Freedman (1971) for details.

9.2 Transition Probabilities

In this section, we study some elementary properties of the conditional probability $P\{X_{n+1} = j | X_n = i\}$, $i, j \in \mathcal{N}$, the right-hand side of (9.1).

Let $\{X_n\}$ be a Markov chain with state space $\mathcal{N}$, and define

$$p_{ij}(n, n+1) \equiv P\{X_{n+1} = j | X_n = i\}, \quad i, j \in \mathcal{N}.$$

The conditional probability $p_{ij}(n, n+1)$ is called the (one-step) *transition* probability from state i to state j at time n; cf. (6.5). The matrix with components $p_{ij}(n, n+1)$, $\mathbf{P}(n, n+1) = (p_{ij}(n, n+1))$ say, is called the (one-step) *transition matrix* at time n. The m-step transition probabilities at time n are defined by

$$p_{ij}(n, n+m) = P\{X_{n+m} = j | X_n = i\}, \quad i, j \in \mathcal{N},$$

and the corresponding m-step transition matrix at time n is denoted by

$$\mathbf{P}(n, n+m) = (p_{ij}(n, n+m)).$$

Although the transition matrices may not be square in general, we assume that they are square in the following.

The one-step transition probabilities determine the m-step transition probabilities. To see this, note that

$$\begin{aligned} p_{ij}(n, n+2) &= P\{X_{n+2} = j | X_n = i\} \\ &= \sum_k P\{X_{n+1} = k, X_{n+2} = j | X_n = i\} \\ &= \sum_k p_{ik}(n, n+1)\, p_{kj}(n+1, n+2), \quad i, j \in \mathcal{N}, \end{aligned}$$

where the sum is taken over all the states possible for X_{n+1}. Here, the third equality follows from the chain rule for conditional probabilities and the Markov property. The above equations can be written in matrix form as

$$\mathbf{P}(n, n+2) = \mathbf{P}(n, n+1)\,\mathbf{P}(n+1, n+2). \tag{9.5}$$

It follows by an induction argument that

$$\begin{aligned} \mathbf{P}(n, n+m) &= \mathbf{P}(n, n+1)\mathbf{P}(n+1, n+2) \\ &\quad \times \cdots \times \mathbf{P}(n+m-1, n+m), \quad m = 0, 1, 2, \ldots, \end{aligned} \tag{9.6}$$

where we use the convention $\mathbf{P}(n, n) = \mathbf{I}$, the identity matrix. See Exercise 9.3 for a related result. It should be noted that matrix multiplication is *not* commutative in general, i.e. $\mathbf{AB} \neq \mathbf{BA}$ for two square matrices $\mathbf{A}$ and $\mathbf{B}$.

Definition 9.1 (Stochastic Matrix) A square matrix $\mathbf{A} = (a_{ij})$ is said to be *stochastic* if

$$a_{ij} \geq 0 \quad \text{and} \quad \sum_j a_{ij} = 1 \quad \text{for all } i, j \in \mathcal{N}.$$

If $\sum_j a_{ij} \leq 1$ then $\mathbf{A}$ is called *substochastic*. A substochastic matrix is called *strictly substochastic* if $\sum_j a_{ij} < 1$ for at least one state i.

The definition states that a stochastic (substochastic, respectively) matrix is a non-negative matrix whose row sums are unity (less than or equal to unity). In matrix notation, $\mathbf{A}$ is stochastic (substochastic) if and only if

$$\mathbf{A} \geq \mathbf{O} \quad \text{and} \quad \mathbf{A1} = \mathbf{1} \quad (\mathbf{A1} \leq \mathbf{1}), \tag{9.7}$$

where $\mathbf{O}$ denotes the *zero matrix*, the matrix whose components are all zero, and $\mathbf{1}$ the column vector whose components are all unity. (9.7) reveals that the

Perron–Frobenius eigenvalue of any finite, square stochastic matrix is unity and the corresponding right eigenvector is **1**. See Kijima (1997) for details. Note that, for each n, the m-step transition matrix $\mathbf{P}(n, n+m)$ is stochastic for all $m = 0, 1, 2, \ldots$.

Let $\alpha_i = P\{X_0 = i\}$ and let $\boldsymbol{\alpha} = (\alpha_i)$. The column vector $\boldsymbol{\alpha}$ is called the *initial distribution* of $\{X_n\}$. Note that $\boldsymbol{\alpha}$ is a *probability vector* in the sense that $\alpha_i \geq 0$ and $\sum_i \alpha_i = 1$ or, in vector notation,

$$\boldsymbol{\alpha} \geq \mathbf{0} \quad \text{and} \quad \boldsymbol{\alpha}^\top \mathbf{1} = 1, \tag{9.8}$$

where **0** denotes the *zero vector*, the column vector whose components are all zero, and $\top$ the transpose. Defining the unconditional probabilities

$$\pi_i(n) = P\{X_n = i\}, \quad n = 0, 1, \ldots,$$

the column vector $\boldsymbol{\pi}(n) = (\pi_i(n))$ is called the *state distribution* of $\{X_n\}$ at time n. Of course, $\boldsymbol{\pi}(0) = \boldsymbol{\alpha}$, the initial distribution. The state distribution can be computed by

$$\boldsymbol{\pi}^\top(n) = \boldsymbol{\alpha}^\top \mathbf{P}(0, n), \quad n = 0, 1, 2, \ldots. \tag{9.9}$$

It follows from (9.6) that

$$\boldsymbol{\pi}^\top(n+1) = \boldsymbol{\pi}^\top(n) \mathbf{P}(n, n+1), \quad n = 0, 1, 2, \ldots. \tag{9.10}$$

Note that, since $\mathbf{P}(0, n)$ is stochastic and $\boldsymbol{\alpha}$ is a probability vector, one has $\boldsymbol{\pi}(n) \geq \mathbf{0}$ and $\boldsymbol{\pi}^\top(n) \mathbf{1} = 1$ from (9.9), whence the state distribution $\boldsymbol{\pi}(n)$ is a probability vector for all $n = 0, 1, \ldots$. See Exercise 9.6 for a numerical example.

If one-step transition probabilities are independent of time n, the Markov chain $\{X_n\}$ is said to be *time-homogeneous* (or *homogeneous* for short), and *non-homogeneous* otherwise. In the homogeneous case, we can define

$$p_{ij} \equiv P\{X_{n+1} = j | X_n = i\}, \quad i, j \in \mathcal{N},$$

for all $n = 1, 2, \ldots$. For a homogeneous Markov chain, the transition matrix does not depend on time, and we denote it by $\mathbf{P} = (p_{ij})$. That is,

$$\mathbf{P} = \mathbf{P}(n, n+1), \quad n = 0, 1, 2, \ldots.$$

The m-step transition matrix in (9.6) becomes

$$\mathbf{P}(n, n+m) = \mathbf{P}^m, \quad m = 0, 1, \ldots,$$

the mth power of $\mathbf{P}$, which is independent of n. The m-step transition probability from state i to state j at time n is therefore independent of n and given by the (i, j)th component of $\mathbf{P}^m$. See Exercise 9.7 for the two-state case with an application to credit risk management. See also Exercise 9.8 for the limiting distribution of a homogeneous Markov chain.

Example 9.1 A p-random walk studied in Chapter 6 is a time-homogeneous Markov chain defined on the integers with upward probability p and downward probability $q = 1 - p$. A simple generalization of the random walk is to assume that a particle, if it is in state i, can in a single transition either stay at i or move to one of the adjacent states $i-1$ or $i+1$. Let $q_i > 0$ be the downward transition probability, $r_i \geq 0$ the probability of no transition (self-transition), and $p_i > 0$ the upward transition probability. That is,

$$\begin{aligned} q_i &= P\{X_{n+1} = i-1 | X_n = i\}, \\ r_i &= P\{X_{n+1} = i | X_n = i\}, \\ p_i &= P\{X_{n+1} = i+1 | X_n = i\}, \end{aligned}$$

and $p_i + r_i + q_i = 1$ for all i. Such a Markov chain is useful, because it is related to a trinomial model and serves as a good discrete approximation to a diffusion process (to be discussed in Chapter 12).

9.3 Absorbing Markov Chains

State i is called *absorbing* if $p_{ii}(t, t+1) = 1$ for all $t = 0, 1, 2, \ldots$. That is, once the Markov chain $\{X_t\}$ enters an absorbing state, it stays there forever. In this section, it is assumed that there is a single absorbing state, N say, where the state space is given by $\mathcal{N} = \{1, 2, \ldots, N\}$. Then, the transition matrix can be written in a canonical form as

$$\mathbf{P}(t, t+1) = \begin{pmatrix} \mathbf{Q}(t, t+1) & \mathbf{r}(t, t+1) \\ \mathbf{0}^\top & 1 \end{pmatrix}. \tag{9.11}$$

The component 1 on the south-east corner implies that the state N is absorbing. Note that

$$\mathbf{Q}(t, t+1)\mathbf{1} + \mathbf{r}(t, t+1) = \mathbf{1},$$

since $\mathbf{P}(t, t+1)$ is stochastic.

In general, it can be shown that

$$\mathbf{P}(0, t) = \begin{pmatrix} \mathbf{Q}(0, t) & \mathbf{r}(0, t) \\ \mathbf{0}^\top & 1 \end{pmatrix}, \tag{9.12}$$

where

$$\mathbf{Q}(0, t) = \mathbf{Q}(0, 1)\mathbf{Q}(1, 2) \cdots \mathbf{Q}(t-1, t), \quad t = 0, 1, 2, \ldots,$$

and

$$\mathbf{r}(0, t) = \mathbf{Q}(0, t-1)\mathbf{r}(t-1, t) + \mathbf{r}(0, t-1), \quad t = 1, 2, \ldots,$$

with $\mathbf{r}(0, 0) = \mathbf{0}$. The proof is left in Exercise 9.9. In particular, when the Markov chain is homogeneous, we have $\mathbf{Q}(0, t) = \mathbf{Q}^t$ and

$$\mathbf{r}(0, t) = (\mathbf{Q}^{t-1} + \cdots + \mathbf{Q} + \mathbf{I})\mathbf{r}, \quad t = 1, 2, \ldots,$$

where $\mathbf{Q} = \mathbf{Q}(t, t+1)$ and $\mathbf{r} = \mathbf{r}(t, t+1)$ for all $t = 0, 1, \ldots$. In the following, we assume that $\mathbf{Q}$ is irreducible and $\mathbf{r}$ is non-zero. These assumptions guarantee that absorption at N occurs in a finite time with probability one, wherever the Markov chain starts from. See, e.g., Kijima (1997) for details.

Example 9.2 Credit ratings are published in a timely manner by rating agencies, and they provide investors invaluable information to assess firms' abilities to meet their debt obligations. For many reasons, however, credit

Table 9.1 *An example of transition matrix for credit ratings (taken from the web site of Standard & Poors)*

Initial	AAA	AA	A	BBB	BB	B	CCC	D	N.R.
AAA	90.34	5.62	0.39	0.08	0.03	0.00	0.00	0.00	3.54
AA	0.64	88.78	6.72	0.47	0.06	0.09	0.02	0.01	3.21
A	0.07	2.16	87.94	4.97	0.47	0.19	0.01	0.04	4.16
BBB	0.03	0.24	4.56	84.26	4.19	0.76	0.15	0.22	5.59
BB	0.03	0.06	0.40	6.09	76.09	6.82	0.96	0.98	8.58
B	0.00	0.09	0.29	0.41	5.11	74.62	3.43	5.30	10.76
CCC	0.13	0.00	0.26	0.77	1.66	8.93	53.19	21.94	13.14

N.R. : Rating withdrawn.

ratings change from time to time to reflect firms' unpredictable credit risk. In recent years, it becomes common to use a Markov chain model to describe the dynamics of a firm's credit rating as an indicator of the likelihood of default. To be more specific, consider a time-homogeneous Markov chain $\{X_t\}$ defined on the state space $\mathcal{N} = \{1, 2, \ldots, N-1, N\}$ with transition matrix $\mathbf{P} = (p_{ij})$. Here, state 1 represents the highest credit class, state 2 the second highest, $\ldots$, state $N-1$ the lowest credit class, and state N designates default. It is usually assumed for simplicity that the default state N is absorbing. The transition probabilities p_{ij} represent the actual probabilities of going from state i to state j in a unit of time. Table 9.1 shows an actual transition matrix for credit ratings published by Standard & Poors.

Let τ represent the time of absorption at state N, called the *absorption time*, in the Markov chain $\{X_t\}$. Since the Markov chain stays there forever once it enters, we have

$$p_{jN}(0,t) = P\{\tau \leq t | X_0 = j\}, \quad t = 1, 2, \ldots.$$

With the convention $P\{\tau \leq 0\} = 0$, it follows that

$$P\{\tau = t | X_0 = j\} = P\{\tau \leq t | X_0 = j\} - P\{\tau \leq t-1 | X_0 = j\}$$

and

$$P\{\tau \geq t | X_0 = j\} = 1 - P\{\tau \leq t-1 | X_0 = j\}, \quad t = 1, 2, \ldots.$$

Hence, if $X_0 = j$, the hazard rate (see Definition 8.1) of the random variable τ is given by

$$\begin{aligned} h_j(t) &\equiv \frac{P\{\tau = t | X_0 = j\}}{P\{\tau \geq t | X_0 = j\}} \\ &= \frac{p_{jN}(0,t) - p_{jN}(0,t-1)}{1 - p_{jN}(0,t-1)} \\ &= 1 - \frac{1 - p_{jN}(0,t)}{1 - p_{jN}(0,t-1)}, \quad t = 1, 2, \ldots, \end{aligned} \qquad (9.13)$$

provided that $p_{jN}(0, t-1) < 1$. Recall that the transition probability $p_{jN}(0, t)$

is the (j, N)th component of the matrix $\mathbf{P}(0, t)$, which can be calculated recursively using (9.5).

Associated with the absorbing Markov chain $\{X_t\}$ is a Markov chain conditional on the event of no absorption. To explain such a Markov chain, let us denote the non-absorbing state space by

$$\hat{\mathcal{N}} = \{1, 2, \ldots, N-1\}$$

and the conditional process by $\{\hat{X}_t\}$. Recall that the submatrix $\mathbf{Q}(t, t+1)$ of the transition matrix $\mathbf{P}(t, t+1)$ in (9.11) is strictly substochastic and associated with the non-absorbing states.

Let $\hat{\pi}_j(t) = P\{\hat{X}_t = j\}$, $j \in \hat{\mathcal{N}}$, be the state probability of the conditional process $\{\hat{X}_t\}$, and define the column vector $\hat{\boldsymbol{\pi}}(t) = (\hat{\pi}_j(t))$. Since $\hat{X}_t$ has not been absorbed until time t, we obtain

$$\hat{\boldsymbol{\pi}}^\top(t+1) = \frac{\hat{\boldsymbol{\pi}}^\top(t)\mathbf{Q}(t, t+1)}{\hat{\boldsymbol{\pi}}^\top(t)\mathbf{Q}(t, t+1)\mathbf{1}}. \tag{9.14}$$

Here, the denominator represents the event of no absorption. Repeated application of (9.14) then yields

$$\hat{\boldsymbol{\pi}}^\top(t) = \frac{\boldsymbol{\alpha}^\top \mathbf{Q}(0, t)}{\boldsymbol{\alpha}^\top \mathbf{Q}(0, t)\mathbf{1}}, \quad t = 0, 1, 2, \ldots, \tag{9.15}$$

where $\boldsymbol{\alpha}$ denotes the initial probability vector defined on $\hat{\mathcal{N}}$. The proof is left in Exercise 9.10. In particular, if the original Markov chain is homogeneous, we have

$$\hat{\boldsymbol{\pi}}^\top(t) = \frac{\boldsymbol{\alpha}^\top \mathbf{Q}^t}{\boldsymbol{\alpha}^\top \mathbf{Q}^t \mathbf{1}}, \quad t = 0, 1, 2, \ldots.$$

See Exercise 9.10 for a numerical example and Exercise 9.11 for the limiting distribution of $\hat{\boldsymbol{\pi}}(t)$ as $t \to \infty$.

Denote the (i, j)th component of $\mathbf{Q}(0, t)$ by $q_{ij}(0, t)$. Then, since the state N is absorbing, we obtain

$$1 - p_{jN}(0, t) = \sum_{k=1}^{N-1} q_{jk}(0, t) \tag{9.16}$$

and

$$P\{\tau = t | X_0 = j\} = \sum_{k=1}^{N-1} P\{X_{t-1} = k, X_t = N | X_0 = j\}$$

$$= \sum_{k=1}^{N-1} q_{jk}(0, t-1) \, p_{kN}(t-1, t).$$

It follows that the hazard rate of τ is given by

$$h_j(t) = \frac{\sum_{k=1}^{N-1} q_{jk}(0, t-1) \, p_{kN}(t-1, t)}{\sum_{k=1}^{N-1} q_{jk}(0, t-1)}$$

$$= \sum_{k=1}^{N-1} \frac{q_{jk}(0,t-1)}{\sum_{\ell=1}^{N-1} q_{j\ell}(0,t-1)} p_{kN}(t-1,t)$$

$$= \sum_{k=1}^{N-1} \hat{p}_{jk}(0,t-1) p_{kN}(t-1,t)$$

$$= E\left[p_{\hat{X}_{t-1},N}(t-1,t)\Big|\hat{X}_0=j\right]. \quad (9.17)$$

Here, from (9.15), the probability

$$\hat{p}_{jk}(0,t-1) \equiv \frac{q_{jk}(0,t-1)}{\sum_{\ell=1}^{N-1} q_{j\ell}(0,t-1)}, \quad k \in \hat{\mathcal{N}},$$

is the transition probability of $\{\hat{X}_t\}$ given $\hat{X}_0 = j$.

For notational simplicity, let us denote

$$g_t(k) = p_{kN}(t,t+1), \quad t = 0,1,2,\ldots.$$

It follows that the process $\{\hat{h}(t); t = 0,1,\ldots\}$ defined by

$$\hat{h}(t) \equiv g_t(\hat{X}_t), \quad t = 0,1,2,\ldots,$$

represents the probability of absorption at time $(t+1)$ given no absorption before that time. Such a process $\{\hat{h}(t)\}$ is called a *hazard process*. We note from (9.17) that

$$h_j(t) = E\left[\hat{h}(t-1)\Big|\hat{X}_0 = j\right], \quad t = 1,2,\ldots.$$

Compare the hazard process with the Cox process formulation.

9.4 Applications to Finance

The theory of Markov chains has been applied to diverse areas of practical problems and enables one to derive many useful results. The field of finance is not an exception. This section provides potential applications of Markov chains to financial engineering.

9.4.1 An Extension of the Binomial Model

In Exercise 6.12, it is shown that the binomial model (6.24) can be expressed as
$$S(t) = u^{(t+W_t)/2} d^{(t-W_t)/2} S, \quad t = 0,1,2,\ldots,T,$$
where $\{W_t\}$ is the associated p-random walk with $x = 1$ and $y = -1$. The random variable W_t represents the difference between the number of upward jumps and that of downward jumps. The random walk $\{W_t\}$ is a Markov chain with state space $\mathcal{N} = \{-T,\ldots,-1,0,1,\ldots,T\}$. Recall that the random walk $\{W_t\}$ is spatially homogeneous.

Following the idea of Example 9.1, we extend the binomial model to include the possibility that the transition probabilities are state-dependent. Let $\{X_t\}$

APPLICATIONS TO FINANCE 149

be a Markov chain defined on $\mathcal{N}$, where X_t represents the difference between the number of upward jumps and that of downward jumps. Suppose that, when $X_n = i \in \mathcal{N}$, the up-factor is u_i while the down-factor is d_i. The risk-free interest rate is assumed to be $R - 1$, independent of the state X_n. The risk-neutral upward probability is then given by

$$p_i^* = \frac{R - d_i}{u_i - d_i}, \quad i \in \mathcal{N},$$

under the assumption of no-arbitrage opportunities. Hence, the process $\{X_n\}$ is a Markov chain with upward transition probabilities p_i^* and downward transition probabilities $q_i^* = 1 - p_i^*$ under the risk-neutral probability measure P^*. The state probabilities $\pi_i^*(T) = P^*\{X_T = i\}$ can be calculated recursively by matrix multiplication (9.6).

In order to obtain the stock price $S(T)$ in a simple form, however, we need an additional assumption, since the jump sizes u_i and d_i depend on the state i of the process $\{X_n\}$. Here, we require the price process $\{S(t)\}$ to be *path-independent*, i.e.

$$u_i d_{i+1} = d_i u_{i-1}, \quad i \in \mathcal{N}. \tag{9.18}$$

The process $\{S(t)\}$ is then recombining (see Definition 7.1) and, at the maturity T, the value $S(T)$ is independent of the sample path that the Markov chain $\{X_n\}$ actually passes through. For example, consider the case $T = 3$. Since

$$u_0 u_1 d_2 = u_0 d_1 u_0 = d_0 u_{-1} u_0$$

as far as $X_3 = 1$, the value of $S(3)$ is the same no matter which path the Markov chain $\{X_n\}$ actually takes (see Exercise 9.12). The call option price, for example, is then given by

$$C = R^{-T} \sum_{j \in \mathcal{N}} \{S_j - K\}_+ \pi_j^*(T),$$

where S_j denotes the time T stock price when $X_T = j$.

9.4.2 Barrier Options

In the above path-independent setting, consider a knock-out call option with knock-out barriers k_u and k_d, where $k_d < 0 < k_u$. That is, it expires worthless if the underlying price level ever hits the specified levels. The option price is therefore given by

$$C_B = R^{-T} E^* \left[\{S(T) - K\}_+ 1_{\{\tau > T\}} \right],$$

where $\tau = \inf\{t > 0 : S(t) \notin (S_{k_d}, S_{k_u})\}$ denotes the first exit time from the interval (S_{k_d}, S_{k_u}).

In order to evaluate the option price, we define an absorbing Markov chain $\{X_n\}$ starting from state 0 with state space

$$\mathcal{N} = \{k_d, \ldots, -1, 0, 1, \ldots, k_u\},$$

where the states k_d and k_u are absorbing. Denoting the state probabilities of

$\{X_n\}$ under the risk-neutral probability measure P^* by $\pi_j^*(T)$, as above, the option value is obtained as

$$C_B = R^{-T} \sum_{j=k_d+1}^{k_u-1} \{S_j - K\}_+ \pi_j^*(T),$$

where $S_{k_d} < K < S_{k_u}$. The state probabilities $\pi_j^*(T)$ can be calculated recursively by matrix multiplication (9.6). We note that other barrier options such as knock-in options can be evaluated similarly.

In particular, when $u_i = u$ and $d_i = d$, i.e. the up-factor as well as the down-factor are the same for all the states, the state probabilities $\pi_j^*(T)$ can be obtained in closed form. To this end, we renumber the states so that the new state space is $\{0, 1, \ldots, N\}$, where $N = k_u - k_d$. Then, the submatrix of the associated transition matrix is given by

$$\mathbf{Q} = \begin{pmatrix} 0 & p & & & \\ q & 0 & p & & \\ & \ddots & \ddots & \ddots & \\ & & q & 0 & p \\ & & & q & 0 \end{pmatrix},$$

where the empty components are all zero, $p = (R - d)/(u - d)$ and $q = 1 - p$. Let

$$\pi_i = \left(\frac{p}{q}\right)^i > 0, \quad i = 1, 2, \ldots, N - 1,$$

and define the diagonal matrix $\mathbf{D}$ with diagonal components $\sqrt{\pi_i}$, i.e.

$$\mathbf{D} = \begin{pmatrix} \sqrt{\pi_1} & & & \\ & \sqrt{\pi_2} & & \\ & & \ddots & \\ & & & \sqrt{\pi_{N-1}} \end{pmatrix}.$$

It is easily seen that

$$\mathbf{DQD}^{-1} = \sqrt{pq} \begin{pmatrix} 0 & 1 & & & \\ 1 & 0 & 1 & & \\ & \ddots & \ddots & \ddots & \\ & & 1 & 0 & 1 \\ & & & 1 & 0 \end{pmatrix}.$$

It follows from well-known results in linear algebra that its spectral decomposition is given by

$$\mathbf{DQD}^{-1} = \sum_{k=1}^{N-1} \lambda_k \mathbf{x}_k \mathbf{x}_k^\top,$$

where the eigenvalues λ_k are obtained as

$$\lambda_k = 2\sqrt{pq} \cos \frac{k\pi}{N}, \quad k = 1, \ldots, N - 1,$$

and the associated eigenvectors $\mathbf{x}_k = (x_{k1},\ldots,x_{k,N-1})^\top$ are given by

$$x_{ki} = C\sin\frac{ki\pi}{N}, \quad i = 1,\ldots,N-1.$$

Here, we take $C = \sqrt{2/N}$, since

$$\sum_{i=1}^{N}\sin^2\frac{ki\pi}{N} = \frac{N}{2}, \quad k = 1,\ldots,N-1,$$

so that $\mathbf{x}_k^\top \mathbf{x}_k = 1$ for all k. But, since

$$(\mathbf{DQD}^{-1})^n = \underbrace{(\mathbf{DQD}^{-1})(\mathbf{DQD}^{-1})\cdots(\mathbf{DQD}^{-1})}_{n} = \mathbf{DQ}^n\mathbf{D}^{-1},$$

we have

$$(\mathbf{DQD}^{-1})^n = \sum_{k=1}^{N-1}\lambda_k^n\,\mathbf{x}_k\mathbf{x}_k^\top.$$

It follows that

$$\mathbf{Q}^n = \sum_{k=1}^{N-1}\lambda_k^n(\mathbf{D}^{-1}\mathbf{x}_k)(\mathbf{D}\mathbf{x}_k)^\top, \quad n = 0,1,2,\ldots.$$

Therefore, we obtain the following transition probabilities:

$$p_{ij}(n) = \frac{2}{N}K_{j-i}(n;p)\sum_{k=1}^{N-1}\cos^n\frac{k\pi}{N}\sin\frac{ki\pi}{N}\sin\frac{kj\pi}{N}, \quad i,j = 1,2,\ldots,N-1, \tag{9.19}$$

where $K_j(n;p) = 2^n p^{(n+j)/2}q^{(n-j)/2}$.

9.4.3 The Jarrow–Lando–Turnbull Model

We have seen in the previous chapter that, in the Jarrow–Turnbull framework (1995), the time t price of a corporate discount bond maturing at time T is given by

$$v^c(t,T) = v(t,T)\left[\delta + (1-\delta)P_t^*\{\tau > T\}\right], \quad t \le T,$$

where τ denotes the default time epoch of the bond, δ is a constant recovery rate, $v(t,T)$ is the time t price of the default-free discount bond maturing at T, and P_t^* is the risk-neutral probability measure given the information $\mathcal{F}_t$. Hence, in this framework, the pricing of a corporate bond is reduced to obtaining the survival probability $P_t^*\{\tau > T\}$ of the corporate bond.

For this purpose, Jarrow, Lando, and Turnbull (1997) developed a Markov chain model to incorporate a firm's credit rating as an indicator of the likelihood of default. It is important to include credit risk information into pricing models, since credit derivatives are sensitive to the firm's credit quality. The Markov chain model explained below is called the *JLT model*. The JLT Markov chain model provides the evolution of an arbitrage-free term structure of credit risk spreads, and it would be appropriate for the pricing and hedging of credit derivatives.

As in Example 9.2, let X_t represent the time t credit rating of a discount bond issued by a corporate firm. We assume that $\{X_t\}$ is a time-homogeneous, absorbing Markov chain defined on the state space $\mathcal{N} = \{1, 2, \ldots, N-1, N\}$, where state N is absorbing (i.e. default). The one-step transition matrix of $\{X_t\}$ is denoted by $\mathbf{P} = (p_{ij})$, where

$$p_{ij} = P\{X_{t+1} = j | X_t = i\}, \quad i, j \in \mathcal{N},$$

and P denotes the actual probability measure. The transition matrix is written in the canonical form (9.11) with $\mathbf{Q} = \mathbf{Q}(t, t+1)$ and $\mathbf{r} = \mathbf{r}(t, t+1)$.

For the pricing of the defaultable discount bond, we need to consider the corresponding stochastic process $\{X_t^*\}$ of credit rating under the risk-neutral probability measure P^*. It is well known that $\{X_t^*\}$ need not be Markovian; however, the JLT model assumes that it is an absorbing Markov chain. The absorbing Markov chain $\{X_t^*\}$ may not be time-homogeneous, so we denote its one-step transition probabilities at time t by

$$p_{ij}^*(t, t+1) = P^*\{X_{t+1}^* = j | X_t^* = i\}, \quad i, j \in \mathcal{N},$$

and the corresponding transition matrix by $\mathbf{P}^*(t, t+1)$ with the canonical partition (9.11). The default-free spot rate under P^* is denoted by $r(t)$. It is assumed throughout that, as in Jarrow and Turnbull (1995), the non-homogeneous Markov chain $\{X_t^*\}$ and the spot rate process $\{r(t)\}$ are independent under the risk-neutral probability measure P^*. According to Jarrow, Lando, and Turnbull (1997), this assumption appears to be a reasonable, first approximation for the actual probabilities in investment grade debts, but the accuracy of this approximation deteriorates for speculative grade debts.

The transition probabilities of $\{X_t^*\}$ conditional on the history up to time t are, in general, given by

$$p_{ij}^*(t, t+1) = \pi_{ij}(t) p_{ij}, \quad i, j \in \mathcal{N}, \tag{9.20}$$

where $\pi_{ij}(t)$ are the *risk-premia adjustments* that may depend on the whole history, and p_{ij} are the actual transition probabilities of the *observed* time-homogeneous Markov chain $\{X_t\}$. If one assumes that

$$\pi_{ij}(t) = \pi_i(t) \quad \text{for} \quad j \neq i, \tag{9.21}$$

and they are deterministic functions of time t, then we have the JLT model, where the risk-neutralized process $\{X_t^*\}$ is a non-homogeneous Markov chain with transition probabilities

$$p_{ij}^*(t, t+1) = \begin{cases} \pi_i(t) p_{ij}, & i \neq j, \\ 1 - \pi_i(t)(1 - p_{ii}), & i = j. \end{cases} \tag{9.22}$$

The transition probability $p_{ii}^*(t, t+1)$ must be so, because

$$\sum_{j=1}^{N} p_{ij}^*(t, t+1) = \sum_{j=1}^{N} p_{ij} = 1.$$

APPLICATIONS TO FINANCE

In the JLT model, the assumption (9.21) is imposed to facilitate statistical estimation.

Note that the risk-adjusted transition probabilities need to satisfy the condition
$$p^*_{ij}(t, t+1) \geq 0, \quad i, j \in \mathcal{N}.$$
From (9.22) with $i = j$, this in turn implies that the risk-premia adjustments $\pi_i(t)$ must satisfy the condition
$$0 < \pi_i(t) < \frac{1}{1 - p_{ii}}, \quad i \neq N. \tag{9.23}$$
The strict inequalities in (9.23) follow since the probability measures P and P^* are equivalent. If (9.23) is violated, we can at least say that the model does not describe the market appropriately.

Now, suppose that we have the market data $v(0, t)$ and $v^c_i(0, t)$ for all $i = 1, 2, \ldots, N-1$ and all $t = 1, 2, \ldots$, where $v^c_i(t, T)$ denotes the time t corporate discount bond maturing at T with the time t credit rating being i. From the pricing formula (8.17), we have
$$P^*\{\tau_i \leq T\} = 1 - P^*\{\tau_i > T\} = \frac{v(0, T) - v^c_i(0, T)}{(1 - \delta)v(0, T)},$$
where τ_i denotes the default time epoch conditional on $X_0 = i$. On the other hand, from the definition of risk-premia adjustments (9.20), it follows that
$$P^*\{\tau_i \leq 1\} = p^*_{iN}(0, 1) = \pi_i(0)p_{iN},$$
whence
$$\pi_i(0) = \frac{v(0, 1) - v^c_i(0, 1)}{(1 - \delta)v(0, 1)p_{iN}}. \tag{9.24}$$
It is easy to see that the condition (9.23) is violated if p_{iN} is sufficiently small compared to the spread $v(0, 1) - v^c_i(0, 1)$. Note that the default probabilities p_{iN} are usually very small especially for high credit ratings i. Notice also that this drawback occurs mainly because the small probability p_{iN} is placed in the denominator of (9.24).

Another possibility to simplify (9.20) is to assume that
$$\pi_{ij}(t) = \ell_i(t) \quad \text{for} \quad j \neq N, \tag{9.25}$$
and they are deterministic functions of time t. Then, we obtain the Kijima–Komoribayashi model, where the risk-neutralized process $\{X^*_t\}$ is a non-homogeneous Markov chain with transition probabilities
$$p^*_{ij}(t, t+1) = \begin{cases} \ell_i(t)p_{ij}, & j \neq N, \\ 1 - \ell_i(t)(1 - p_{iN}), & j = N. \end{cases} \tag{9.26}$$
Note the difference between (9.21) and (9.25). The risk-premia adjustments $\ell_i(t)$ given in (9.25) has been proposed by Kijima and Komoribayashi (1998).

Notice that the risk-premia adjustments $\ell_i(t)$ must satisfy the condition
$$0 < \ell_i(t) \leq \frac{1}{1 - p_{iN}}, \quad i \neq N.$$

The estimation of $\ell_i(t)$ is stable because the non-default probability $1 - p_{iN}$ rather than the default probability p_{iN} is placed in the denominator; see (9.29) below. In the following, we assume that $r_i = 1 - p_{iN} > 0$ for all $i \neq N$. This assumption precludes the possibility that all the bonds in credit class i will default within the next period. Note that, in the actual market, the non-default probabilities r_i are large compared to the default probabilities p_{iN} (see Table 9.1).

Recall that the transition matrix $\mathbf{P}^*(t, t+1)$ with transition probabilities given by (9.26) has the canonical form (9.11), where the submatrix $\mathbf{Q}^*(t, t+1)$ and the column vector $\mathbf{r}^*(t, t+1)$ are obtained, respectively, as

$$\mathbf{Q}^*(t, t+1) = \mathbf{L}_D(t)\mathbf{Q}, \quad \mathbf{r}^*(t, t+1) = \mathbf{1} - \mathbf{L}_D(t)\mathbf{Q}\mathbf{1}. \qquad (9.27)$$

Here, $\mathbf{Q}$ is the submatrix of $\mathbf{P}$, $\mathbf{1}$ is the column vector with all components being unity, and $\mathbf{L}_D(t)$ is the diagonal matrix with diagonal components being the risk-premia adjustments $\ell_i(t)$, $i \neq N$.

The risk-premia adjustments $\ell_i(t)$ can be calculated from the market data $v(0, t)$ and $v_i^c(0, t)$, $i \neq N$. Let us denote the transition probabilities of $\{X_t^*\}$ at time t by

$$p_{ij}^*(t, T) = P_t^*\{X_T^* = j | X_t^* = i\}, \quad i, j \in \mathcal{N},$$

and define the matrices $\mathbf{Q}^*(0, t)$ successively by

$$\mathbf{Q}^*(0, t+1) \equiv \mathbf{Q}^*(0, t)\mathbf{Q}^*(t, t+1), \quad t = 0, 1, 2, \ldots, \qquad (9.28)$$

where $\mathbf{Q}^*(0, 0) = \mathbf{I}$ is the identity matrix. It can be shown that, if $\mathbf{Q}^*(0, t)$ is invertible, then

$$\ell_i(t) = \frac{1}{1 - p_{iN}} \sum_{k=1}^{N-1} q_{ik}^{-1}(0, t) \frac{v_k^c(0, t+1) - \delta v(0, t+1)}{(1 - \delta)v(0, t+1)}, \quad i \neq N, \qquad (9.29)$$

where $q_{ij}^{-1}(0, t)$ denotes the (i, j)th component of the inverse matrix of $\mathbf{Q}^*(0, t)$. The proof is left in Exercise 9.13. In particular, for $t = 0$, we obtain

$$\ell_i(0) = \frac{1}{1 - p_{iN}} \frac{v_i^c(0, 1) - \delta v(0, 1)}{(1 - \delta)v(0, 1)}, \quad i \neq N.$$

Note that, as far as $v_i^c(0, 1) > \delta v(0, 1)$, the risk-premia adjustments $\ell_i(0)$ are well defined.

9.5 Exercises

Exercise 9.1 For the Markov chain $\{X_n\}$, prove that Equation (9.2) holds.

Exercise 9.2 Prove the equivalence between the conditional independence (9.3) and the Markov property (9.1).

Exercise 9.3 (Chapman–Kolmogorov Equation) For any Markov chain $\{X_n\}$, denote its transition probabilities by $p_{ij}(m, n)$. Prove that

$$p_{ij}(m, n) = \sum_k p_{ik}(m, \ell) p_{kj}(\ell, n), \quad m \leq \ell \leq n.$$

EXERCISES

This identity can be compactly written in matrix form as

$$\mathbf{P}(m,n) = \mathbf{P}(m,\ell)\,\mathbf{P}(\ell,n), \quad m \le \ell \le n.$$

Exercise 9.4 Suppose that square matrices $\mathbf{A}$ and $\mathbf{B}$ are stochastic. Show that the matrices $\mathbf{AB}$ and $\alpha\mathbf{A} + (1-\alpha)\mathbf{B}$, $0 < \alpha < 1$, are also stochastic.

Exercise 9.5 A financial institution succeeded to estimate the transition matrix $\mathbf{P}$ for credit ratings, where the unit time is taken to be a year. However, according to a change of risk management strategy, the institution needs a transition matrix $\mathbf{Q}$ for credit ratings with the unit time being a half-year. Confirm that there may not exist a stochastic matrix $\mathbf{Q}$ such that $\mathbf{Q}^2 = \mathbf{P}$. See Israel, Rosenthal, and Wei (2001) for related results.

Exercise 9.6 Suppose that a Markov chain $\{X_n\}$ is governed by the transition matrix

$$\mathbf{P} = \begin{pmatrix} 0.2 & 0.4 & 0.4 \\ 0.4 & 0.1 & 0.5 \\ 0.3 & 0.3 & 0.4 \end{pmatrix}$$

with the initial distribution $\boldsymbol{\alpha} = (0.5, 0.5, 0)^\top$. Calculate the state distributions for X_3 and X_5.

Exercise 9.7 Consider a corporate firm that is influenced by an economic condition, and suppose that there are two states, good (state 1) and bad (state 2), for the condition. If the state is good, the hazard rate for default of the firm is λ, while it is given by μ, $\mu > \lambda$, when the state is bad. Suppose that the economic condition is currently good, and it evolves according to a homogeneous Markov chain with transition probabilities $p_{11} = a$ and $p_{22} = b$, where $0 < a, b < 1$. Determine the survival probability of the firm. *Hint*: The state probabilities at time t of the two-state Markov chain can be obtained in closed form by using the spectral decomposition of the transition matrix.

Exercise 9.8 Let $\{X_n\}$ be a homogeneous Markov chain defined on a finite state space $\mathcal{N}$ with transition matrix $\mathbf{P}$. From (9.9), the state distribution $\boldsymbol{\pi}(n)$ is given by $\boldsymbol{\pi}^\top(n) = \boldsymbol{\alpha}^\top \mathbf{P}^n$. Show that the *limiting distribution* $\boldsymbol{\pi} = \lim_{n\to\infty} \boldsymbol{\pi}(n)$ exists, if $\mathbf{P}$ is *primitive* (see Kijima (1997) for the definition). Also, using (9.10), show that the limiting distribution is a *stationary distribution* in the sense that it satisfies $\boldsymbol{\pi}^\top = \boldsymbol{\pi}^\top \mathbf{P}$. Calculate the stationary distribution for the Markov chain given in Exercise 9.6. *Hint*: By the Perron–Frobenius theorem, we have $\mathbf{P}^n = \mathbf{1}\boldsymbol{\pi}^\top + \boldsymbol{\Delta}^n$, $n = 1, 2, \ldots$, where the spectral radius of $\boldsymbol{\Delta}$ is strictly less than unity.

Exercise 9.9 Prove that the canonical form of $\mathbf{P}(0,t)$ is given by (9.12).

Exercise 9.10 Prove that Equation (9.15) holds. Also, calculate the state distribution $\hat{\boldsymbol{\pi}}(3)$ of the conditional Markov chain governed by

$$\mathbf{P} = \begin{pmatrix} 0.2 & 0.4 & 0.4 \\ 0.4 & 0.1 & 0.5 \\ 0 & 0 & 1 \end{pmatrix}$$

with the initial distribution $\boldsymbol{\alpha} = (0.5, 0.5, 0)^\top$.

Exercise 9.11 In Equation (9.15), suppose that the underlying Markov chain is finite and homogeneous, i.e. $\mathbf{Q}(t, t+1) = \mathbf{Q}$ for all t. Show that the *quasi-limiting distribution* $\hat{\boldsymbol{\pi}} = \lim_{n \to \infty} \hat{\boldsymbol{\pi}}(n)$ exists, if $\mathbf{Q}$ is primitive. Also, using (9.14), show that the quasi-limiting distribution is a *quasi-stationary distribution* in the sense that it satisfies $\hat{\boldsymbol{\pi}}^\top = \gamma \hat{\boldsymbol{\pi}}^\top \mathbf{Q}$ for some $\gamma > 0$. Calculate the quasi-stationary distribution for the Markov chain given in Exercise 9.10. *Hint*: Use the Perron–Frobenius theorem.

Exercise 9.12 Write down the conditions of path-independence (9.18) for the case $T = 4$.

Exercise 9.13 Assuming that the matrix $\mathbf{Q}^*(0, t)$ is invertible, prove that Equation (9.29) holds.

CHAPTER 10

Monte Carlo Simulation

Monte Carlo simulation produces results using a stochastic model, rather than an actual experiment with a real system under study. It is regarded as mathematical experimentation and best fit to modern computer. It is one of the easiest things we can do with a stochastic model. Since finance models have become more complicated than ever, practitioners want to use Monte Carlo simulation more and more. This chapter is intended to serve as an introduction of Monte Carlo simulation with emphasis on how to use it in financial engineering. See, e.g., Ripley (1987) and Ross (1990) for general introduction to stochastic simulation.

10.1 Mathematical Backgrounds

For a family of IID random variables $X_1, X_2, \ldots$, define the partial-sum process $\{S_n\}$ by

$$S_n \equiv X_1 + X_2 + \cdots + X_n, \quad n = 1, 2, \ldots, \tag{10.1}$$

with $S_0 = 0$. Associated with S_n is the *sample mean* defined by $\bar{X}_n \equiv S_n/n$. The IID assumption leads to such classic limit theorems as the strong law of large numbers and the central limit theorem. This section provides these two limit theorems crucial for Monte Carlo simulation.

Proposition 10.1 (Strong Law of Large Numbers) *For a family of IID random variables* $X_1, X_2, \ldots$, *suppose that the mean* $\mu = E[X_1]$ *exists. Then,*

$$\lim_{n \to \infty} \frac{X_1 + X_2 + \cdots + X_n}{n} = \mu \tag{10.2}$$

with probability one.

The strong law of large numbers ensures that the sample mean $\bar{X}_n$ converges to the (unknown) population mean μ almost surely as $n \to \infty$. This is the reason why the sample mean is used as an estimate of the population mean for large samples. The proof is beyond the scope of this book and omitted. See, e.g., Feller (1971) for detailed discussions about the strong law of large numbers. See also Exercise 10.1 for a related problem.

For a family of IID random variables $X_1, X_2, \ldots$, suppose that $E[X_1^2] < \infty$. We denote $E[X_1] = \mu$ and $V[X_1] = \sigma^2$. Then, for S_n defined by (10.1), we have

$$E[S_n] = n\mu, \quad V[S_n] = n\sigma^2.$$

The next result is the classic central limit theorem.

Proposition 10.2 (Central Limit Theorem) *For a family of IID random variables $X_1, X_2, \ldots$ with finite mean μ and finite variance $\sigma^2 > 0$, define*

$$Z_n = \frac{X_1 + X_2 + \cdots + X_n - n\mu}{\sigma\sqrt{n}}, \quad n = 1, 2, \ldots.$$

Then,

$$\lim_{n\to\infty} P\{Z_n \leq x\} = \Phi(x), \quad x \in \mathbf{R}, \tag{10.3}$$

where $\Phi(x)$ is the standard normal distribution function (3.14).

Proof. We prove the proposition for the case that X has the moment generating function (MGF). The idea to prove the central limit theorem for the general case is similar. Let $Y_i \equiv (X_i - \mu)/\sigma$, and denote the MGF of Y_i by $m_Y(t)$. Since Y_i are independent and since

$$Z_n = \frac{1}{\sqrt{n}} \sum_{i=1}^{n} Y_i,$$

the MGF of Z_n is given by

$$m_n(t) = E\left[\exp\left\{t \sum_{i=1}^{n} \frac{Y_i}{\sqrt{n}}\right\}\right] = [m_Y(t/\sqrt{n})]^n.$$

Note that $m_Y(0) = 1$ and

$$m_Y'(0) = E[Y_i] = 0, \quad m_Y''(0) = E[Y_i^2] = 1,$$

whence, from Taylor's expansion (1.16), we obtain

$$m_Y(t) = 1 + \frac{t^2}{2} + \frac{t^3}{3!} m_Y'''(0) + \cdots.$$

It follows that

$$m_n(t) = \left[1 + \frac{t^2}{2n} + \frac{t^3 m_Y'''(0)}{3! \, n\sqrt{n}} + \cdots\right]^n.$$

Now, due to the well-known result

$$\lim_{n\to\infty} a_n = b \quad \Rightarrow \quad \lim_{n\to\infty} \left[1 + \frac{a_n}{n}\right]^n = e^b,$$

we conclude that

$$\lim_{n\to\infty} m_n(t) = e^{t^2/2},$$

which is the MGF of the standard normal distribution. The result follows from Proposition 2.12. □

According to the classic central limit theorem, if we sample X_i more and more from an infinite population with mean μ and variance σ^2, then the probability distribution of the partial sum S_n can be approximated by the normal distribution $N(n\mu, n\sigma^2)$ for sufficiently large n (see Exercise 10.2). This property makes the normal distribution special in stochastic models. We shall present a more general form of the central limit theorem in the next chapter.

Example 10.1 Consider a financial security whose price at the end of day n is denoted by S_n. Since

$$\frac{S_T}{S_0} = \frac{S_1}{S_0}\frac{S_2}{S_1}\cdots\frac{S_T}{S_{T-1}}, \quad T \geq 1,$$

defining $X_t = \log[S_t/S_{t-1}]$, we have

$$S_T = S_0 \exp\{X_1 + X_2 + \cdots + X_T\}.$$

Suppose that $X_1, X_2, \ldots, X_T$ are IID with mean $\mu = E[X_1]$ and variance $\sigma^2 = V[X_1] > 0$. Then, from the central limit theorem, the probability distribution of $\log[S_T/S_0]$ can be approximated by the normal distribution $N(T\mu, T\sigma^2)$ for sufficiently large T, no matter what distribution X_t actually follows. Remark that, in this story, the IID assumption is crucial. The IID assumption usually fails in the real market.

10.2 The Idea of Monte Carlo

For any random variable X and a real-valued function $h(x)$, the composition $h(X)$ is again a random variable under a regularity condition. Also, if $X_1, X_2, \ldots$ are IID, then so are the random variables $h(X_1), h(X_2), \ldots$. Hence, for a family of IID random variables $X_1, X_2, \ldots$, the strong law of large numbers ensures that

$$\lim_{n\to\infty} \frac{h(X_1) + h(X_2) + \cdots + h(X_n)}{n} = E[h(X_1)] \quad (10.4)$$

with probability one, provided that the expectation $E[h(X_1)]$ exists.

To explain the idea of the *Monte Carlo simulation*, consider the definite integral $I \equiv \int_0^1 g(x)\mathrm{d}x$. Suppose that you know the function $g(x)$, but the integration is difficult to calculate by some reasons. Let $X_1, X_2, \ldots$ be a family of IID random variables with density function $f(x)$ over $[0,1]$. Then, since

$$I = \int_0^1 \frac{g(x)}{f(x)} f(x)\mathrm{d}x = E[h(X)], \quad h(x) = \frac{g(x)}{f(x)},$$

we can estimate the integral I based on (10.4) for sufficiently large n. That is, for IID random variables $X_1, X_2, \ldots, X_n$ generated from $f(x)$, the integral I is approximated by the sample mean,

$$I \approx \bar{Z}_n = \frac{h(X_1) + h(X_2) + \cdots + h(X_n)}{n},$$

for sufficiently large n.

Moreover, we can calculate the confidence interval of the value I based on the central limit theorem. To this end, consider the probability

$$P\{|\bar{Z}_n - I| \leq \varepsilon\} = \alpha, \quad \varepsilon > 0,$$

for a given confidence coefficient α. By the central limit theorem, $\bar{Z}_n$ is approximately normally distributed with mean $I = E[h(X_1)]$ and variance σ^2/n,

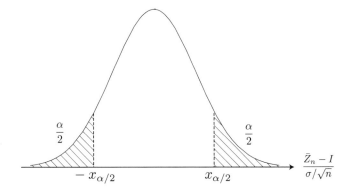

Figure 10.1 *The confidence interval for the estimate I*

where $\sigma^2 = V[h(X_1)]$. It follows that

$$P\left\{-\frac{\varepsilon}{\sigma/\sqrt{n}} \leq \frac{\bar{Z}_n - I}{\sigma/\sqrt{n}} \leq \frac{\varepsilon}{\sigma/\sqrt{n}}\right\} = \alpha$$

and the normalized random variable $\sqrt{n}(\bar{Z}_n - I)/\sigma$ is approximated by the standard normal variable $N(0,1)$. Hence, denoting the $100(1-\alpha)$-percentile of the standard normal distribution by x_α, we have $\sqrt{n}\varepsilon/\sigma = x_{\alpha/2}$ (see Figure 10.1). It follows that

$$\varepsilon = x_{\alpha/2}\frac{\sigma}{\sqrt{n}}, \qquad (10.5)$$

and the confidence interval with the confidence α is given by

$$\left[\bar{Z}_n - x_{\alpha/2}\frac{\sigma}{\sqrt{n}},\ \bar{Z}_n + x_{\alpha/2}\frac{\sigma}{\sqrt{n}}\right].$$

When the variance σ^2 is unknown, σ^2 is approximated by the sample variance

$$S^2 = \frac{1}{n-1}\sum_{i=1}^{n}\left(h(X_i) - \bar{Z}_n\right)^2,$$

in practice.

The result (10.5) can be seen in another way. Suppose that the confidence coefficient α as well as the confidence interval $\varepsilon > 0$ are given. Then, the problem is how many samples are enough to guarantee the confidence level. The result (10.5) tells us that

$$n = \left(\frac{\sigma x_{\alpha/2}}{\varepsilon}\right)^2.$$

See Exercise 10.3 for an example.

Consider a probability space $(\Omega, \mathcal{F}, P)$, and let $X_1, X_2, \ldots$ be a family of IID random variables with finite mean, $\mu = E[X_1]$ say, defined on it. A realization of X_i is a sample $x_i = X_i(\omega)$ for some $\omega \in \Omega$. According to the strong law of large numbers, the limit (10.2) holds for almost all $\omega \in \Omega$. It follows that, for

THE IDEA OF MONTE CARLO   161

almost all realizations of $x_i = X_i(\omega)$, we have

$$\lim_{n\to\infty} \frac{x_1 + x_2 + \cdots + x_n}{n} = \mu. \qquad (10.6)$$

That is, the mean $\mu = E[X_1]$ can be estimated by almost all realizations x_i of IID random variables X_i.

Definition 10.1 (Random Number) A realization of a random variable is called a *random number*, and a realization of a multivariate random variable is called a *random vector*. Random numbers (or vectors) are called IID, if they are realizations of IID (multivariate) random variables.

For IID random variables $X_1, X_2, \ldots$, the sample mean $\bar{X}_n$ is an *estimator* of the population mean $\mu = E[X_1]$. For realizations $x_1, x_2, \ldots$ of those random variables, the realization of $\bar{X}_n$, i.e. $(x_1 + \cdots + x_n)/n$, is called an *estimate* of the mean. Hence, in statistics, an estimate is a realization of an estimator; but, those terms are often used interchangeably. In the case of Monte Carlo simulation, we will also use the terms 'random number' and 'random variable' interchangeably. The strong law of large numbers is then stated as follows: The sample mean of any IID random numbers converges almost surely to the population mean, if it exists, as the number of random numbers tends to infinity.

Suppose that a random variable X is defined by

$$X = f(Z_1, Z_2, \ldots, Z_n),$$

for some random variables $Z_1, Z_2, \ldots, Z_n$ and some function $f(z_1, z_2, \ldots, z_n)$, and suppose that we like to know the expected value $E[X]$. To this end, let $(z_1^i, z_2^i, \ldots, z_n^i)$ be the ith random vector of the multivariate random variable $(Z_1, Z_2, \ldots, Z_n)$. Realizations of X are generated by

$$x^i = f(z_1^i, z_2^i, \ldots, z_n^i), \quad i = 1, 2, \ldots.$$

If the random vectors $(z_1^i, z_2^i, \ldots, z_n^i)$ are IID, the values x^i are IID random numbers of X. Hence, according to the strong law of large numbers, we can evaluate the mean $E[X]$ based on (10.6).

In many practical situations, the function $f(z_1, z_2, \ldots, z_n)$ cannot be expressed in closed form, but the values x^i can be generated numerically from the underlying random numbers z_t^i. The next example highlights this point.

Example 10.2 Let S_t be the time t price of a financial security, and consider a *stochastic difference equation* defined by

$$\frac{S_{t+1} - S_t}{S_t} = \mu(S_t, t) + \sigma(S_t, t)\Delta Z_t, \quad t = 0, 1, \ldots, T-1, \qquad (10.7)$$

where $S_0 = S$ and $\Delta Z_t = Z_{t+1} - Z_t$ with $Z_0 = 0$. This equation describes the situation that the time t rate of return of the security (the left-hand side) is expressed by a sum of a trend $\mu(S_t, t)$ and a stochastic quantity $\sigma(S_t, t)\Delta Z_t$. See Exercise 10.4 for a related problem.

Note that, if $\sigma(x, t) = 0$ in (10.7), then the equation is an approximation of the ordinary *differential equation*

$$dS(t) = S(t)\mu(S(t), t)dt, \quad t \geq 0,$$

which has no uncertainty. However, in many real situations, the increment $S_{t+1} - S_t$ may be influenced by noise, and we cannot completely determine its future behavior. For example, no one believes that the future price S_{t+1} can be completely predicted today. For such a case, we need to consider the stochastic term $\sigma(S_t, t)\Delta Z_t$ as in (10.7).

Now, suppose that we like to know the expected payoff of a European call option with strike price K and maturity T written on the security. That is, we like to evaluate the mean $E[\{S_T - K\}_+]$, where $\{x\}_+ = \max\{x, 0\}$. To this end, we note that the stochastic difference equation (10.7) can be rewritten as the stochastic recursive equation

$$S_{t+1} = S_t \left[1 + \mu(S_t, t) + \sigma(S_t, t)(Z_{t+1} - Z_t) \right], \quad t = 0, 1, \ldots, T-1, \quad (10.8)$$

starting with $S_0 = S$. Hence, even if the functions $\mu(s, t)$ and $\sigma(s, t)$ are complicated, we can calculate the realizations of S_T numerically based on the random vectors $(z_1^i, z_2^i, \ldots, z_T^i)$ and the recursive relation (10.8). Denoting the ith realization of S_T by s_T^i, the expected payoff from the call option can be evaluated as

$$E\left[\{S_T - K\}_+\right] \approx \frac{1}{N} \sum_{i=1}^{N} \{s_T^i - K\}_+ \quad (10.9)$$

for sufficiently large N.

10.3 Generation of Random Numbers

We have seen in the previous section that, in order to evaluate the expected value of a random variable, we need to generate sufficiently many IID random numbers (or random vectors). This section briefly describes how to generate such random numbers.

10.3.1 Uniform Random Numbers

A random number generated from the standard uniform distribution $U(0, 1)$ is called a *uniform random number*. We are interested in how to generate IID uniform random numbers $u_1, u_2, \ldots$ in computer.

Typical modern computers generate the so-called pseudo random numbers. A sequence of real numbers $\{x_n\}$ is said to be *pseudo random numbers*, if they are generated by the recursive relation

$$x_n = ax_{n-1} + c \pmod{m}, \quad n = 1, 2, \ldots, \quad (10.10)$$

where $y \pmod{m} = y - m \times [y/m]$ and $[x]$ denotes the largest integer not greater than x. Hence, x_n takes values on $\{0, 1, \ldots, m-1\}$ (see Exercise 10.5). This method to generate the pseudo random numbers is called the *linear congruential method*.

GENERATION OF RANDOM NUMBERS

The sequence $\{x_n\}$ is called pseudo random, because they are *not* random. It is well known that $x_{n+\ell} = x_n$ for some $\ell \leq m$, so that the same value appears periodically. But, if we take m appropriately, the sequence $\{u_n\}$ obtained by

$$u_n = \frac{x_n}{m}, \quad n = 1, 2, \ldots,$$

looks very similar to IID uniform random numbers. It is a very important subject how to set the values of a and m in (10.10). See, e.g., Tezuka (1995) for details of uniform random numbers.

10.3.2 Normal Random Numbers

Undoubtedly, one of the most important probability distributions in financial engineering is the normal distribution $N(\mu, \sigma^2)$. In this subsection, we describe how to generate normally distributed random numbers, called *normal random numbers*, from IID uniform random numbers.

Let $F(x)$ be the distribution function of a random variable X defined on $\mathbf{R}$, and suppose that the inverse function $F^{-1}(x)$ exists. We have already seen in (3.33) that

$$X \stackrel{\mathrm{d}}{=} F^{-1}(U), \quad U \sim U(0, 1).$$

Hence, given a sequence of IID uniform random numbers $\{u_n\}$ and the inverse function $F^{-1}(x)$, we can generate a sequence of IID random numbers generated from X according to the equation

$$x_n = F^{-1}(u_n), \quad n = 1, 2, \ldots. \tag{10.11}$$

This method to generate the random numbers is called the *inverse transform method* and useful, in particular, for the case that the inverse function $F^{-1}(x)$ is known in closed form.

Recall that the standard normal distribution function $\Phi(x)$ given in (3.14) cannot be represented in closed form. Accordingly, an analytical form of its inverse function is not known. Hence, when applying the inverse transform method to the normal random numbers, we need either to employ a numerical technique to calculate the equation (10.11), or to develop an approximation of the inverse function $\Phi^{-1}(x)$ (see, e.g., Moro (1995) for an approximation formula). See Exercises 10.6 and 10.7 for other methods to generate normal random numbers.

As we have seen in Chapter 3, we often use multivariate normal distributions in financial engineering. In order to treat such a case by Monte Carlo simulation, we need to generate *random vectors* (not merely random numbers) from a multivariate normal distribution.

Let $\mathbf{X} = (X_1, X_2, \ldots, X_n)$ follow an n-variate normal distribution with mean vector

$$\boldsymbol{\mu} = (\mu_1, \mu_2, \ldots, \mu_n)^\top, \quad \mu_i = E[X_i],$$

and covariance matrix

$$\mathbf{\Sigma} = \begin{pmatrix} \sigma_{11} & \sigma_{12} & \cdots & \sigma_{1n} \\ \sigma_{21} & \sigma_{22} & \cdots & \sigma_{2n} \\ \vdots & \vdots & \ddots & \vdots \\ \sigma_{n1} & \sigma_{n2} & \cdots & \sigma_{nn} \end{pmatrix}, \quad \sigma_{ij} = C[X_i, X_j].$$

Here, $\sigma_{ii} = \sigma_i^2 = V[X_i]$ and $\sigma_{ij} = \sigma_{ji}$. It is assumed for simplicity that the covariance matrix $\mathbf{\Sigma}$ is positive definite and so, the inverse $\mathbf{\Sigma}^{-1}$ exists.

Suppose that we are given IID normal random numbers $(x_1, x_2, \ldots, x_n)$. This means that the associated random variable $(X_1, X_2, \ldots, X_n)$ follows the n-variate standard normal distribution $N_n(\mathbf{0}, \mathbf{I})$, where $\mathbf{0}$ denotes the zero vector, and $\mathbf{I}$ is the identity matrix. The n-variate random variable distributed by $N_n(\boldsymbol{\mu}, \mathbf{\Sigma})$ can be constructed from $(X_1, X_2, \ldots, X_n) \sim N_n(\mathbf{0}, \mathbf{I})$. For this purpose, we decompose the given covariance matrix $\mathbf{\Sigma}$ as

$$\mathbf{\Sigma} = \mathbf{C}\mathbf{C}^\top \tag{10.12}$$

for some non-singular matrix $\mathbf{C}$. The decomposition (10.12) is not unique; however, for our purpose, the next decomposition is useful. The proof is left to the reader.

Proposition 10.3 (Cholesky Decomposition) *Given the covariance matrix $\mathbf{\Sigma} = (\sigma_{ij})$, the matrix $\mathbf{C} = (c_{ij})$ in (10.12) can be defined as follows:*

$$\begin{aligned}
c_{11} &= \sqrt{\sigma_{11}}, \\
c_{i1} &= \frac{\sigma_{i1}}{\sqrt{\sigma_{11}}}, \quad i = 2, \ldots, n, \\
c_{jj} &= \sqrt{\sigma_{jj} - \sum_{k=1}^{j-1} c_{jk}^2}, \quad j = 2, \ldots, n, \\
c_{ij} &= \frac{1}{c_{jj}}\left(\sigma_{ij} - \sum_{k=1}^{j-1} c_{ik}c_{jk}\right), \quad j < i, \; i = 2, \ldots, n-1.
\end{aligned}$$

Thus, the matrix $\mathbf{C}$ is lower triangular.

The next result suggests how to obtain random vectors distributed by $N_n(\boldsymbol{\mu}, \mathbf{\Sigma})$ from IID normal random numbers. See Exercise 10.8 for the bivariate case.

Proposition 10.4 *Suppose that $\mathbf{X} \sim N_n(\mathbf{0}, \mathbf{I})$ and that a positive definite covariance matrix $\mathbf{\Sigma}$ is decomposed as in (10.12). Then, we have*

$$\mathbf{Y} = \mathbf{C}\mathbf{X} + \boldsymbol{\mu} \sim N_n(\boldsymbol{\mu}, \mathbf{\Sigma}).$$

Conversely, for $\mathbf{Y} \sim N_n(\boldsymbol{\mu}, \mathbf{\Sigma})$, we have

$$\mathbf{X} = \mathbf{C}^{-1}(\mathbf{Y} - \boldsymbol{\mu}) \sim N_n(\mathbf{0}, \mathbf{I}).$$

Based on Proposition 10.4, IID random vectors distributed by $N_n(\boldsymbol{\mu}, \mathbf{\Sigma})$ can be generated as follows:

(a) Generate IID normal random numbers $x_1, x_2, \ldots, x_n \sim N(0,1)$, and define
$$\mathbf{x} = (x_1, x_2, \ldots, x_n)^\top.$$

(b) Define the random vector $\mathbf{y} = (y_1, y_2, \ldots, y_n)^\top$ by
$$\mathbf{y} = \mathbf{C}\mathbf{x} + \boldsymbol{\mu}.$$

(c) Repeat (a) and (b) until enough random vectors are generated.

10.3.3 Discrete Random Numbers

Finally, we consider a discrete random variable X and its random numbers. Suppose that we are given a uniform random number u, and like to generate a random number x from the probability distribution

$$P\{X = y_i\} = p_i, \quad i = 1, \ldots, n, \tag{10.13}$$

where $p_i > 0$ and $\sum_{i=1}^n p_i = 1$. Given the distribution, we define

$$x = \begin{cases} y_1, & \text{if } u \leq p_1, \\ y_i, & \text{if } p_1 + \cdots + p_{i-1} < u \leq p_1 + \cdots + p_i, \quad i = 2, \ldots, n. \end{cases} \tag{10.14}$$

Then, we claim that x is a random number generated from X. For example, suppose that $X \sim Be(p)$, the Bernoulli distribution with success probability p. From the criteria (10.14), we define

$$x = 0 \text{ if } u \leq 1 - p \text{ while } x = 1 \text{ if } u > 1 - p. \tag{10.15}$$

Then, x is a random number generated from $X \sim Be(p)$.

To prove the claim, we let $U \sim U(0,1)$ and define a random variable X by

$$X = \begin{cases} y_1, & 0 \leq U \leq p_1, \\ y_i, & p_1 + \cdots + p_{i-1} < U \leq p_1 + \cdots + p_i, \quad i = 2, \ldots, n. \end{cases}$$

It is readily seen that

$$\begin{aligned} P\{X = y_i\} &= P\{p_1 + \cdots + p_{i-1} < U \leq p_1 + \cdots + p_i\} \\ &= P\{U \leq p_1 + \cdots + p_i\} - P\{U \leq p_1 + \cdots + p_{i-1}\}. \end{aligned}$$

But, since $U \sim U(0,1)$ so that

$$P\{U \leq x\} = x, \quad 0 < x < 1,$$

it follows that

$$P\{X = y_i\} = (p_1 + \cdots + p_i) - (p_1 + \cdots + p_{i-1}) = p_i,$$

whence X is distributed by (10.13) and the claim follows.

10.4 Some Examples from Financial Engineering

This section considers typical examples of Monte Carlo simulation in financial engineering.

Example 10.3 Consider the binomial model (6.24), i.e.
$$S(t) = u^{W_t} d^{t-W_t} S, \quad t = 0, 1, 2, \ldots, T,$$
where $\{W_t\}$ is the random walk with the underlying IID Bernoulli random variables X_i defined by (6.3). The default-free interest rate is assumed to be constant and given by $r = R - 1$. Then, the binomial model is complete and the upward probability p^* under the risk-neutral probability measure P^* is given by (7.3). That is,
$$P^*\{X_i = 1\} = 1 - P^*\{X_i = 0\} = p^* \equiv \frac{R - d}{u - d}.$$
Since $W_t = W_{t-1} + X_t$, we obtain
$$S(t) = u^{X_t} d^{1-X_t} S(t-1), \quad t = 1, 2, \ldots, T, \quad (10.16)$$
with $S(0) = S$.

Given a family of IID uniform random numbers $\{u_n\}$, we generate a family of IID Bernoulli random numbers $\{x_t^i\}$ according to (10.15) with $p = p^*$. The ith realization of the stock price $s^i(T)$ is generated by the recursion (10.16). Hence,
$$s^i(t) = u^{x_t^i} d^{1-x_t^i} s^i(t-1), \quad t = 1, 2, \ldots, T,$$
with $s^i(0) = S$. The price of a derivative with payoff function $h(x)$ is then approximated based on (10.4) as
$$C \approx R^{-T} \frac{1}{N} \sum_{i=1}^{N} h(s^i(T)) \quad (10.17)$$
for sufficiently large N; cf. (10.9).

Monte Carlo simulation can treat more complicated models such as the one in Example 7.1. That is, consider a binomial model in which the up-factor and down-factor are dependent on the current price and time and given by $u(S,t)$ and $d(S,t)$, respectively, when $S(t) = S$. If the interest rate is also dependent only on the current price and time, $R(S,t)$ say, then the model remains complete and the upward probability, say $p^*(S,t)$ when $S(t) = S$, under the risk-neutral probability measure P^* is given by (7.15). Hence, we need to generate random numbers $\{x_t^i\}$ according to
$$P^*\{X_t = 1|S(t) = S\} = 1 - P^*\{X_t = 0|S(t) = S\} = p^*(S,t),$$
and the stock price $s^i(T)$ is generated by
$$s^i(t) = \begin{cases} u(s^i(t-1), t-1) s^i(t-1), & \text{if } x_t^i = 1, \\ d(s^i(t-1), t-1) s^i(t-1), & \text{if } x_t^i = 0, \end{cases}$$
where $s^i(0) = S$ and $t = 1, 2, \ldots, T$. The price of a derivative with payoff function $h(x)$ is calculated by (10.17) with an appropriate modification. An actual calculation is left in Exercise 10.10.

Example 10.4 Consider the stochastic difference equation (10.7) given in Example 10.2; but, in this example, we assume that ΔZ_t are IID and follow the standard normal distribution $N(0,1)$. Suppose that we divide each year into n equally spaced subintervals with length $h = 1/n$, and let $S(t)$ be the time t price of a financial security, $t = 0, h, 2h, \ldots$. Denote $S_t = S(th)$. In the stochastic difference equation (10.7), the term $\mu(S_t, t)$ represents the mean rate of return per unit of interval. In finance literature, however, it is common to measure the rate of return in terms of year. Hence, it would be better to write the term as $\mu(S_t, t)h$, where $\mu(S_t, t)$ in turn represents the mean rate of return per year. Similarly, in (10.7), the term $\sigma^2(S_t, t)$ represents the variance for the rate of return per unit of interval. Since ΔZ_t are IID and the variance of a sum of independent random variables is a sum of the variances of the random variables, we write the term as $\sigma(S_t, t)\sqrt{h}$, where $\sigma^2(S_t, t)$ in turn represents the variance for the rate of return per year. The standard deviation $\sigma(S, t)$ is called the *volatility* of the price process. In summary, the price process $\{S_t\}$ follows the stochastic difference equation

$$\frac{S_{t+1} - S_t}{S_t} = \mu(S_t, t)h + \sigma(S_t, t)\sqrt{h}\Delta Z_t, \quad t = 0, 1, 2, \ldots, \qquad (10.18)$$

with $S_0 = S$, under the physical probability measure P. It follows that the ith realization $\{s_t^i\}$ of the price process can be generated by the recursion

$$s_{t+1}^i = s_t^i \left[1 + \mu(s_t^i, t)h + \sigma(s_t^i, t)\sqrt{h}\Delta Z_t\right], \quad t = 0, 1, 2, \ldots,$$

with $s_0^i = S$. The expected payoff from a derivative with payoff function $h(x)$ is calculated accordingly. See Exercises 10.11 and 10.12 for actual simulation experiments.

Example 10.5 Let X_t represent the time t credit rating of a corporate bond, and suppose that the process $\{X_t\}$ is a homogeneous Markov chain defined on the state space $\mathcal{N} = \{1, 2, \cdots, K-1, K\}$ with transition matrix $\mathbf{P} = (p_{ij})$. Here, state K is assumed to be absorbing. For the purpose of credit risk management, we simulate $\{X_t\}$ in order to obtain the future distribution of X_T at the risk horizon T. Let $X_0 = i$, and consider the one-step transition probabilities

$$p_{ij} = P\{X_1 = j | X_0 = i\}, \quad j = 1, 2, \ldots, K.$$

Given a uniform random number u, a realization of X_1 can be simulated according to (10.14). Suppose $X_1 = x_1$. If $x_1 = K$, then we put $X_T = K$, since K is absorbing. Otherwise, since

$$p_{x_1, j} = P\{X_2 = j | X_1 = x_1\}, \quad j = 1, 2, \ldots, K,$$

a realization of X_2 can be generated according to (10.14) again. This procedure is repeated until time T to obtain a realization of X_T. Let x_T^i be the ith realization of X_T. Gathering enough realizations, we can discuss the distribution of X_T statistically.

Example 10.6 Recall that the Markov chain $\{X_t\}$ considered in Example 10.5 represents the dynamics of the credit rating of an individual bond. For the valuation of credit risk of a *portfolio*, however, the above model cannot be used directly except in the independent case. The very reason is lack of consideration of correlation effects on the dynamics of credit ratings. That is, let $\{X_t^n\}$, $n = 1, 2, \ldots, N$, be a homogeneous Markov chain describing the dynamics of the credit rating of the nth bond, and consider the multivariate Markov chain

$$\{\mathbf{X}_t = (X_t^1, X_t^2, \ldots, X_t^N); t = 0, 1, \ldots, T\}.$$

We are interested in the *joint* distribution of $\mathbf{X}_T$ at the risk horizon T. As such, a candidate is a state-expanding, one-dimensional Markov chain with K^N states. However, it leads to an exponential growth in the size of the state space as N gets large. Duffie and Singleton (1998) commented a more tractable version of this approach by assuming symmetry conditions among entities.

Recently, Kijima, Komoribayashi, and Suzuki (2002) proposed a multivariate Markov model in which each Markov chain $\{X_t^n\}$ follows

$$X_{t+1}^n = \begin{cases} \xi\left(X_t^n + Z_{t+1}^n + (-1)^{c_n} B_{t+1}^n Y_{t+1}\right), & X_t^n \neq K, \\ X_t^n, & X_t^n = K. \end{cases} \quad (10.19)$$

Here, c_n is either 0 or 1, B_t^n are IID Bernoulli random variables with parameter α_n, i.e.

$$P\{B_t^n = 1\} = 1 - P\{B_t^n = 0\} = \alpha_n, \quad 0 \leq \alpha_n \leq 1,$$

Y_t are defined by

$$P\{Y_t = 1\} = P\{Y_t = -1\} = r, \quad P\{Y_t = 0\} = 1 - 2r,$$

where $0 < r \leq 1/2$ and $P\{|Y_t| > 1\} = 0$, and each Z_{t+1}^n denotes an incremental factor depending on X_t^n whose conditional distribution is given by

$$P\{Z_{t+1}^n = j - i | X_t^n = i\} = q^n(i, j), \quad i, j \in \mathcal{N}. \quad (10.20)$$

The function $\xi(x)$ determines the boundary behavior so that $\xi(x) = 1$ if $x \leq 1$, $\xi(x) = x$ if $1 < x < K$, and $\xi(x) = K$ if $x \geq K$. This is needed, because the highest state 1 (default state K, respectively) is a retaining (absorbing) boundary, meaning that a transition to $x \leq 1$ ($x \geq K$) is pulled back to (put into) state 1 (K).

The nth set of random variables $\{B_t^n\}$ and $\{Z_t^n\}$ is independent of $\{Y_t\}$ and the other processes. Hence, the Markov chains $\{X_t^n\}$ are correlated only through the random variables Y_t which are the common factor to all the Markov chains. The multivariate Markov model (10.19) can be evaluated by a simple Monte Carlo simulation. See Exercise 10.13 for an actual simulation experiment.

The multivariate Markov model can be seen as follows. Suppose $\xi(x) = x$. Then, (10.19) reveals that the increment $\Delta_{t+1}^n \equiv X_{t+1}^n - X_t^n$ is explained by a single common factor Y_{t+1} and a specific term Z_{t+1}^n to the Markov chain $\{X_t^n\}$. The common random variable Y_{t+1} is considered to represent the *systematic risk* of the market, while Z_{t+1}^n is *specific risk* of firm n. The factor $(-1)^{c_n} B_{t+1}^n$ corresponds to the "beta" in the single-index model of modern portfolio theory, although it is a random variable, and measures the change in Δ_{t+1}^n given a change in Y_{t+1}. Note that, when $B_t^n = 0$ (or $Y_t = 0$) with probability 1 for all t and n, the processes $\{X_t^n\}$ constitute independent, time-homogeneous Markov chains, each having one-step transition probabilities $q^n(i,j)$ given by (10.20). The value of c_n determines the sign of correlation coefficient. See Kijima, Komoribayashi, and Suzuki (2002) for more details.

10.5 Variance Reduction Methods

We have already seen in (10.5) that the accuracy of Monte Carlo simulation is proportional to $1/N^\alpha$ with $\alpha = 1/2$. That is, more precisely, the accuracy is given by $\varepsilon = C/\sqrt{N}$, where C is the constant of proportionality. Unfortunately, we cannot improve the power α in the stochastic simulation framework.* However, we *can* improve the constant C by reducing the variance of the estimate in Monte Carlo simulation. Any methods to reduce the variance of an estimate are called *variance reduction methods*. In this section, we explain some of these important methods.

10.5.1 Importance Sampling

In order to explain the idea of importance sampling, consider the definite integral $I = \int_a^b f(x) \mathrm{d}x$, where $f(x) > 0$. Let Y be a random variable with density function $g(x)$ defined on $[a, b]$. Then, since

$$I = \int_a^b \frac{f(x)}{g(x)} g(x) \mathrm{d}x,$$

we obtain

$$I = E\left[\frac{f(Y)}{g(Y)}\right] \approx I_n \equiv \frac{1}{n} \sum_{i=1}^n \frac{f(y_i)}{g(y_i)}$$

for sufficiently large n, where y_i are IID random numbers taken from $g(x)$.

Now, try to choose $g(x)$ so that the variance of the estimate is minimized. The variance of I_n is given by

$$\begin{aligned} V[I_n] &= V\left[\frac{1}{n}\sum_{i=1}^n \frac{f(Y_i)}{g(Y_i)}\right] \\ &= \frac{1}{n^2} \sum_{i=1}^n \left\{ E\left[\left(\frac{f(Y_i)}{g(Y_i)}\right)^2\right] - I^2 \right\} \\ &= \frac{1}{n}\left\{ \int_a^b \frac{f^2(x)}{g(x)} \mathrm{d}x - I^2 \right\}, \end{aligned}$$

where Y_i denotes the random variable associated with the random number y_i.

---

* The power α can be improved by using quasi-random numbers. More specifically, the accuracy is proportional to $N^{-1}(\log N)^k$ in the k-dimensional quasi-Monte Carlo simulation. See, e.g., Niederreiter (1992) for details of quasi-Monte Carlo methods.

It follows that, if we set
$$g(x) = \frac{f(x)}{\int_a^b f(x)\,\mathrm{d}x}, \qquad (10.21)$$
then the variance is minimized and equal to $V[I_n] = 0$. Of course, since $I = \int_a^b f(x)\,\mathrm{d}x$ is unknown, we cannot choose $g(x)$ as in (10.21). In practice, we try to choose $g(x)$ proportional to $f(x)$.

10.5.2 Antithetic Variable Methods

Suppose that we have generated X_1 and X_2 by Monte Carlo simulation, and suppose that our estimate for $E[X]$ is given by $\bar{X} = (X_1 + X_2)/2$. The variance of the estimate is equal to
$$V[\bar{X}] = \frac{1}{4}(V[X_1] + V[X_2] + 2C[X_1, X_2]).$$
Hence, the variance becomes smaller if X_1 and X_2 are negatively correlated. This is the idea of antithetic variable methods.

To be more specific, let $X \sim U(0, \pi/2)$, and consider the definite integral
$$I \equiv \int_0^{\pi/2} \sin x\,\mathrm{d}x = \frac{\pi}{2} E[\sin X].$$
Let $X_1 = X$ and define
$$X_2 = \frac{\pi}{2} - X_1.$$
Then, after some algebra, we obtain
$$C[\sin X_1, \sin X_2] = \frac{\pi - 4}{\pi^2} < 0$$
and
$$V[\sin X_i] = \frac{\pi^2 - 8}{2\pi^2} \approx 0.0947, \quad i = 1, 2,$$
since X_1 and X_2 are equal in law. It follows that
$$V\left[\frac{\sin X_1 + \sin X_2}{2}\right] = \frac{\pi^2 + 2\pi - 16}{4\pi^2} \approx 0.00387.$$
Hence, about 92% of the variance can be reduced, compared to the independent case.

10.5.3 Control Variate Methods

Suppose that two estimates X and Y are obtained by the same simulation experiment, and the mean μ_Y of Y is known while that of X is *unknown*. For some constant α, let
$$Z = X + \alpha(Y - \mu_Y). \qquad (10.22)$$
Then, since $E[Z] = E[X]$, Z is an unbiased estimator for X. The variance of Z is given by
$$V[Z] = V[X] + \alpha^2 V[Y] + 2\alpha C[X, Y].$$

VARIANCE REDUCTION METHODS

Hence, when $\alpha = -C[X,Y]/V[Y]$, the variance $V[Z]$ is minimized and equal to

$$V[Z] = V[X] - \frac{(C[X,Y])^2}{V[Y]}.$$

The random variable Y is called a *control variate* for the estimation of $E[X]$. Since the covariance $C[X,Y]$ is usually unknown, we cannot choose the optimal α (see Exercise 10.15). However, roughly speaking, if X and Y are positively correlated, the optimal α should be negative, and a large value of X is compensated by $\alpha(Y - \mu_Y)$, so that the variance of Z becomes small. If X and Y are negatively correlated, the optimal α should be positive.

Example 10.7 (Average Option) In this example, we consider the binomial model (6.24), i.e.

$$S(t) = u^{W_t} d^{t-W_t} S, \quad t = 0, 1, 2, \ldots, T.$$

The default-free interest rate is assumed to be constant and given by $r = R - 1$. As in Example 10.3, the time t price $S(t)$ can be simulated according to

$$S(t) = u^{X_t} d^{1-X_t} S(t-1), \quad t = 1, 2, \ldots, T,$$

with $S(0) = S$, where the risk-neutral probability is given by

$$P^*\{X_i = 1\} = 1 - P^*\{X_i = 0\} = p^* \equiv \frac{R-d}{u-d}.$$

Let $A(T) = \sum_{t=0}^{T} S(t)/T$ denote the time T average of the price process. An option written on the average is called an *average option* or an *Asian option*. In particular, the payoff function of an average call option with strike price K and maturity T is given by $\{A(T) - K\}_+$.

We estimate the premium of the average option by Monte Carlo simulation. To this end, let $s^i(T)$ denote the ith realization of $S(T)$ in the Monte Carlo simulation described in Example 10.3. Then, from (10.17), we have, for sufficiently large N,

$$Y = R^{-T} \frac{1}{N} \sum_{i=1}^{N} \{s^i(T) - K\}_+$$

as an estimate of the ordinary European call option premium written on $S(T)$. At the same time, we can obtain the ith realization of $A(T)$, denoted by $a^i(T)$, in the same simulation run, and an estimate of the average option premium is given by

$$X = R^{-T} \frac{1}{N} \sum_{i=1}^{N} \{a^i(T) - K\}_+.$$

Note here that X and Y should be positively correlated, since they use the same simulation runs and both are the estimates of call options. Hence, using Y as a control variate, we can define from (10.22) a better estimate of the premium as

$$Z = R^{-T}\frac{1}{N}\sum_{i=1}^{N}\{a^i(T) - K\}_+ - R^{-T}\frac{1}{N}\sum_{i=1}^{N}\{s^i(T) - K\}_+ + C,$$

where $\alpha = -1$ and C denotes the ordinary European call option premium calculated by the Cox, Ross, and Rubinstein formula (see Theorem 7.2).

10.6 Exercises

Exercise 10.1 (Weak Law of Large Numbers) For a family of random variables $X_1, X_2, \ldots$, let

$$m_n = E[X_1 + \cdots + X_n], \quad \sigma_n^2 = V[X_1 + \cdots + X_n].$$

Suppose that $\sigma_n/n \to 0$ as $n \to \infty$. Prove that, for any $\varepsilon > 0$, we have

$$\lim_{n\to\infty} P\left\{\left|\frac{X_1 + \cdots + X_n - m_n}{n}\right| > \varepsilon\right\} = 0.$$

Hint: Use Chebyshev's inequality (Exercise 2.14). Note that we do not assume that X_i are independent nor identical.

Exercise 10.2 Let X be the number of times that a coin, flipped 100 times, lands heads. Assuming the probability that the coin lands a head is 0.49, use the normal approximation to find the probability that $X \geq 60$.

Exercise 10.3 Let $X_1, X_2, \ldots$ be a family of IID random variables that follow the uniform distribution $U(0, \pi/2)$. In order to evaluate the integral $I = \int_0^{\pi/2} \sin x \, dx$, let $\bar{Z}_n = \sum_{i=1}^{n} \sin X_i / n$. Determine the number of samples necessary to guarantee the confidence $\alpha = 95\%$ with the confidence interval $\varepsilon = 0.1\%$, i.e. $P\{|\bar{Z}_n - I| \leq 0.001\} = 0.95$. *Hint*: First show that $I = \pi E[\sin X_1]/2$ and $V[\sin X_1] = (\pi^2 - 8)/2\pi^2$.

Exercise 10.4 In the binomial model (6.24), determine the trend μ and the stochastic term $\sigma \Delta Z_t$ with $E[\Delta Z_t] = 0$ and $V[\Delta Z_t] = 1$ so as to represent the stock price $S(t)$ in terms of (10.7).

Exercise 10.5 In the linear congruential method (10.10) with $c = 1$, $a = 6$, and $m = 5$, generate pseudo random numbers $x_1, \ldots, x_7$ starting with $x_0 = 2$.

Exercise 10.6 Let $U_1, U_2, \ldots, U_{12}$ be IID random variables that follow the standard uniform distribution $U(0, 1)$, and define

$$X = U_1 + U_2 + \cdots + U_{12} - 6.$$

Many practitioners use X as a normal random variable. Discuss how this approximation is justified.

Exercise 10.7 (Box–Müller) For independent random variables U_1 and U_2 that follow $U(0, 1)$, define

$$Z_1 \equiv \sqrt{-2\log U_1}\sin(2\pi U_2), \quad Z_2 \equiv \sqrt{-2\log U_1}\cos(2\pi U_2).$$

Prove that $E[Z_1] = E[Z_2] = 0$, $V[Z_1] = V[Z_2] = 1$ and $E[Z_1 Z_2] = 0$. *Note*: It is known that Z_1 and Z_2 are independent and follow $N(0, 1)$.

Exercise 10.8 In Proposition 10.4, consider the bivariate case with

$$\Sigma = \begin{pmatrix} \sigma_1^2 & \rho\sigma_1\sigma_2 \\ \rho\sigma_1\sigma_2 & \sigma_2^2 \end{pmatrix}, \quad \sigma_1, \sigma_2 > 0.$$

Obtain the Cholesky decomposition of Σ, and state (Y_1, Y_2) explicitly.

Exercise 10.9 Suppose that you have bought a call option with strike price 100 yen, and want to evaluate its expected return. Suppose that you obtain somehow random numbers of the underlying stock price at the maturity, and they are given by

$$\begin{array}{cccccccccc} 100 & 110 & 95 & 108 & 97 & 101 & 96 & 118 & 97 & 91 \\ 93 & 128 & 97 & 111 & 109 & 108 & 93 & 110 & 95 & 106 \end{array}$$

The option premium was 5 yen. Estimate the mean rate of return of the option.

Exercise 10.10 In the extended binomial model considered in Example 10.3, suppose that the up-factor is given by $u(S, t) = uS^\alpha$ when $S(t) = S$ and the down-factor $d(S, t) = dS^{-\alpha}$ (cf. Example 7.1). Assuming that $R(S, t) = R > 1$, calculate by Monte Carlo simulation the call option premium written on this security with parameters $u = 1.1$, $d = 0.9$, $R = 1.05$, $S = K = 100$, $T = 10$, and $\alpha = 0.1$.

Exercise 10.11 In the model considered in Example 10.4, suppose $\mu(S, t) = c(\mu - S)$ and $\sigma(S, t) = \sigma\sqrt{S}$. Calculate by Monte Carlo simulation the expected payoff from the call option written on this security with parameters $c = 0.1$, $\mu = 110$, $\sigma = 0.2$, $S = K = 100$, $T = 1$, and $h = 1/250$.

Exercise 10.12 In the same setting as Exercise 10.11, obtain the probability distribution of $M_T = \max_{0 \le t \le T} S_t$ by Monte Carlo simulation.

Exercise 10.13 In the multivariate Markov model (10.19) introduced in Example 10.6, suppose that $X_0^1 = X_0^2 = A$ as the initial rating, $a_n = c_n = 1$ for $n = 1, 2$, and the conditional transition probabilities $q^n(i, j)$, $n = 1, 2$, are given by Table 9.1. Obtain the joint distribution (X_T^1, X_T^2) with $T = 10$ for the cases $r = 0.5$ and $r = 0.1$ by Monte Carlo simulation. Simulate the same multivariate Markov model except that $c_1 = 1$ and $c_2 = 0$, i.e. the negatively correlated case.

Exercise 10.14 In the framework of Exercise 9.7, suppose that the hazard rate for default of firm i is $\lambda_i = (20 - i)^{-1}$ when the economic condition is good, while it is given by $\mu_i = (10 + i)^{-1}$ when the condition is bad, where $i = 1, 2, \ldots, 5$. By Monte Carlo simulation with $a = 1 - b = 0.7$, obtain the probability that no firms will default within 10 periods.

Exercise 10.15 Consider a control variate method to evaluate the integral

$$I = \frac{\pi}{2} E[\sin X] = \int_0^{\pi/2} \sin x \, dx.$$

We take $X \sim U(0, \pi/2)$ as a control variate. Calculate the variance of the estimate $Z = \sin X - \alpha(X - E[X])$, and draw its graph with respect to α.

CHAPTER 11

From Discrete to Continuous: Towards the Black–Scholes

Random walks and Brownian motions are related through the central limit theorem. Accordingly, the CRR binomial model and the Black–Scholes model are connected in an obvious way. This chapter begins by the formal definition of Brownian motions and then introduces geometric Brownian motions used in the Black–Scholes model to describe the dynamics of a risky security price. It is demonstrated that the Black–Scholes formula is obtained as a limit of the CRR binomial formula. As an alternative security price model, a Poisson process model is also described. A technique to transfer results in random walks to results in the corresponding continuous-time setting will be presented.

11.1 Brownian Motions

We begin this chapter by the formal definition of Brownian motions. Recall that a *stochastic process* $\{X(t),\, t \geq 0\}$ is a collection of random variables $X(t)$ with index set $\{t \geq 0\}$ representing time. Throughout this chapter, we denote the probability space by $(\Omega, \mathcal{F}, P)$.

For two distinct time epochs $t > s$, the random variable $X(t) - X(s)$ is called an *increment* of $\{X(t)\}$. In the case that $s = t + \Delta t$, $\Delta t > 0$, the increment is denoted by $\Delta X(t) = X(t + \Delta t) - X(t)$, as before. The time intervals $(s_i, t_i]$, $i = 1, 2, \ldots$, are called *non-overlapping* if

$$s_1 < t_1 \leq s_2 < t_2 \leq \cdots \leq s_i < t_i \leq \cdots.$$

The process $\{X(t),\, t \geq 0\}$ is said to have *independent increments* if the increments $X(t_i) - X(s_i)$ over non-overlapping intervals $(s_i, t_i]$ are independent.

Definition 11.1 (Brownian Motion) Let $\{z(t),\, t \geq 0\}$ be a stochastic process defined on the probability space $(\Omega, \mathcal{F}, P)$. The process $\{z(t)\}$ is called a *standard Brownian motion* if

(1) it has independent increments,

(2) the increment $z(t+s) - z(t)$ is normally distributed with mean 0 and variance s, independently of time t, and

(3) it has continuous sample paths and $z(0) = 0$.

In Definition 11.1, the second property reveals that the increments are *stationary* in the sense that they are independent of time t. Hence, the stochastic process $\{z(t)\}$ is a standard Brownian motion if it has the independent and stationary (normally distributed) increments with continuous sample paths (see Exercise 11.1).

In what follows, the *information* $\mathcal{F}_t$ up to time t is defined as a σ-field generated from the history $\{z(s), s \leq t\}$. Hence, $\mathcal{F}_t$ can be thought of as the information obtained by knowing all possible outcomes of the standard Brownian motion $\{z(t)\}$ up to time t. The information $\mathcal{F}_t$ increases as time passes, i.e. $\mathcal{F}_s \subset \mathcal{F}_t \subset \mathcal{F}$ for $s < t$. The first property in Definition 11.1 implies that the increment $z(t+s) - z(t)$ is independent of the history $\mathcal{F}_t$.

Proposition 11.1 *For a standard Brownian motion $\{z(t)\}$, the correlation coefficient between $z(t)$ and $z(t+s)$ is given by*

$$\rho(t, t+s) = \sqrt{\frac{t}{t+s}}, \quad t \geq 0, \ s > 0.$$

Proof. Since $E[z(t)] = E[z(t+s)] = 0$, the covariance is given by

$$C[z(t), z(t+s)] = E[z(t)z(t+s)].$$

In order to calculate the covariance, we use the property of independent increments. Namely,

$$z(t)z(t+s) = z(t)\{z(t+s) - z(t)\} + z^2(t).$$

Since the increment $z(t+s) - z(t)$ is independent of $z(t)$, we obtain

$$E[z(t)z(t+s)] = E[z(t)]E[z(t+s) - z(t)] + E[z^2(t)] = t.$$

Here, we have used the fact that

$$V[z(t)] = E[z^2(t)] = t.$$

The proposition now follows at once. □

The algebra used in the above proof is standard when dealing with Brownian motions; that is, we must use the properties (1) and (2) in Definition 11.1 appropriately. See Exercise 11.2 for a related problem.

For a standard Brownian motion $\{z(t)\}$, define the *transition probability function*

$$P(y, s|x, t) \equiv P\{z(s) \leq y | z(t) = x\}, \quad t < s,$$

where $x, y \in \mathbf{R}$. From Definition 11.1, we have

$$P(y, s|x, t) = P\{z(s) - z(t) \leq y - x | z(t) = x\} = \Phi\left(\frac{y-x}{\sqrt{s-t}}\right), \quad t < s.$$

Hence, standard Brownian motions are homogeneous in time as well as in space in the sense that

$$P(y, s|x, t) = P(y - x, s - t|0, 0), \quad t < s,$$

for all $x, y \in \mathbf{R}$. That is, the transition probability function depends only on the differences $s - t$ and $y - x$. Hence, we shall denote it simply by $P(y, t)$ with the understanding that $z(0) = 0$. Note that

$$P(y, t) = \Phi\left(y/\sqrt{t}\right), \quad t > 0, \ y \in \mathbf{R}.$$

BROWNIAN MOTIONS

The *transition density function* is given by

$$p(y,t) \equiv \frac{\partial}{\partial y} P(y,t) = \frac{1}{\sqrt{2\pi t}} e^{-y^2/2t}, \quad t > 0,$$

for all $y \in \mathbf{R}$.

For real numbers μ and σ, define

$$X(t) = X(0) + \mu t + \sigma z(t), \quad t \geq 0. \tag{11.1}$$

The process $\{X(t)\}$ is called a Brownian motion with *drift* μ and *diffusion coefficient* σ. Since $z(t) \sim N(0,t)$, we know from (3.16) that the random variable $X(t)$ is distributed by $N(\mu t, \sigma^2 t)$. Therefore, the diffusion coefficient σ is understood to be the standard deviation of an increment over the unit of time interval. Moreover, its transition probability function is given by

$$P(x,y,t) = \Phi\left(\frac{y - x - \mu t}{\sigma \sqrt{t}}\right), \quad t > 0, \tag{11.2}$$

where we in turn denote $P(x,y,t) = P(y,t|x,0)$, $x, y \in \mathbf{R}$. The proof is left to the reader.

As is shown in Exercise 11.3, the function $P(x,y,t)$ satisfies the *partial differential equation* (abbreviated PDE)

$$\frac{\partial}{\partial t} P = \mu \frac{\partial}{\partial x} P + \frac{\sigma^2}{2} \frac{\partial^2}{\partial x^2} P. \tag{11.3}$$

The PDE (11.3) is called the *Kolmogorov backward equation*; cf. (6.10). The reason of the term 'backward' becomes apparent later. The transition density function is given by

$$p(x,y,t) = \frac{1}{\sigma\sqrt{2\pi t}} \exp\left\{-\frac{(y - x - \mu t)^2}{2t\sigma^2}\right\}, \quad t > 0,$$

for all $x, y \in \mathbf{R}$. The density function $p(x,y,t)$ also satisfies the PDE (11.3). See Exercise 11.4 for a related problem.

Brownian motions can be seen as a special case of diffusion processes (to be defined in the next chapter). For a diffusion process $\{X(t)\}$, let

$$\mu(x,t) = \lim_{\Delta t \to 0} \frac{1}{\Delta t} E[\Delta X(t)|X(t) = x] \tag{11.4}$$

and

$$\sigma^2(x,t) = \lim_{\Delta t \to 0} \frac{1}{\Delta t} E\left[\{\Delta X(t)\}^2 | X(t) = x\right], \tag{11.5}$$

where $\Delta X(t) = X(t + \Delta t) - X(t)$. The quantity $\mu(x,t)$ represents the *infinitesimal mean* of the increment $\Delta X(t)$ conditional on $X(t) = x$, whereas $\sigma^2(x,t)$ represents its conditional *infinitesimal variance*. In the case of Brownian motion $\{X(t)\}$ defined by (11.1), it is readily verified that

$$\mu(x,t) = \mu, \quad \sigma^2(x,t) = \sigma^2. \tag{11.6}$$

That is, the drift μ is equal to the conditional infinitesimal mean, whereas σ^2 is the conditional infinitesimal variance.

Example 11.1 Let $S(t)$ denote the time t price of a risky security, and suppose that $\{X(t)\}$ defined by $X(t) = \log[S(t)/S(0)]$ is a Brownian motion with drift μ and diffusion coefficient σ. The process $\{S(t)\}$ defined by

$$S(t) = S(0)\, e^{X(t)}, \quad t \geq 0, \tag{11.7}$$

is called a *geometric Brownian motion*. The Black–Scholes model assumes that the dynamics of a risky security price follows a geometric Brownian motion. The infinitesimal mean (11.4) of the geometric Brownian motion is given by

$$\mu(x,t) = \left(\mu + \frac{\sigma^2}{2}\right)x. \tag{11.8}$$

To see this, let $S(t) = x$ and consider

$$\Delta S(t) = x\left(e^{\Delta X(t)} - 1\right),$$

where $\Delta S(t) = S(t + \Delta t) - S(t)$. Since $\Delta X(t) \sim N(\mu \Delta t, \sigma^2 \Delta t)$, it follows from (3.17) that

$$E[\Delta S(t) | S(t) = x] = x\left(e^{(\mu + \sigma^2/2)\Delta t} - 1\right).$$

The result (11.8) is obtained by applying L'Hospital's rule. Similarly, it can be shown that the infinitesimal variance (11.5) of the geometric Brownian motion is given by

$$\sigma^2(x,t) = \sigma^2 x^2. \tag{11.9}$$

The proof is left in Exercise 11.5.

11.2 The Central Limit Theorem Revisited

Let $\{W_n\}$ be a random walk defined in Definition 6.1, where it is assumed that the increments X_n take values of either x or y and the time step is unity. In contrast, this section assumes that X_n take values of either $\pm \Delta x$ and the time step is Δt, where

$$\sigma^2 \Delta t = (\Delta x)^2 \tag{11.10}$$

for some σ. It will be shown that the random walk $\{W_n\}$ converges in law to a Brownian motion as $\Delta t \to 0$ with keeping the relation (11.10).

For sufficiently small $\Delta x > 0$, let $\{X_n\}$ be a sequence of IID random variables with probability distribution

$$P\{X_n = \Delta x\} = 1 - P\{X_n = -\Delta x\} = \frac{1}{2} + \frac{\mu}{2\sigma^2}\Delta x. \tag{11.11}$$

Let $W_0 = 0$, and define

$$W_n \equiv \sum_{i=1}^{n} X_i, \quad n = 1, 2, \ldots.$$

From (11.10) and (11.11), we obtain

$$E[X_n] = \frac{\mu \Delta x}{\sigma^2}\Delta x = \mu \Delta t \tag{11.12}$$

and
$$V[X_n] = (\Delta x)^2 - \mu^2(\Delta t)^2 = \sigma^2 \Delta t - \mu^2(\Delta t)^2. \tag{11.13}$$
Supposing that $n = t/\Delta t$, it then follows that
$$E[W_n] = \mu t, \quad V[W_n] = \sigma^2 t - \mu^2 t \Delta t. \tag{11.14}$$

Since W_n follows a binomial distribution, it is expected from the central limit theorem that W_n converges in law to a normally distributed random variable with mean μt and variance $\sigma^2 t$ as $\Delta t \to 0$. However, the classic central limit theorem (Proposition 10.2) is not exactly what we want, since X_n defined by (11.11) changes as $\Delta t \to 0$.

The following extension of the classic central limit theorem is useful in financial engineering. The proof is beyond the scope of this book and omitted. See, e.g., Gnedenko and Kolmogorov (1954) for a proof. In the next result, we consider *triangular arrays* of random variables X_{n,k_n}, where k_n increases to infinity as $n \to \infty$ and $X_{n1}, X_{n2}, \ldots, X_{n,k_n}$ are independent for every fixed n. The array $\{X_{n,k_n}\}$ is said to be *infinitesimal* if
$$\lim_{n \to \infty} \max_{1 \le j \le k_n} P\{|X_{nj}| \ge \varepsilon\} = 0$$
for any $\varepsilon > 0$.

Proposition 11.2 (Central Limit Theorem) *Suppose that the array $\{X_{n,k_n}\}$ is infinitesimal, and define*
$$Y_n = X_{n1} + \cdots + X_{n,k_n} - A_n$$
for some real A_n. Then, Y_n converges in law to a normal variable with mean 0 and variance σ^2 if and only if, for any $\varepsilon > 0$,
$$\lim_{n \to \infty} \sum_{j=1}^{k_n} P\{|X_{nj}| \ge \varepsilon\} = 0$$
and
$$\lim_{n \to \infty} \sum_{j=1}^{k_n} \left[\int_{|x|<\varepsilon} x^2 \mathrm{d}F_{nj}(x) - \left(\int_{|x|<\varepsilon} x \mathrm{d}F_{nj}(x) \right)^2 \right] = \sigma^2,$$
where $F_{nj}(x)$ is the distribution function of X_{nj}.

We can now prove the following important result. See Exercise 11.6 for another verification of this result when the MGF of each X_n exists.

Corollary 11.1 *The random walk defined above converges in law to a Brownian motion with drift μ and diffusion coefficient σ.*

Proof. Fix t, and let $X(t)$ denote the limiting random variable to which the random variable W_n converges in law. The proof of the convergence is left in Exercise 11.7. The independent and stationary increment properties obviously hold for the process $\{X(t)\}$. Also, the sample paths of $\{X(t)\}$ is continuous, since
$$P\{|X(h)| > \delta\} = o(h), \quad \delta > 0,$$

where $o(h)$ denotes the small order of h (see Definition 1.1). Therefore, the process $\{X(t)\}$ is a Brownian motion with drift μ and diffusion coefficient σ due to Definition 11.1 and (11.14). □

Another yet informal justification of Corollary 11.1 is as follows. Let $u_{ij}(n)$ be the transition probabilities of random walk $\{W_n\}$ with no restriction on the movement, and define

$$i = \frac{x}{\Delta x}, \quad j = \frac{y}{\Delta x}, \quad n = \frac{t}{\Delta t}. \tag{11.15}$$

Let

$$p(x, y, t) = \frac{1}{\Delta x} u_{x/\Delta x, y/\Delta x}(t/\Delta t). \tag{11.16}$$

Note that the denominator Δx is needed for (11.16), because we intend to obtain the transition *density* function in the limit. From the *backward* equation (6.10), we obtain

$$p(x, y, t + \Delta t) = \left(\frac{1}{2} + \frac{\mu}{2\sigma^2}\Delta x\right) p(x + \Delta x, y, t) \tag{11.17}$$
$$+ \left(\frac{1}{2} - \frac{\mu}{2\sigma^2}\Delta x\right) p(x - \Delta x, y, t).$$

Assuming that $p(x, y, t)$ is sufficiently smooth, we apply Taylor's expansion to obtain

$$p(x, y, t + \Delta t) = p(x, y, t) + \Delta t \frac{\partial p}{\partial t} + \cdots.$$

Similarly, we obtain

$$p(x \pm \Delta x, y, t) = p(x, y, t) \pm \Delta x \frac{\partial p}{\partial x} + \frac{(\Delta x)^2}{2} \frac{\partial^2 p}{\partial x^2} + \cdots.$$

Substitution of these terms into (11.17) yields

$$\Delta t \frac{\partial p}{\partial t} = \frac{\mu}{\sigma^2}(\Delta x)^2 \frac{\partial p}{\partial x} + \frac{(\Delta x)^2}{2} \frac{\partial^2 p}{\partial x^2} + o(\Delta t). \tag{11.18}$$

Substitute $(\Delta x)^2 = \sigma^2 \Delta t$ into (11.18), divide the resulting equation by Δt, and then take the limit $\Delta t \to 0$. It follows that

$$\frac{\partial p}{\partial t} = \mu \frac{\partial p}{\partial x} + \frac{\sigma^2}{2} \frac{\partial^2 p}{\partial x^2},$$

the Kolmogorov *backward* equation (11.3) satisfied by the Brownian motion with drift μ and diffusion coefficient σ. Since $u_{ij}(n)$ satisfies the initial condition $u_{ij}(0) = \delta_{ij}$, the initial condition for $p(x, y, t)$ is given by

$$p(x, y, 0+) = 0, \quad x \neq y.$$

Hence, we conclude that $p(x, y, t)$ is the transition density function of a Brownian motion with no restriction on the movement.

11.3 The Black–Scholes Formula

Cox, Ross, and Rubinstein (1979) showed that the binomial pricing formula (Theorem 7.2) converges to the Black–Scholes formula (4.21) using the central limit theorem (Proposition 11.2). Recall that, for the pricing of a derivative security, only the distribution under the risk-neutral probability measure is relevant. This section demonstrates their idea by a slightly different limiting argument.

Suppose that the current time is 0, and consider a European call option written on a stock with strike price K and maturity T. The stock price process $\{S(t)\}$ is modeled by the geometric Brownian motion (11.7) and the money-market account is given by

$$B(t) = e^{rt}, \quad t \geq 0,$$

where $r > 0$ is a constant. This is the setting considered by Black and Scholes (1973). It will be shown that the geometric Brownian motion is obtained as a limit of the binomial model.

We divide the time interval $[0, T]$ into n equally spaced subintervals. Let $h = T/n$, and consider the binomial model with n periods. For the IID Bernoulli random variables $X_1, X_2, \ldots$, let

$$Y_t = X_t \log u + (1 - X_t) \log d, \quad t = 1, 2, \ldots.$$

Then,

$$P\{Y_t = \log u\} = 1 - P\{Y_t = \log d\} = p, \quad (11.19)$$

where $0 < p < 1$, and

$$\log \frac{S_t}{S} = Y_1 + Y_2 + \cdots + Y_t, \quad t = 1, 2, \ldots, n, \quad (11.20)$$

with $S_0 = S$. Here, S_t denotes the stock price at period t in the binomial model. In the rest of this section, we denote $x = \log u$, $y = \log d$ and

$$D_t = \sum_{k=1}^{t} Y_k, \quad t = 1, 2, \ldots, n.$$

Since Y_t are IID, it follows that

$$E[D_n] = n(px + qy), \quad V[D_n] = npq(x - y)^2, \quad (11.21)$$

where $q = 1 - p$.

In the continuous-time Black–Scholes setting, we have from (11.1) and (11.7) that*

$$\log \frac{S(t)}{S} = \mu t + \sigma z(t), \quad 0 \leq t \leq T. \quad (11.22)$$

Since we make the binomial model (11.20) converge to the Black–Scholes

---

* To be consistent with what follows (cf. Example 13.1), the parameter μ in (11.22) should be understood to be equal to $\mu' - \sigma^2/2$, where μ' denotes the instantaneous mean rate of return of the stock.

model (11.22), we require that
$$E[D_n] = \mu T, \quad V[D_n] = \sigma^2 T.$$

Note that there are three parameters (p, x, and y) for two equations in (11.21). In this section, we fix p and set two equations for two parameters x and y. That is,
$$\mu T = n(px + qy), \quad \sigma^2 T = npq(x-y)^2. \tag{11.23}$$

These equations are solved in terms of x and y as
$$x = \frac{\mu T}{n} + \sigma\sqrt{\frac{1-p}{p}}\sqrt{\frac{T}{n}}, \quad y = \frac{\mu T}{n} - \sigma\sqrt{\frac{p}{1-p}}\sqrt{\frac{T}{n}}. \tag{11.24}$$

The proof is left to the reader. See Exercise 11.8 for another treatment of the three parameters in (11.21).

First, we obtain the limit of S_n as $n \to \infty$. Since
$$nE[Y_1] = \mu T, \quad nV[Y_1] = \sigma^2 T,$$
it follows from Proposition 11.2 that
$$\lim_{n\to\infty} P\left\{\frac{Y_1 + \cdots + Y_n - nE[Y_1]}{\sqrt{nV[Y_1]}} \leq x\right\}$$
$$= P\left\{\frac{\log[S(T)/S] - \mu T}{\sigma\sqrt{T}} \leq x\right\} = \Phi(x).$$

That is, $\log[S(T)/S]$ follows the normal distribution with mean μT and variance $\sigma^2 T$.

Next, we derive the risk-neutral upward probability in the binomial model. To this end, let $R-1$ be the default-free interest rate. Then, we want to have $R^n = e^{rT}$, so that $R = e^{rT/n}$. From (7.3) and the definition of u and d, the risk-neutral upward probability is given by
$$p_n^* = \frac{e^{rT/n} - \exp\left\{\mu\frac{T}{n} - \sigma\sqrt{\frac{p}{1-p}\frac{T}{n}}\right\}}{\exp\left\{\mu\frac{T}{n} + \sigma\sqrt{\frac{1-p}{p}\frac{T}{n}}\right\} - \exp\left\{\mu\frac{T}{n} - \sigma\sqrt{\frac{p}{1-p}\frac{T}{n}}\right\}}$$
$$= \frac{\exp\left\{(r-\mu)\frac{T}{n}\right\} - \exp\left\{-\sigma\sqrt{\frac{p}{1-p}\frac{T}{n}}\right\}}{\exp\left\{\sigma\sqrt{\frac{1-p}{p}\frac{T}{n}}\right\} - \exp\left\{-\sigma\sqrt{\frac{p}{1-p}\frac{T}{n}}\right\}}.$$

An application of L'Hospital's rule yields
$$\lim_{n\to\infty} p_n^* = p. \tag{11.25}$$

The proof is left to the reader.

As was shown in Theorem 7.2, the option premium C_n is determined through the risk-neutral probability p_n^*. Hence, instead of assuming (11.19), we need to assume
$$P^*\{Y_t = x\} = 1 - P^*\{Y_t = y\} = p_n^*, \quad t = 1, 2, \ldots, n,$$

where P^* denotes the risk-neutral probability measure. Let $D_n = Y_1 + \cdots + Y_n$, as before. Denoting $\mu_n = E^*[Y_1]$ and $\sigma_n^2 = V^*[Y_1]$, we obtain

$$\mu_n = p_n^* x + q_n^* y, \quad \sigma_n^2 = p_n^* q_n^* (x-y)^2,$$

where $q_n^* = 1 - p_n^*$. Since Y_t are IID, it follows from (11.24) and (11.25) that

$$\lim_{n \to \infty} V^*[D_n] = \lim_{n \to \infty} n\sigma_n^2 = \sigma^2 T \qquad (11.26)$$

under P^*. The limit of the mean is obtained by applying L'Hospital's rule twice as

$$\lim_{n \to \infty} E^*[D_n] = \lim_{n \to \infty} n\mu_n = \left(r - \frac{\sigma^2}{2}\right)T. \qquad (11.27)$$

The proof is left to the reader. It follows from Proposition 11.2 that

$$\lim_{n \to \infty} P^* \left\{ \frac{D_n - n\mu_n}{\sigma_n \sqrt{n}} \leq x \right\} = P^* \left\{ \frac{\log[S(T)/S] - (r - \sigma^2/2)T}{\sigma \sqrt{T}} \leq x \right\} = \Phi(x). \qquad (11.28)$$

That is, D_n converges in law to the normally distributed random variable $\log[S(T)/S]$ with mean $(r - \sigma^2/2)T$ and variance $\sigma^2 T$ under the risk-neutral probability measure P^*.

Finally, from (7.13) with $h(x) = \{x - K\}_+$ and (11.20), we observe that

$$C_n = R^{-n} E^* \left[\{S e^{D_n} - K\}_+ \right].$$

Recall from Proposition 2.12 that D_n converges in law to $X \equiv \log[S(T)/S]$ if and only if $E[e^{tD_n}]$ converges to $E[e^{tX}]$ for which the MGFs exist. Also, from (2.33), we have

$$\lim_{n \to \infty} E[f(D_n)] = E[f(X)]$$

for any bounded, continuous function $f(x)$, for which the expectations exist. Since, from the put–call parity (see Exercise 4.9),

$$\{S e^x - K\}_+ = S e^x - K + \{K - S e^x\}_+,$$

we conclude that the option premium C_n converges as $n \to \infty$ to

$$C = e^{-rT} E^* \left[\{S e^{\log[S(T)/S]} - K\}_+ \right], \qquad (11.29)$$

where $\log[S(T)/S] \sim N((r - \sigma^2/2)T, \sigma^2 T)$ under P^*. Evaluation of the expectation in (11.29) then leads to the Black–Scholes formula (4.21). We prove this result in the next section by applying the change of measure (see Example 11.2 below).

11.4 More on Brownian Motions

Results in random walks can be transferred to the corresponding results in Brownian motions by a similar manner to the above limiting arguments. This section demonstrates how to obtain results in Brownian motions from the discrete counterparts.

As the first example, consider a standard Brownian motion $\{z(t)\}$, and let
$$P^*(A) \equiv E[1_A Y(T)], \quad Y(t) = e^{\mu z(t) - \mu^2 t/2}. \tag{11.30}$$
It is readily seen that P^* defines a new probability measure. As a special case of the well-known Girsanov theorem (see Chapter 13), the process $\{z^*(t)\}$ defined by
$$z^*(t) \equiv z(t) - \mu t, \quad t \geq 0, \tag{11.31}$$
is a standard Brownian motion under the new probability measure P^*.

To prove the result, let X_n be defined by
$$P^*\{X_n = \Delta x\} = 1 - P^*\{X_n = -\Delta x\} = p,$$
where $p \equiv (1 + \mu \Delta x)/2$ and P^* is defined in (6.19). That is, X_n follows the same distribution as in (11.11) with $\sigma = 1$. Also, as in Corollary 6.1(1), let $\theta \Delta x = \log \sqrt{p/(1-p)}$. Then,
$$\theta = \frac{1}{2\Delta x} \log\left(\frac{1 + \mu \Delta x}{1 - \mu \Delta x}\right). \tag{11.32}$$

From Corollary 6.1(1), a symmetric random walk $\{W_n\}$ under the original probability measure P is transferred to a p-random walk under P^*. Also, by the transformation (11.15), the symmetric random walk converges in law to a standard Brownian motion under P, while it converges to a Brownian motion $\{z(t)\}$ with drift μ under P^* as $\Delta x \to 0$. Hence, the process $\{z^*(t)\}$ defined by (11.31) is a standard Brownian motion under P^*.

We will obtain the explicit form of the transformed measure P^*. The MGF for the symmetric X_n is given by
$$m(\theta) = \frac{1}{2}\left(e^{\theta \Delta x} + e^{-\theta \Delta x}\right) = \left(1 - \mu^2 \Delta t\right)^{-1/2}.$$
Since $\Delta t = (\Delta x)^2$ and $n = t/\Delta t$, it follows that
$$\lim_{\Delta t \to 0} m^{-n}(\theta) = \lim_{\Delta t \to 0} \left(1 - \mu^2 \Delta t\right)^{t/2\Delta t} = e^{-\mu^2 t/2}. \tag{11.33}$$
From (11.32), an application of L'Hospital's rule yields
$$\lim_{\Delta x \to 0} \theta = \frac{1}{2} \lim_{\Delta x \to 0} \left\{\frac{\mu}{1 + \mu \Delta x} + \frac{\mu}{1 - \mu \Delta x}\right\} = \mu. \tag{11.34}$$
On the other hand, from Corollary 11.1, W_n converges in law to a random variable $z(t) \sim N(0, t)$ as $\Delta t \to 0$. Hence, as $\Delta t \to 0$, the random variable defined by
$$Y_n = m^{-n}(\theta) e^{\theta(W_n - W_0)}, \quad n = 1, 2, \cdots,$$
converges in law to
$$Y(t) = \exp\left\{-\frac{\mu^2 t}{2} + \mu z(t)\right\}$$
under the original measure P, whence we obtain (11.30).

Example 11.2 This example calculates the expectation in (11.29) using the

change of measure formula (11.30). As in Example 6.1, let $A = \{S(T) \geq K\}$, so that
$$E^* \left[\{S(T) - K\}_+ \right] = E^* \left[S(T) 1_A \right] - K E^* \left[1_A \right].$$
Note that
$$E^* \left[1_A \right] = P^* \{S(T) \geq K\} = P^* \{\log[S(T)/S] \geq \log[K/S]\}.$$
Since $\log[S(T)/S] \sim N((r - \sigma^2/2)T, \sigma^2 T)$ under the risk-neutral probability measure P^*, we obtain
$$P^* \{\log[S(T)/S] \geq \log[K/S]\} = \Phi(\xi - \sigma\sqrt{T})$$
after standardization, where
$$\xi = \frac{\log[S/K] + rT}{\sigma\sqrt{T}} + \frac{\sigma\sqrt{T}}{2}.$$
On the other hand, let $\mu = \sigma$ in (11.30) with P^* being replaced by $\widetilde{P}$. Then,
$$\widetilde{P}(A) = E^* \left[e^{\log[S(T)/S] - rT} 1_A \right],$$
since $\log[S(T)/S] = (r - \sigma^2/2)T + \sigma z^*(T)$ under P^*. It follows that
$$E^* \left[S(T) 1_A \right] = S \, e^{rT} \widetilde{P} \{S(T) \geq K\}.$$
But, from (11.31), $\widetilde{z}(t) = z^*(t) - \sigma t$ is a standard Brownian motion under $\widetilde{P}$. Also, $\log[S(T)/S] = (r + \sigma^2/2)T + \sigma \widetilde{z}(T)$. Hence, we obtain
$$\widetilde{P} \{\log[S(T)/S] \geq \log[K/S]\} = \Phi(\xi).$$
In summary, we have obtained that
$$E^* \left[\{S(T) - K\}_+ \right] = S \, e^{rT} \Phi(\xi) - K \Phi(\xi - \sigma\sqrt{T}).$$
The Black–Scholes formula (4.21) is now derived at once.

Next, we consider a driftless Brownian motion $\{X(t)\}$. Such a Brownian motion is related to a symmetric random walk in an obvious manner. Let
$$M(t) \equiv \min_{0 \leq u \leq t} X(u), \quad t > 0,$$
where $X(t) = x + \sigma z(t)$ and $\{z(t)\}$ denotes a standard Brownian motion. For $t > 0$, define
$$a(x, y, t) = -\frac{\partial}{\partial y} P\{X(t) > y, M(t) > 0 | X(0) = x\}, \quad x, y > 0.$$
Hence, $a(x, y, t)$ denotes the transition density function of the driftless Brownian motion $\{X(t)\}$ with state 0 being absorbing.

In order to obtain the transition density function, we recall the transition probability of a symmetric random walk with state 0 being absorbing. From Proposition 6.3, the transition probability is given by
$$a_{ij}(n) = v_{j-i}(n) - v_{j+i}(n), \quad n = 0, 1, \ldots,$$

where $i, j > 0$ and $v_j(n)$ denotes the transition probability of a symmetric random walk with no restriction on the movement. The symmetric random walk converges in law to a standard Brownian motion by the transformation (11.15) with $\sigma = 1$. That is,

$$\lim_{\Delta t \to 0} \frac{1}{\Delta x} v_{x/\Delta x}(t/\Delta t) = p(x,t),$$

where $p(x,t)$ denotes the transition density function of a standard Brownian motion. It follows that

$$a(x,y,t) = p(y-x,t) - p(y+x,t), \quad x, y > 0, \tag{11.35}$$

whence we obtain

$$a(x,y,t) = \frac{1}{\sqrt{2\pi t}} \exp\left\{-\frac{(y-x)^2}{2t}\right\} - \frac{1}{\sqrt{2\pi t}} \exp\left\{-\frac{(y+x)^2}{2t}\right\}, \quad x, y > 0, \tag{11.36}$$

where $t > 0$. Alternatively, this result can be obtained using the reflection principle for the standard Brownian motion (see Exercise 11.9).

We note that, since $p(y-x,t)$ satisfies the backward equation (11.3), the transition density function $a(x,y,t)$ also satisfies the backward equation. Also, they satisfy the same initial condition. But, $a(x,y,t)$ is the transition density function of a driftless Brownian motion with state 0 being absorbing and differs from $p(y-x,t)$. Hence, the PDE (11.3) and the initial condition alone do not determine the solution of the backward equation. The property of the boundary state is needed to determine the solution uniquely.

Finally, we consider a Brownian motion with drift μ. Recall from (6.21) that the transition probability of a p-random walk with state 0 being absorbing is given by

$$a_{ij}(n) = u_{j-i}(n) - \left(\frac{q}{p}\right)^i u_{j+i}(n), \quad n = 0, 1, 2, \ldots,$$

where $i, j > 0$. Since $u_j(n)$ is the transition probability of a p-random walk with no restriction on the movement, the transformation (11.15) with $\sigma = 1$ yields

$$\lim_{\Delta t \to 0} \frac{1}{\Delta x} u_{x/\Delta x}(t/\Delta t) = p(x,t),$$

where $p(x,t)$ in turn denotes the transition density function of a Brownian motion with drift μ. Also, as in (11.34), we obtain

$$\left(\frac{q}{p}\right)^i = \left(\frac{1-\mu\Delta x}{1+\mu\Delta x}\right)^{x/\Delta x} \to e^{-2\mu x}$$

as $\Delta x \to 0$. It follows that, for $t > 0$,

$$a(x,y,t) = p(y-x,t) - e^{-2\mu x} p(y+x,t), \quad x, y > 0, \tag{11.37}$$

whence we obtain

$$a(x,y,t) = \frac{1}{\sqrt{2\pi t}} \exp\left\{-\frac{(y-x-\mu t)^2}{2t}\right\}$$

$$-e^{-2\mu x}\frac{1}{\sqrt{2\pi t}}\exp\left\{-\frac{(y+x-\mu t)^2}{2t}\right\}, \quad x, y > 0.$$

See Exercise 11.10 for alternative derivation of (11.37) and Exercises 11.11 and 11.12 for related problems.

11.5 Poisson Processes

A Poisson process is a Markov process that counts the number of randomly occurring events in time. Let $N(t)$ denote the number of events having occurred before time t. Such a stochastic process $\{N(t), t \geq 0\}$ is called a *counting process*. It takes values on non-negative integers and is non-decreasing in time t. The increment $N(t) - N(s)$, $s < t$, counts the number of events having occurred during the time interval $(s, t]$. It is usually assumed that $N(0) = 0$.

Definition 11.2 (Poisson Process) A counting process $\{N(t)\}$ is called a *Poisson process* if

(1) it has independent increments,

(2) the increment is stationary in time in the sense that the distribution of $N(t+s) - N(t)$ depends only on s, and

(3) the process admits at most single jump with unit magnitude during a small interval, i.e.

$$P\{N(t+h) - N(t) \geq 2|N(t)\} = o(h),$$

where $o(h)$ denotes the small order of h, and $N(0) = 0$.

As for the case of Brownian motions (see Definition 11.1), Poisson processes have independent and stationary increments. From the second and third properties, we assume that

$$P\{N(h) = 1|N(0) = 0\} = \lambda h + o(h) \tag{11.38}$$

for some $\lambda > 0$. The parameter λ is referred to as the *intensity* of the process. It is readily seen that the Poisson process is a Markov process. The stationarity assumption can be dropped by assuming the intensity to be time-dependent (see Exercise 11.13).

Let $\{N(t)\}$ be a Poisson process with intensity $\lambda > 0$, and define

$$p_n(t) = P\{N(t) = n\}, \quad n = 0, 1, \ldots.$$

For sufficiently small $h > 0$, consider the event $\{N(t+h) = 0\}$. Since

$$\{N(t+h) = 0\} = \{N(t) = 0\} \cap \{N(t+h) - N(t) = 0\},$$

it follows from (11.38) that

$$\begin{aligned} p_0(t+h) &= P\{N(t) = 0, N(t+h) - N(t) = 0\} \\ &= P\{N(t) = 0\}P\{N(h) = 0\} \\ &= p_0(t)\{1 - \lambda h + o(h)\}. \end{aligned}$$

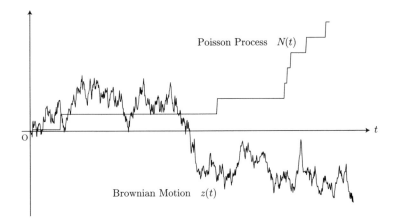

Figure 11.1 *Sample paths of a Brownian motion and a Poisson process*

Here, we have used the properties (1) and (2) in Definition 11.2 for the second equality. Hence,
$$\frac{p_0(t+h)-p_0(t)}{h} = \lambda p_0(t) + \frac{o(h)}{h},$$
so that we obtain
$$p_0'(t) = -\lambda p_0(t), \quad t \geq 0.$$
Since $p_0(0) = 1$ (recall that $N(0) = 0$), it follows that
$$p_0(t) = e^{-\lambda t}, \quad t \geq 0. \tag{11.39}$$
Similarly, for any $n = 1, 2, \ldots$, we obtain
$$p_n'(t) = -\lambda p_n(t) + \lambda p_{n-1}(t), \quad t > 0. \tag{11.40}$$
Since $p_n(0) = 0$, $n = 1, 2, \ldots$, it can be shown that
$$p_n(t) = \frac{(\lambda t)^n}{n!} e^{-\lambda t}, \quad t \geq 0. \tag{11.41}$$
The proof is left in Exercise 11.14. Note that (11.41) holds also for $n = 0$. The distribution (11.41) is a Poisson distribution with mean λt, whence we call the counting process $\{N(t)\}$ a *Poisson* process. Note that $E[N(t)] = \lambda t$. The parameter λ thus represents the mean number of events occurring in the unit of time, whence it is called the intensity of the process.

Figure 11.1 depicts sample paths of a Brownian motion and a Poisson process. These sample paths look very different; however, as we have seen in the definitions, they are very close in their mathematical properties, i.e., they both possess the property of independent and stationary increments. Also, as the next example reveals, they are characterized as limits of random walks.

Example 11.3 (Alternative Price Process) In the binomial model, suppose that
$$d = e^{-\zeta T/n}, \quad p = \frac{\lambda T}{n}, \quad R = \hat{R}^{T/n}$$

with $u > 1$ being fixed, where $\zeta > 0$. For sufficiently large n, this correspondence captures the essence of a pure jump process of Cox and Ross (1976) in which each successive stock price is almost always close to the previous price, but occasionally, with very low probability, significantly different changes occur. Note that, as $n \to \infty$, the probability of a change by d becomes larger and larger, while the probability of a change by u approaches zero.

With these specifications, since the binomial distribution converges to a Poisson distribution (see Section 3.2), it is expected that the stock price process converges in law to a log-Poisson model as $n \to \infty$. Indeed, define

$$\overline{\Psi}(n;\lambda) = \sum_{k=n}^{\infty} \frac{\lambda^k}{k!} e^{-\lambda}, \quad n = 0, 1, 2, \ldots,$$

as the survival probability of a Poisson distribution with parameter λ. From Theorem 7.2, the limiting call option pricing formula is then given by

$$C = S\overline{\Psi}(k;y) - K\hat{R}^{-T}\overline{\Psi}(k;y/u), \quad y = (\log \hat{R} + \zeta)\frac{uT}{u-1},$$

where S is the current price of the underlying stock and

$$k = \min\{a : a > (\log[K/S] + \zeta T)/\log u\}.$$

The proof is left to the reader. Note that a very similar formula holds if we set $u = e^{\zeta T/n}$ and $1 - p = \lambda T/n$ with d being fixed (see Exercise 11.15).

Let X_n denote the length of the time interval between the $(n-1)$th jump and the nth jump in a Poisson process $\{N(t)\}$ with intensity λ. Since the event $\{X_1 > t\}$ is equivalent to the event that no jump has occurred before time t, we obtain from (11.39) that

$$P\{X_1 > t\} = P\{N(t) = 0\} = e^{-\lambda t}, \quad t \geq 0. \tag{11.42}$$

Hence, X_1 follows an exponential distribution with mean $1/\lambda$. Similarly, it can be shown that X_n are IID and follow the same exponential distribution. The proof is left to the reader.

In this regard, the Poisson process $\{N(t)\}$ is defined by

$$N(t) = \max\{n \geq 0 : X_1 + \cdots + X_n \leq t\}, \quad t \geq 0,$$

since $N(t)$ counts the number of events having occurred before time t; see (3.29). Notice that a simple generalization of Poisson processes is to assume a general distribution for the intervals X_n. Such a process is called a *renewal process*, and has many applications in reliability models, queueing models, and others. Note that the general renewal process does not have the property of independent increments in the continuous-time setting.

Example 11.4 (Collective Risk Process) Let ξ_n, $n = 1, 2, \ldots$, be a sequence of IID random variables, and let $\{N(t), t \geq 0\}$ be a Poisson process with intensity λ, which is independent of ξ_n, $n = 1, 2, \ldots$. Define

$$\eta(t) = \sum_{i=1}^{N(t)} \xi_i, \quad t \geq 0, \tag{11.43}$$

where $\eta(t) = 0$ if $N(t) = 0$. The process $\{\eta(t), t \geq 0\}$ is called a *compound Poisson process* and has apparent applications in, e.g., an actuarial model.

Recall that a Poisson process counts the number of events randomly occurring in time. Hence, it is appropriate to describe the arrival of claims on an insurance company due to unanticipated accidents, e.g. traffic accidents, by the Poisson process. Upon the nth arrival of claims, the company has to pay a (positive) random amount of money, ξ_n say, according to a contract. Instead, the company receives c units of money from the customers, called the *risk premium* rate, per unit of time. The amount of money that the company holds at time t is therefore given by

$$Y(t) = u + ct - \sum_{i=1}^{N(t)} \xi_i, \quad t \geq 0,$$

where $Y(0) = u$ denotes the given initial capital. The process $\{Y(t), t \geq 0\}$ is called the *risk process* in collective risk theory, and the problem is to calculate the probability of *ruin*, i.e. $P\{Y(t) \leq 0 \text{ for some time } t\}$. Since $\{N(t)\}$ and ξ_n are independent, we obtain

$$E[\eta(t)|N(t) = n] = E[\xi_1 + \cdots + \xi_n|N(t) = n] = nE[\xi_1]. \quad (11.44)$$

It follows from (11.43) and (11.44) that

$$E[\eta(t)] = E[E[\eta(t)|N(t)]] = E[\xi_1]E[N(t)].$$

This equation is called the *Wald identity* (cf. Exercise 2.17).

If the intensity of the Poisson process is λ, i.e. $E[N(t)] = \lambda t$, the *safety loading*, ρ say, is defined by

$$\rho = \frac{c}{E[\xi_1]\lambda} - 1.$$

Note that, even if the risk process has a positive safety loading, i.e. $c > \lambda E[\xi_1]$, there is a positive ruin probability for the insurance company. See, e.g., Grandell (1991) for details.

11.6 Exercises

Exercise 11.1 For $\alpha \neq 0$, let $X(t) = \alpha z(t/\alpha^2)$, where $\{z(t)\}$ is a standard Brownian motion. Prove that the process $\{X(t)\}$ satisfies the three conditions given in Definition 11.1 (hence, it is also a standard Brownian motion).

Exercise 11.2 For a standard Brownian motion $\{z(t)\}$, let us define $I(t) = \int_0^t z(u)du$. Then,

$$I^2(t) = 2\int_0^t \int_0^u z(u)z(v)dvdu.$$

Using this and the independent increment property of the Brownian motion, prove that $V[I(t)] = t^3/3$.

Exercise 11.3 For a Brownian motion $\{X(t)\}$, prove that the transition probability function $P(x, y, t)$ given by (11.2) as well as the density function $p(x, y, t)$ satisfy the Kolmogorov backward equation (11.3).

Exercise 11.4 (Fokker–Planck Equation) For a Brownian motion, prove that the transition probability function $P(x, y, t)$ as well as the density function $p(x, y, t)$ satisfy the PDE

$$\frac{\partial}{\partial t}P = -\mu \frac{\partial}{\partial y}P + \frac{\sigma^2}{2}\frac{\partial^2}{\partial y^2}P.$$

This PDE is called the *Fokker–Planck equation* or the *Kolmogorov forward equation* by an obvious reason; cf. (6.7).

Exercise 11.5 In Example 11.1, show that the infinitesimal variance of the geometric Brownian motion conditional on $S(t) = x$ is given by (11.9).

Exercise 11.6 Let $\{X_n\}$ be given by (11.11) and define $W_n = \sum_{i=1}^n X_i$. Under the relations $\sigma^2 \Delta t = (\Delta x)^2$ and $t = n\Delta t$, show that the MGF of W_n is given by

$$m_n(s) = \left[1 + \frac{\mu t s + \sigma^2 t s^2/2 + a_n}{n}\right]^n$$

for some a_n such that $\lim_{n\to\infty} a_n = 0$, whence

$$\lim_{n\to\infty} m_n(s) = \exp\left\{\mu t s + \frac{\sigma^2 t s^2}{2}\right\}.$$

It follows that W_n converges in law to a random variable $X(t) \sim N(\mu t, \sigma^2 t)$.

Exercise 11.7 In the proof of Corollary 11.1, confirm that the conditions of Proposition 11.2 are satisfied.

Exercise 11.8 As in Cox, Ross, and Rubinstein (1979), let

$$x = \sigma\sqrt{T/n}, \quad y = -\sigma\sqrt{T/n}, \quad p = \frac{1}{2} + \frac{\mu}{2\sigma}\sqrt{T/n}.$$

Confirm that they satisfy (11.23) as $n \to \infty$. Also, show that the central limit result (11.28) holds even in this case.

Exercise 11.9 First, confirm that the reflection principle (Proposition 6.2) holds for any standard Brownian motion. Next, for a standard Brownian motion $\{z(t)\}$ with $z(0) = x > 0$, let $M(t) = \min_{0 \le u \le t} z(u)$. Using the reflection principle, prove that

$$P\{z(t) > y, \, M(t) \le 0 | z(0) = x\} = P\{z(t) < -y | z(0) = x\}, \quad y > 0.$$

Finally, use this equation to obtain (11.35) and (11.36).

Exercise 11.10 Applying the change of measure formula (11.30) to (11.36), derive the transition density function (11.37) for a Brownian motion with drift μ and diffusion coefficient $\sigma = 1$.

Exercise 11.11 Consider a Brownian motion $\{X(t)\}$ with drift μ and diffusion coefficient σ. Suppose that $X(0) = 0$, and let τ denote the first time that the Brownian motion reaches the state $x > 0$. Prove that the density function for τ is given by

$$f(t) = \frac{x}{\sqrt{2\pi t^3}\sigma} \exp\left\{-\frac{(x-\mu t)^2}{2\sigma^2 t}\right\}, \quad t > 0.$$

Note: For a standard Brownian motion $\{z(t)\}$, let τ be the first passage time to the boundary $a(t) > 0$. If the derivative $a'(t)$ exists and is continuous, then the density function $f(t)$ for τ satisfies the equation

$$\left\{\frac{a(t)}{t} - a'(t)\right\} p(a(t), t)$$
$$= f(t) + \int_0^t f(u) \left\{\frac{a(t) - a(u)}{t - u} - a'(t)\right\} p(a(t) - a(u), t - u) du,$$

where $p(x, t)$ is the transition density function of $\{z(t)\}$. See Williams (1992) for details.

Exercise 11.12 Consider a Brownian motion $\{X(t)\}$ with drift μ and diffusion coefficient σ defined on the interval $[0, a]$, where the boundaries 0 and $a > 0$ are absorbing. Using the transformation (11.15) and the result (9.19) with $N = a/\Delta x$, show that the transition density function $a(x, y, t)$ of the Brownian motion is given by

$$a(x, y, t) = \frac{2e^{\mu(y-x)/\sigma^2 - \mu^2 t/2\sigma^2}}{a} \sum_{k=1}^{\infty} \exp\left\{-\frac{\sigma^2 t}{2}\left(\frac{k\pi}{a}\right)^2\right\} \sin\frac{kx\pi}{a} \sin\frac{ky\pi}{a}.$$

Exercise 11.13 (Non-Stationary Poisson Process) Consider a counting process $\{N(t)\}$ satisfying the properties (1) and (3) in Definition 11.2; but, in this exercise, let $\lambda(t)$ be a non-negative, deterministic function of time t. Instead of (11.38), suppose that

$$P\{N(t+h) = k+1 | N(t) = k\} = \lambda(t)h + o(h), \quad t \geq 0,$$

for all $k = 0, 1, \ldots$. Show that

$$P\{N(t) = n\} = \frac{\Lambda^n(t)}{n!} e^{-\Lambda(t)}, \quad t \geq 0,$$

where $\Lambda(t) = \int_0^t \lambda(u) du$.

Exercise 11.14 Prove that Equation (11.40) holds for a Poisson process $\{N(t)\}$ with intensity $\lambda > 0$. Also, derive Equation (11.41).

Exercise 11.15 In the binomial model, suppose instead that $u = e^{\zeta T/n}$, $1 - p = \lambda T/n$ and $R = \hat{R}^{T/n}$ with $d < 1$ being fixed, where $\zeta > \log \hat{R} > 0$. Obtain the limiting call option pricing formula for this specification.

CHAPTER 12

Basic Stochastic Processes in Continuous Time

This chapter introduces stochastic processes necessary for the development of continuous-time securities market models provided in the next chapter. Namely, we discuss diffusion processes and stochastic integrals with respect to Brownian motions. A diffusion process is a natural extension of a Brownian motion and has many applications in biology, physics, operations research, and economics. Also, it is a solution of a stochastic differential equation, a key tool to describe the stochastic behavior of security prices in finance. Our arguments in this chapter are not rigorous, but work for almost all practical situations. The reader should consult any textbooks of mathematical finance such as Elliott and Kopp (1999), Klebaner (1998), Neftci (1996), and Shiryaev (1999) for discussions of more comprehensive results in continuous-time stochastic processes.

12.1 Diffusion Processes

Let $\{X(t)\}$ be a continuous-time Markov process defined on the real line R, and consider the increment $\Delta X(t) = X(t + \Delta t) - X(t)$, where $\Delta t > 0$ is sufficiently small. Because of the Markov property, the distribution of $\Delta X(t)$ depends only on the value of $X(t)$ and time t. In particular, if $\{X(t)\}$ is a Brownian motion with drift μ and diffusion coefficient σ, the increment $\Delta X(t)$ follows the normal distribution $N(\mu \Delta t, \sigma^2 \Delta t)$. Hence, we conclude that, given $X(t) = x$,

(1) the mean of the increment $\Delta X(t)$ is $\mu \Delta t$,

(2) the variance of the increment is $\sigma^2 \Delta t$, and

(3) for any $\delta > 0$, the probability that the increment is greater than δ in the magnitude is the small order of Δt.

The proof of the third statement is left in Exercise 12.1. The first and second statements were verified in the previous chapter.

A natural extension of the Brownian motion is to assume that the mean and the variance of the increment $\Delta X(t)$ are dependent on time t and state $X(t)$. That is, we assume that the limits

$$\mu(x, t) = \lim_{\Delta t \to 0} \frac{1}{\Delta t} E[\Delta X(t) | X(t) = x] \qquad (12.1)$$

as well as
$$\sigma^2(x,t) = \lim_{\Delta t \to 0} \frac{1}{\Delta t} E\left[\{\Delta X(t)\}^2 | X(t) = x\right] \quad (12.2)$$
exist with $\sigma(x,t) \neq 0$; see (11.4) and (11.5). The above property (3) remains valid and is stated formally as
$$\lim_{\Delta t \to 0} \frac{1}{\Delta t} P\{|\Delta X(t)| > \delta | X(t) = x\} = 0 \quad (12.3)$$
for any $\delta > 0$. The condition (12.3) can be viewed as a formalization of the property that the sample paths of the process $\{X(t)\}$ are continuous.

Definition 12.1 (Diffusion Process) A Markov process $\{X(t)\}$ in continuous time is called a *diffusion process* (or diffusion for short), if it has continuous sample paths and the limits (12.1) as well as (12.2) exist with $\sigma(x,t) \neq 0$. The function $\mu(x,t)$ is called the *drift function*, while $\sigma(x,t)$ is called the *diffusion function*. If the two functions do not depend on time t, then the diffusion is called *time-homogeneous* (or homogeneous for short).

Diffusion processes satisfy the strong Markov property (9.4) and, conversely, a diffusion process is characterized as a strong Markov process with continuous sample paths.

Alternatively, suppose that the limits (12.1) and (12.2) exist for a Markov process $\{X(t)\}$ and the limiting functions are sufficiently smooth. Further suppose that
$$\lim_{\Delta t \to 0} \frac{1}{\Delta t} E[\{\Delta X(t)\}^r | X(t) = x] = 0 \quad (12.4)$$
for some $r > 2$. Then, the process $\{X(t)\}$ is a diffusion. Hence, a diffusion process is characterized as a Markov process with the three limits (12.1), (12.2) with $\sigma(x,t) \neq 0$, and (12.4) for some $r > 2$. We note that it is often easier to check the case that $r = 4$ for (12.4) in many practical situations (see Exercise 12.1).

Example 12.1 Let $\{X(t)\}$ be a Brownian motion with drift μ and diffusion coefficient σ, and consider a process $\{Y(t)\}$ defined by
$$Y(t) = f(X(t)), \quad t \geq 0,$$
where the function $f(x)$ is sufficiently smooth and strictly increasing in x. It is evident that the process $\{Y(t)\}$ is a strong Markov process with continuous sample paths, whence it is a time-homogeneous diffusion. The drift and diffusion functions of $\{Y(t)\}$ are obtained as follows. Let $X(t) = x$. By Taylor's expansion, we obtain
$$f(X(t + \Delta t)) - f(x) = \Delta X(t) f'(x) + \frac{\{\Delta X(t)\}^2}{2} f''(\xi),$$
for some ξ such that $|\xi - x| < |X(t + \Delta t) - x|$. Note that, due to the continuity of sample paths of the Brownian motion, $f''(\xi)$ converges to $f''(x)$ as $\Delta t \to 0$ almost surely. Denoting $Y(t) = y = f(x)$, we obtain
$$\mu_Y(y) \equiv \lim_{\Delta t \to 0} \frac{1}{\Delta t} E[\Delta Y(t) | Y(t) = y] = \mu f'(x) + \frac{\sigma^2}{2} f''(x)$$

DIFFUSION PROCESSES

and
$$\sigma_Y^2(y) \equiv \lim_{\Delta t \to 0} \frac{1}{\Delta t} E\left[\{\Delta Y(t)\}^2 | Y(t) = y\right] = \sigma^2 \{f'(x)\}^2,$$
where $\Delta Y(t) = Y(t + \Delta t) - y$. In particular, when $f(x) = e^x$, we have
$$\mu_Y(y) = \left(\mu + \frac{\sigma^2}{2}\right) y, \quad \sigma_Y^2(y) = \sigma^2 y^2.$$

Hence, we recovered the drift function $(\mu + \sigma^2/2)y$ and the diffusion function σy of a geometric Brownian motion (see Example 11.1).

Let $\{X(t)\}$ be a time-homogeneous diffusion, for simplicity, with drift function $\mu(x)$ and diffusion function $\sigma(x)$. In the following, we shall prove that the transition probability function
$$u(x, t) = P\{X(t) \leq y | X(0) = x\}, \quad t \geq 0, \quad x, y \in \mathbf{R}, \tag{12.5}$$
satisfies the *partial differential equation* (PDE)
$$u_t(x, t) = \mu(x) u_x(x, t) + \frac{1}{2} \sigma^2(x) u_{xx}(x, t) \tag{12.6}$$
with the boundary condition
$$u(x, 0+) \equiv \lim_{h \downarrow 0} u(x, h) = \begin{cases} 1, & x \leq y, \\ 0, & x > y. \end{cases}$$
Here, u_x denotes the partial derivative of $u(x, t)$ with respect to x, u_{xx} denotes the second-order partial derivative, etc.

To that end, it suffices to show the following. The desired result follows by a simple limiting argument, since the function defined by
$$g_n(x) = \begin{cases} 1, & x \leq y, \\ e^{-n(x-y)}, & x > y, \end{cases}$$
is bounded and continuous, and converges to the indicator function $1_{\{x \leq y\}}$ as $n \to \infty$.

Proposition 12.1 *Let $g(x)$ be a bounded and continuous function defined on $\mathbf{R}$. Then, the function defined by*
$$u(x, t) = E[g(X(t)) | X(0) = x], \quad t \geq 0, \tag{12.7}$$
satisfies the PDE (12.6) with the boundary condition
$$u(x, 0+) = g(x), \quad x \in \mathbf{R}. \tag{12.8}$$

Outline of the proof. Consider $u(X(\Delta t), t)$ for sufficiently small $\Delta t > 0$. Since $X(0) = x$, assuming that the function $u(x, t)$ is sufficiently smooth, Taylor's expansion (1.16) yields
$$u(X(\Delta t), t) = u(x, t) + u_x(x, t) \Delta X(t) + \frac{\{\Delta X(t)\}^2}{2} u_{xx}(x, t) + o(\Delta t).$$

Note that
$$\begin{aligned}u(x,t+\Delta t) &= E[E[g(X(t+\Delta t))|X(\Delta t)]|X(0)=x] \\ &= E[u(X(\Delta t),t)|X(0)=x],\end{aligned}$$

since the process is time-homogeneous. It follows from (12.1) and (12.2) that

$$u(x,t+\Delta t) = u(x,t) + u_x(x,t)\mu(x)\Delta t + \frac{\sigma^2(x)}{2}u_{xx}(x,t)\Delta t + o(\Delta t).$$

The proposition now follows at once. □

We note that the above proof is not mathematically rigorous. In particular, we left the proof for differentiability of the function $u(x,t)$. However, this kind of argument works and, in fact, is enough for many practical instances. See Exercises 12.2 and 12.3 for other examples.

Before proceeding, one remark should be in order. Note that the function $g(x)$ in (12.7) does not appear in the PDE (12.6), but does appear only in the boundary condition (12.8). This observation is very important for the pricing of derivative securities, since the price of a contingent claim is in general given by the form of (12.7), as we shall prove later.

Next, consider a time-homogeneous diffusion $\{X(t)\}$ with state space $\boldsymbol{R} \cup \{\kappa\}$, where $\kappa \notin \boldsymbol{R}$ denotes an absorbing state. The diffusion enters the state κ only if it is *killed*. The diffusion is killed according to the rate $k(x)$ when $X(t) = x$. That is, denoting the *killing time* by ζ, the killing rate is formally defined by

$$k(x) = \lim_{h \to 0} \frac{1}{h} P\{t < \zeta \le t+h | X(t) = x\}, \quad x \in \boldsymbol{R}.$$

Thus, the killing rate is the same as the hazard rate of ζ, since $X(t) \in \boldsymbol{R}$ implies that $\zeta > t$; see Definition 8.1.

Now, let $g(x)$ be a bounded and continuous function defined on $\boldsymbol{R} \cup \{\kappa\}$, and consider

$$u(x,t) = E\left[g(X(t))1_{\{\zeta > t\}}|X(0) = x\right], \quad t \ge 0, \tag{12.9}$$

for $x \in \boldsymbol{R}$. If $g(x)$ represents the cashflow when $X(t) = x$, the function $u(x,t)$ is the expected cashflow at time t, conditional on $\{\zeta > t\}$, when the process starts from $X(0) = x$. It can be shown that the function $u(x,t)$ satisfies the PDE

$$u_t(x,t) = \mu(x)u_x(x,t) + \frac{1}{2}\sigma^2(x)u_{xx}(x,t) - k(x)u(x,t), \tag{12.10}$$

where $\mu(x)$ is the drift function and $\sigma(x)$ is the diffusion function of the diffusion $\{X(t)\}$. The proof is left in Exercise 12.4.

Alternatively, the function $u(x,t)$ in (12.9) can be interpreted as follows. Let $k(x)$ and $g(x)$ denote an instantaneous interest rate and the cashflow, respectively, when $X(t) = x$. For the diffusion process $\{X(t)\}$ without killing, consider

$$u(x,t) = E\left[e^{-\int_0^t k(X(s))ds} g(X(t)) \Big| X(0) = x\right], \quad t \geq 0. \quad (12.11)$$

The function $u(x,t)$ represents the expected, *discounted* cashflow at time t when $X(0) = x$. It can be shown (see Example 12.5 for a more general result) that the function $u(x,t)$ satisfies the PDE (12.10). This means that the interest rate plays the same role as the killing rate. This observation becomes significantly important for the pricing of defaultable discount bonds.

Example 12.2 Suppose that the state space of a diffusion $\{X(t)\}$ without killing is positive, and consider the case that $g(x) = 1$ and $k(x) = x$ in (12.11). That is, we have

$$u(x,t) = E\left[\exp\left\{-\int_0^t X(s)ds\right\} \Big| X(0) = x\right], \quad t \geq 0,$$

where $x > 0$. If $X(t)$ represents the instantaneous spot rate, then $u(x,t)$ is the price of the default-free discount bond with maturity t when $X(0) = x$. From (12.10), the bond price function $u(x,t)$ satisfies the PDE

$$u_t(x,t) = \mu(x)u_x(x,t) + \frac{1}{2}\sigma^2(x)u_{xx}(x,t) - xu(x,t) \quad (12.12)$$

with the boundary condition $u(x,0+) = 1$, since the default-free discount bond pays the face value $F = 1$ at the maturity. Alternatively, let ζ be the killing time with killing rate $X(t)$, and consider

$$u(x,t) = E\left[1_{\{\zeta > t\}} | X(0) = x\right], \quad t \geq 0.$$

That is, the default-free discount bond pays 1 dollar at the maturity t if $\zeta > t$ and nothing if $\zeta \leq t$ under the economy with zero interest rates.

Example 12.3 Let r be a positive constant, and consider a Brownian motion $\{X(t)\}$ with drift $r - \sigma^2/2$ and diffusion coefficient σ. Let $S(t) = Se^{X(t)}, t \geq 0$, where $S(0) = S$, and suppose that $g(x) = \{x - K\}_+$ and $k(x) = r$. Consider in turn

$$u(S,t) = E\left[e^{-rt}\{S(t) - K\}_+ \Big| S(0) = S\right], \quad t > 0.$$

Recall from Example 12.1 that the process $\{S(t)\}$ is a diffusion with $\mu(S,t) = rS$ and $\sigma^2(S,t) = \sigma^2 S^2$. It follows from (12.10) that the function $u(x,t)$ satisfies the PDE

$$u_t(S,t) = rSu_S(S,t) + \frac{\sigma^2 S^2}{2}u_{SS}(S,t) - ru(S,t) \quad (12.13)$$

with the boundary condition $u(x,0+) = \{x - K\}_+$. The representation of $u(S,t)$ in terms of the killing time ζ is obvious and left to the reader. Note that ζ is exponentially distributed with mean r^{-1}.

12.2 Sample Paths of Brownian Motions

An occurrence of a Brownian motion $\{z(t)\}$ observed from time 0 to time T is called a *realization* or a *sample path* of the process on the time interval

[0, T]. Paths as functions of t, $z(t,\omega)$ for $\omega \in \Omega$, have the following properties: Almost every sample path

(1) is a continuous function of time t,
(2) is not differentiable at any point of time,
(3) has infinite variation on any interval, no matter how small the interval is, and
(4) has quadratic variation on $[0,t]$ equal to t, for any $t \in [0,T]$.

Properties (1) and (2) state that, although any sample path of $\{z(t)\}$ is continuous, it has increments $\Delta z(t)$ much longer than Δt as $\Delta t \to 0$. Since $V[\Delta z(t)] = \Delta t$, this suggests that the increment is roughly equal to $\sqrt{\Delta t}$ as $\Delta t \to 0$. As we will show later, this property can be made precise by the quadratic variation Property (4). Note that, since a monotone function has finite variation, almost every sample path of the Brownian motion is not monotone in any interval, no matter how small the interval is. Also, since a continuous function with a bounded derivative is of finite variation, $z(t,\omega)$ cannot have a bounded derivative on any interval. In this section, we prove Property (4) of Brownian motion paths.

Definition 12.2 The *quadratic variation* of a stochastic process $\{X(t)\}$ is defined, if it exists, as

$$[X,X](T) = \lim_{N \to \infty} \sum_{i=1}^{N} |X(t_i^N) - X(t_{i-1}^N)|^2, \quad (12.14)$$

where the limit is taken over all partitions such that, for each N,

$$0 = t_0^N < t_1^N < \cdots < t_{N-1}^N < t_N^N = T$$

and $\delta_N = \max_i \{t_{i+1}^N - t_i^N\} \to 0$ as $N \to \infty$.

It can be shown (see, e.g., Loeve (1978) for details) that the quadratic variation of the standard Brownian motion $\{z(t)\}$ is given by

$$[z,z](t) = t, \quad t \geq 0. \quad (12.15)$$

Hence, although the sum in the right-hand side of (12.14) is random, the limit is *not* random in the case of Brownian motions.

The above result tells us more. That is, loosely speaking, Equation (12.14) is equal to $\int_0^t (dz(s))^2$. It follows from (12.15) that

$$\int_0^t (dz(s))^2 = t \text{ or, equivalently, } (dz(t))^2 = dt, \quad (12.16)$$

which has been assumed when we derived Ito's formula (see Theorem 1.1). More generally, for Brownian motions $\{z_1(t)\}$ and $\{z_2(t)\}$ with correlation coefficient $\rho(t)$, we have

$$dz_1(t)dz_2(t) = \rho(t)dt. \quad (12.17)$$

The proof is left in Exercise 12.5.

Example 12.4 In Definition 12.2, take $N = 2^n$ and define

$$W_n(t) \equiv \sum_{k=1}^{2^n} \Delta_{nk}^2, \qquad (12.18)$$

where

$$\Delta_{nk} = z(kt/2^n) - z((k-1)t/2^n), \quad k = 1, 2, \ldots, 2^n.$$

$W_n(t)$ is a sum of quadratic increments with time interval 2^{-n}. Since the increments Δ_{nk} are IID and follow the normal distribution $N(0, t/2^n)$, we obtain

$$E[W_n(t)] = \sum_{k=1}^{2^n} E\left[\Delta_{nk}^2\right] = 2^n \frac{t}{2^n} = t. \qquad (12.19)$$

Also, $E[\Delta_{nk}^4] = 3t^2/2^{2n}$, since $E[(X-\mu)^4] = 3\sigma^4$ when $X \sim N(\mu, \sigma^2)$, so that

$$E\left[W_n^2(t)\right] = \sum_{k=1}^{2^n} E\left[\Delta_{nk}^4\right] + \sum_{i \neq j} E\left[\Delta_{ni}^2\right] E\left[\Delta_{nj}^2\right] = t^2 + \frac{2t^2}{2^n}. \qquad (12.20)$$

It follows from (12.19) and (12.20) that, as $n \to \infty$, we obtain

$$E\left[|W_n(t) - t|^2\right] = \frac{2t^2}{2^n} \to 0. \qquad (12.21)$$

Hence, $W_n(t)$ converges to the constant t as $n \to \infty$ in the sense of mean square.

Consider next the *total variation* of the standard Brownian motion $\{z(t)\}$. Since

$$\max_{1 \leq j \leq 2^n} \{|\Delta_{nj}|\} \geq |\Delta_{nk}|, \quad k = 1, 2, \ldots, 2^n,$$

we obtain

$$\sum_{k=1}^{2^n} |\Delta_{nk}| \geq \sum_{k=1}^{2^n} \frac{\{\Delta_{nk}\}^2}{\max_{1 \leq j \leq 2^n}\{|\Delta_{nj}|\}} = \frac{W_n(t)}{\max_{1 \leq j \leq 2^n}\{|\Delta_{nj}|\}}. \qquad (12.22)$$

From (12.14) and (12.15), we know that $W_n(t)$ converges to t as $n \to \infty$ a.s. Also, since $\{z(t)\}$ has continuous sample paths, the term $\max_{1 \leq j \leq 2^n}\{|\Delta_{nj}|\}$ converges to 0 as $n \to \infty$. Hence, the value in (12.22) diverges as $n \to \infty$. This implies that the Brownian motion is *not* of finite variation, and so it is nowhere differentiable. (*Note*: If a function is continuous and of finite variation, then its quadratic variation is zero.)

12.3 Martingales

This section formally defines continuous-time martingales. To this end, we fix a probability space $(\Omega, \mathcal{F}, P)$ equipped with filtration $\mathcal{F} = \{\mathcal{F}_t; 0 \leq t \leq T\}$. Here and hereafter, the time horizon is always finite and given by $[0, T]$ for some $T > 0$. See Definition 5.7 for discrete-time martingales.

While the Markov property (9.1) is concerned with the future probability distribution, the martingale property states its future mean. The next definition introduces continuous-time martingales.

Definition 12.3 (Martingale) A continuous-time stochastic process $\{X(t)\}$ is called a *martingale* with respect to $\mathcal{F}$, if

(M1) $E[|X(t)|] < \infty$ for each $t \in [0,T]$, and

(M2) $E_t[X(s)] = X(t)$, $t < s \leq T$,

where $E_t[X(s)]$ denotes the conditional expectation of $X(s)$ given the information $\mathcal{F}_t$.

Note that the condition (M2) implies that, given the information $\mathcal{F}_t$, the expectation of the future value $X(s)$ is the same as the current value $X(t)$. On the other hand, (M1) states the integrability condition, which is needed because the martingale compares the conditional expectation with the current value.

Before proceeding, we provide some examples of continuous-time martingales. Obviously, the standard Brownian motion $\{z(t)\}$ is a martingale with respect to $\mathcal{F}_t = \sigma(z(s), s \leq t)$, since the increment $z(t+s) - z(t)$, $s > 0$, is independent of $\mathcal{F}_t$ and its mean is zero. Hence,

$$E_t[z(t+s)] = E_t[z(t+s) - z(t)] + E_t[z(t)] = z(t), \quad s > 0.$$

Also, the process $\{X(t)\}$ defined by

$$X(t) = z^2(t) - t, \quad t \geq 0, \tag{12.23}$$

is a martingale, since we have

$$z^2(t+s) = (\Delta_s z(t))^2 + 2z(t)\Delta_s z(t) + z^2(t),$$

where $\Delta_s z(t) = z(t+s) - z(t)$. See Exercise 12.6 for other martingales.

Definition 12.4 Let $\{X(t); 0 \leq t \leq T\}$ be a martingale.

(1) $\{X(t)\}$ is called *square integrable* if

$$\sup_{0 \leq t \leq T} E\left[X^2(t)\right] < \infty.$$

(2) $\{X(t)\}$ is called *uniformly integrable* if

$$\lim_{n \to \infty} \sup_{0 \leq t \leq T} E\left[|X(t)| 1_{\{|X(t)| > n\}}\right] = 0.$$

The standard Brownian motion as well as the process defined by (12.23) are square integrable martingales, provided that they are defined on the finite time interval.

The next proposition provides uniformly integrable martingales. The proof is left in Exercise 12.7.

Proposition 12.2 (1) *Any square integrable martingale defined on a finite time interval is uniformly integrable.*

(2) *For any integrable random variable Y, define*

$$M(t) = E_t[Y], \quad 0 \le t \le T. \tag{12.24}$$

Then, the process $\{M(t)\}$ is uniformly integrable.

The martingale defined by (12.24) is called a Doob's martingale (see Example 5.5). Its immediate consequence is that, if $\{M(t)\}$ is a martingale defined on a finite time interval, then it is uniformly integrable and given by the form (12.24) with Y being replaced by $M(T)$.

Recall that a random time τ is called a *stopping time* if

$$\{\tau \le t\} \in \mathcal{F}_t \text{ for all } t \in [0, T].$$

A martingale stopped at time τ is the process $\{M(t \wedge \tau); 0 \le t \le T\}$, where $t \wedge \tau = \min\{t, \tau\}$. It is readily verified that if $\{M(t)\}$ is a martingale then so is the stopped process $\{M(t \wedge \tau)\}$. The converse statement is not true, however.

Definition 12.5 (Local Martingale) An adapted process $\{X(t)\}$ is called a *local* martingale, if there exists a sequence of stopping times $\{\tau_n\}$ such that $\tau_n \to \infty$ as $n \to \infty$ and that, for each n, the stopped process $\{M(t \wedge \tau_n)\}$ is a uniformly integrable martingale. The sequence of stopping times $\{\tau_n\}$ is called a *localizing sequence*.

Any martingale is a local martingale, but not vice versa (see Exercise 12.8). A natural question is then under what conditions a local martingale becomes a martingale. For example, suppose that $\{M(t)\}$ is a local martingale such that $|M(t)| \le Y$ for some integrable random variable Y. For a localizing sequence $\{\tau_n\}$, the stopped process $\{M(t \wedge \tau_n)\}$ is a martingale, i.e.

$$E_t[M(s \wedge \tau_n)] = M(t \wedge \tau_n), \quad t < s \le T.$$

Since $\lim_{n \to \infty} M(t \wedge \tau_n) = M(t)$, we have $E_t[M(s)] = M(t)$ by the dominated convergence theorem (Proposition 1.11). Since $M(t)$ is integrable, the local martingale $\{M(t)\}$ is a martingale.

In financial engineering applications, we often encounter non-negative local martingales.

Proposition 12.3 *A non-negative local martingale $\{M(t); 0 \le t \le T\}$ is a supermartingale. That is, $E[M(t)] < \infty$ for all t and $E_t[M(s)] \le M(t)$ for any $t < s \le T$. It is a martingale if and only if $E[M(T)] = E[M(0)]$.*

Proof. Let $\{\tau_n\}$ be a localizing sequence. For any n and $t < s$, we have

$$E_t\left[\inf_{m \ge n} M(s \wedge \tau_m)\right] \le E_t[M(s \wedge \tau_n)] = M(t \wedge \tau_n).$$

Since $M(t) \ge 0$, we obtain

$$E_t[M(s)] = E_t\left[\lim_{n \to \infty} M(s \wedge \tau_n)\right] \le \lim_{n \to \infty} M(t \wedge \tau_n) = M(t),$$

whence the result. □

12.4 Stochastic Integrals

In the following, we denote the Brownian filtration by $\{\mathcal{F}_t; 0 \le t \le T\}$. Let $\{\psi(t)\}$ be a stochastic process with continuous sample paths, and consider a stochastic integral

$$I(t) = \int_0^t \psi(u) \mathrm{d}z(u), \quad 0 \le t \le T. \tag{12.25}$$

In order to calculate $I(t)$, we divide the time interval $[0, t]$ into n equally spaced subintervals, i.e.

$$0 = t_0 < t_1 < \cdots < t_n = t, \quad t_i \equiv \frac{t}{n} i.$$

Although $\{z(t)\}$ is not of finite variation, we approximate $I(t)$ by the Riemann–Stieltjes sum (1.27) as

$$J_n \equiv \sum_{i=0}^{n-1} \frac{\psi(t_i) + \psi(t_{i+1})}{2} \{z(t_{i+1}) - z(t_i)\}. \tag{12.26}$$

Alternatively, we may define the Riemann–Stieltjes sum by

$$I_n \equiv \sum_{i=0}^{n-1} \psi(t_i) \{z(t_{i+1}) - z(t_i)\}. \tag{12.27}$$

In the ordinary Riemann–Stieltjes integral, the integral exists if any Riemann–Stieltjes sum converges to the same limit regardless of the approximation.

Since the integrator $\{z(t)\}$ in (12.25) is of infinite variation, the Riemann–Stieltjes sums may converge to different limits. To see this, consider the case that $\psi(t) = z(t)$. Then, from (12.26), we obtain

$$J_n = \frac{1}{2} \sum_{i=0}^{n-1} \{z^2(t_{i+1}) - z^2(t_i)\} = \frac{1}{2} z^2(t).$$

Hence, the limit of the Riemann–Stieltjes sum (12.26) is given by

$$\lim_{n \to \infty} J_n = \int_0^t z(u) \mathrm{d}z(u) = \frac{z^2(t)}{2}, \tag{12.28}$$

which agrees with the ordinary Riemann–Stieltjes integral.

Next, in order to calculate the limit of I_n, we use the identity

$$2J_n - \sum_{i=0}^{n-1} \{z(t_{i+1}) - z(t_i)\}^2 = 2I_n.$$

It follows from (12.15) that

$$\lim_{n \to \infty} I_n = \frac{z^2(t)}{2} - \frac{t}{2}, \tag{12.29}$$

which is different from the limit (12.28).

Now, we are facing the problem of which approximation should be used in finance. As we shall see later, the approximation (12.27) is more appropriate for our purpose, because it is related directly to martingales, one of the key notions in the theory of no-arbitrage pricing (see Theorem 5.3). The stochastic integral resulted from the approximation (12.27) is called the *Ito integral*, while the one based on the approximation (12.26) is called the *Stratonovich integral*. See, e.g., Øksendal (1998) for the relationship between the Ito integral and the Stratonovich integral.

Definition 12.6 (Simple Process) An adapted process $\{\psi(t)\}$ is said to be *simple*, if there exist time epochs such that

$$0 = t_0 < t_1 < \cdots < t_n = T$$

and random variables $\xi_1, \xi_2 \ldots, \xi_{n-1}$ such that ξ_i is $\mathcal{F}_{t_i}$-measurable and

$$\psi(t) = \psi(0)1_{\{0\}}(t) + \sum_{i=1}^{n-1} \xi_i 1_{(t_i, t_{i+1}]}(t), \quad 0 \leq t \leq T, \tag{12.30}$$

where $1_A(t) = 1$ if $t \in A$ and $1_A(t) = 0$ otherwise.

Referring to (12.27), the *Ito integral* for a simple process $\{\psi(t)\}$ is defined by

$$I(t) = \sum_{k=0}^{i-1} \psi(t_k)[z(t_{k+1}) - z(t_k)] + \psi(t_i)[z(t) - z(t_i)], \tag{12.31}$$

where $t_i < t \leq t_{i+1} \leq T$. Since $\{z(t)\}$ has continuous sample paths, the sample paths of the Ito integral process $\{I(t)\}$ defined by (12.31) are continuous. Also, it is readily seen that the process $\{I(t)\}$ is a martingale (see Proposition 5.3), if it is integrable. The proof is left to the reader.

We next calculate the variance of the Ito integral (12.31) under the assumption that

$$\int_0^T E\left[\psi^2(t)\right] \mathrm{d}t < \infty. \tag{12.32}$$

By the chain rule (5.18) of the conditional expectation, we have

$$E\left[I^2(t)\right] = E\left[E_{t_i}\left[I^2(t)\right]\right], \quad t_i < t.$$

Also, from (12.31), we obtain

$$\begin{aligned} E_{t_i}\left[I^2(t)\right] &= I^2(t_i) + 2I(t_i)E_{t_i}\left[\psi(t_i)\{z(t) - z(t_i)\}\right] \\ &\quad + E_{t_i}\left[\psi^2(t_i)\{z(t) - z(t_i)\}^2\right]. \end{aligned}$$

Since $\psi(t_i)$ is $\mathcal{F}_{t_i}$-measurable, it follows that

$$E_{t_i}\left[I^2(t)\right] = I^2(t_i) + \psi^2(t_i)(t - t_i),$$

whence we obtain

$$E\left[I^2(t)\right] = E\left[I^2(t_i)\right] + E\left[\psi^2(t_i)\right](t - t_i).$$

Repeating the same argument, we then have

$$E\left[I^2(t)\right] = \sum_{k=0}^{i-1} E\left[\psi^2(t_k)\right](t_{k+1} - t_k) + E\left[\psi^2(t_i)\right](t - t_i),$$

which is rewritten as

$$E\left[\left\{\int_0^t \psi(s)\mathrm{d}z(s)\right\}^2\right] = \int_0^t E\left[\psi^2(s)\right]\mathrm{d}s. \qquad (12.33)$$

More generally, it can be shown that, for two simple processes $\{f(t)\}$ and $\{g(t)\}$, we have

$$E\left[\left\{\int_0^t f(s)\mathrm{d}z(s)\right\}\left\{\int_0^t g(s)\mathrm{d}z(s)\right\}\right] = \int_0^t E[f(s)g(s)]\mathrm{d}s, \qquad (12.34)$$

provided that the integral in the right-hand side is well defined. The proof is left in Exercise 12.9.

The idea to define the Ito integral for a general integrand $X(t)$ is to approximate the process $\{X(t)\}$ by a sequence of simple processes $\{\psi_n(t)\}$. The problem is then what class of stochastic processes can be approximated in such a way. Recall that, in the definition of discrete stochastic integral (5.10), the integrand is a predictable process. Hence, a plausible choice of such a class is the class of predictable processes. However, the definition of predictable processes in continuous time is rather technical. Intuitively, an adapted process $\{X(t)\}$ is *predictable* if, for any t, the value $X(t)$ is determined by the values of $z(s)$ for $s < t$. For our purposes, the following definition for predictability is enough.

Definition 12.7 (Predictable Process) An adapted process $\{X(t)\}$ is *predictable*, if it is left-continuous in time t, a limit of left-continuous processes, or a measurable function of a left-continuous process. In particular, any adapted and continuous process is predictable.

We are now in a position to define the Ito integral. The proof of the next result is beyond the scope of this book and omitted. See, e.g., Liptser and Shiryaev (1989) or Karatzas and Shreve (1988) for the proof.

Proposition 12.4 Let $\{X(t)\}$ be a predictable process such that $\int_0^T X^2(t)\mathrm{d}t < \infty$ almost surely. Then, the Ito integral $\int_0^T X(t)\mathrm{d}z(t)$ is well defined and a local martingale. Conversely, any local martingale can be represented as an Ito integral process with some predictable process $\{\psi(t)\}$ such that $\int_0^T \psi^2(t)\mathrm{d}t < \infty$ almost surely.

It should be noted that, under the conditions of Proposition 12.4, the Ito integral is well defined, but its mean and variance may not exist when the stronger condition (12.32) fails. That is, if $\int_0^T E\left[X^2(t)\right]\mathrm{d}t = \infty$ while $\int_0^T X^2(t)\mathrm{d}t < \infty$ almost surely, then the Ito integral process $\{I(t)\}$ defined by $I(t) = \int_0^t X(u)\mathrm{d}z(u)$ may not be a martingale; but, it is always a local martingale with continuous sample paths. If the condition (12.32) holds, then from (12.33) the Ito integral has a finite variance and the process $\{I(t)\}$ is a square integrable, zero mean martingale with continuous sample paths. The next result summarizes.

Proposition 12.5 *Let $\{X(t)\}$ be a predictable process such that*

$$\int_0^T E\left[X^2(t)\right] dt < \infty.$$

Then, the Ito integral process $\{I(t)\}$ is a square integrable, zero mean martingale with continuous sample paths. Conversely, any square integrable martingale can be represented as an Ito integral process with some predictable process $\{\psi(t)\}$ such that $\int_0^T E\left[\psi^2(t)\right] dt < \infty$.

The converse statement in each of the above propositions is known as the (local) *martingale representation theorem*, which is frequently used in mathematical finance.

Definition 12.8 (Ito Differential) The Ito integral

$$I(t) = \int_0^t X(s) dz(s),$$

if it exists, is alternatively expressed by the differential form

$$dI(t) = X(t) dz(t),$$

called the *Ito differential*.

12.5 Stochastic Differential Equations

In the literature of financial engineering, it is common to model a continuous-time price process in terms of a *stochastic differential equation* (SDE). This section explains the idea of SDEs as a limit of stochastic difference equations in the discrete-time setting.

Let $\Delta t > 0$ be sufficiently small, and consider the stochastic difference equation

$$\Delta X(t) = \mu(X(t),t)\Delta t + \sigma(X(t),t)\Delta z(t), \quad t \geq 0, \tag{12.35}$$

where $\mu(x,t)$ and $\sigma(x,t)$ are given functions with enough smoothness and where $\Delta z(t) \equiv z(t + \Delta t) - z(t)$ is the increment of the standard Brownian motion $\{z(t)\}$. Note that, at time t, the state $X(t)$ is observed and the values of $\mu(X(t),t)$ and $\sigma(X(t),t)$ are known. Suppose that $X(t) = x$. Then, the limiting process $\{X(t)\}$ as $\Delta t \to 0$, if it exists, is a strong Markov process with continuous sample paths, since so is the Brownian motion $\{z(t)\}$. Moreover, we obtain from (12.35) that

$$\lim_{\Delta t \to 0} \frac{E[\Delta X(t) | X(t) = x]}{\Delta t} = \mu(x,t)$$

and

$$\lim_{\Delta t \to 0} \frac{E[\{\Delta X(t)\}^2 | X(t) = x]}{\Delta t} = \sigma^2(x,t).$$

Hence, the limiting process is expected to be a diffusion with drift function $\mu(x,t)$ and diffusion function $\sigma(x,t)$. However, we cannot take the limit in the ordinary sense, since the Brownian motion $\{z(t)\}$ is *not* differentiable with respect to t. Nevertheless, from (12.35), we can formally obtain the SDE

$$dX(t) = \mu(X(t),t)dt + \sigma(X(t),t)dz(t), \quad 0 \le t \le T, \tag{12.36}$$

under some regularity conditions, where $T > 0$ is a given time horizon.

The SDE (12.36) is understood to be the differential form of the integral equation

$$X(t) - X(0) = \int_0^t \mu(X(u),u)du + \int_0^t \sigma(X(u),u)dz(u), \tag{12.37}$$

where the first term on the right-hand side is the ordinary (path-by-path) integral and the second term is the Ito integral. In the rest of this book, we denote (12.36) as

$$dX = \mu(X,t)dt + \sigma(X,t)dz, \quad 0 \le t \le T,$$

in order to simplify the notation. The solution $\{X(t)\}$ to the SDE (12.36), if it exists, is called the *Ito process*.

Note that the solution to the SDE (12.36), if it exists, is a stochastic process adapted to the Brownian filtration $\{\mathcal{F}_t\}$, where $\mathcal{F}_t = \sigma(z(s), s \le t)$. To see this, given the information $\mathcal{F}_t$, uncertainty in (12.35) is due to $\Delta z(t)$, since $z(t+\Delta t)$ is not yet realized. Once the outcome of $z(t+\Delta t)$ is observed, the value $X(t+\Delta t)$ is determined through (12.35). This means that the limiting process, if it exists, is adapted to the Brownian filtration.

Before proceeding, we confirm that the SDE (12.36) has a meaning at least for some special case. Let $\{X(t)\}$ be the Brownian motion defined in (11.1). For $\Delta t > 0$, consider the increment $\Delta X(t) = X(t+\Delta t) - X(t)$. Then, from (11.1), we obtain

$$\Delta X(t) = \mu \Delta t + \sigma \Delta z(t), \quad \Delta t > 0,$$

whence, letting $\Delta t \to 0$, we have

$$dX = \mu dt + \sigma dz, \quad t \ge 0. \tag{12.38}$$

Comparing (12.38) with (12.36), we observe that $\mu(x,t)$ in (12.36) corresponds to the drift while $\sigma(x,t)$ corresponds to the diffusion coefficient.

We turn to consider the existence of a solution to the SDE (12.36). Suppose that there exist some constants $K, L > 0$ such that

(C1) $\mu^2(x,t) + \sigma^2(x,t) \le K(1+x^2)$, and

(C2) $|\mu(x,t) - \mu(y,t)| + |\sigma(x,t) - \sigma(y,t)| \le L|x-y|$.

Then, it can be shown that the SDE (12.36) has a unique (path-by-path) solution. For example, when the drift μ and diffusion coefficient σ are constants, these conditions are obviously satisfied (see also Exercise 12.10). Indeed, the unique solution to the SDE (12.38) is the Brownian motion defined in (11.1). The condition (C1) is called the *growth condition*, while (C2) is known as the *Lipschitz condition* (see Exercise 1.4).

These conditions are, however, often too strong to be used for practical purposes. For example, consider the case

$$\mu(x,t) = \mu, \quad \sigma(x,t) = \sigma\sqrt{x}.$$

The Lipschitz condition (C2) is not satisfied, since

$$|\sigma(x,t) - \sigma(y,t)| = \sigma \frac{|x-y|}{\sqrt{x}+\sqrt{y}},$$

which cannot be bounded by $L|x-y|$ from above. This kind of model is known to have a solution (see Exercise 12.11) and has many applications in practice. Notice that the conditions (C1) and (C2) are only sufficient for a solution to exist in (12.36).

On the other hand, when estimating the parameters of the SDE from actual security price data, it often happens that the diffusion function is estimated as

$$\sigma(x,t) = \sigma x^\alpha, \quad \alpha > 1.$$

Such a model may not have a solution in any sense. This occurs, since we try to determine the function $\sigma(x,t)$ for *all* $x > 0$ so as to explain the actual data sampled from a *finite* region.

In order for such a model to make sense, we may assume that

$$\sigma(x,t) = \min\{\sigma x^\alpha, M\}, \quad \alpha > 1,$$

for a sufficiently large constant $M > 0$. Then, it can be shown that the corresponding SDE has a solution, since the model satisfies the conditions (C1) and (C2). The proof is left in Exercise 12.10.

Recall that a solution to the SDE (12.36), if it exists, is an adapted process to the Brownian filtration $\{\mathcal{F}_t\}$. It is well known and very important in financial engineering practices that solutions of some SDEs are martingales. We provide such important results without proof. For the proof, see any textbooks of mathematical finance.

Proposition 12.6 *Suppose that the process $\{X(t)\}$ is a solution to the SDE*

$$dX = \mu(X,t)dt + \sigma(X,t)dz, \quad 0 \le t \le T,$$

where $\sigma(x,t)$ is continuous and satisfies $E\left[\int_0^T \sigma^2(X,t)dt\right] < \infty$. Then, the process $\{X(t)\}$ is a martingale if and only if the drift is zero, i.e. $\mu(x,t) = 0$.

Proposition 12.7 *Suppose that the process $\{X(t)\}$ is a solution to the SDE*

$$\frac{dX}{X} = \sigma(X,t)dz, \quad 0 \le t \le T,$$

where $\sigma(x,t)$ is continuous and satisfies $E\left[\exp\left\{\frac{1}{2}\int_0^T \sigma^2(X,t)dt\right\}\right] < \infty$.[*]
Then, the process $\{X(t)\}$ is a martingale.

---

[*] This condition is called the *Novikov condition*.

As we shall show in the next chapter, the process $\{X(t)\}$ defined in Proposition 12.7 satisfies the equation

$$X(t) = X(0) \exp\left\{\int_0^t \sigma(X,s)\mathrm{d}z(s) - \frac{1}{2}\int_0^t \sigma^2(X,s)\mathrm{d}s\right\}, \quad (12.39)$$

whence it is called an *exponential martingale*.

Finally, there is a class of SDEs that can be solved analytically. Consider the *linear* equation

$$\mathrm{d}X = (b(t) + \mu(t)X)\mathrm{d}t + \sigma(t)\mathrm{d}z, \quad 0 \le t \le T, \quad (12.40)$$

where $b(t)$, $\mu(t)$, and $\sigma(t)$ are deterministic functions of time t. Then, as is proved in Exercise 12.12, the solution to the SDE (12.40) is given by

$$X(t) = \mathrm{e}^{\int_0^t \mu(s)\mathrm{d}s}\left[x + \int_0^t b(s)\,\mathrm{e}^{-\int_0^s \mu(u)\mathrm{d}u}\mathrm{d}s + \int_0^t \sigma(s)\,\mathrm{e}^{-\int_0^s \mu(u)\mathrm{d}u}\mathrm{d}z(s)\right], \quad (12.41)$$

where $X(0) = x$. The solution $\{X(t)\}$ is a Gaussian process in the sense that the joint distribution of $\{X(t)\}$ is a multivariate normal. The normality follows from the fact that a linear combination of normally distributed random variables is also normally distributed (cf. Exercise 3.10). The mean of $X(t)$ is given by

$$E[X(t)] = \mathrm{e}^{\int_0^t \mu(s)\mathrm{d}s}\left[x + \int_0^t b(s)\,\mathrm{e}^{-\int_0^s \mu(u)\mathrm{d}u}\mathrm{d}s\right],$$

while its variance is obtained from (12.33) as

$$V[X(t)] = \int_0^t \sigma^2(s)\,\mathrm{e}^{-2\int_0^s \mu(u)\mathrm{d}u}\mathrm{d}s.$$

12.6 Ito's Formula Revisited

Let $\{X(t)\}$ be a solution to the SDE (12.36). In many applications, we often consider a stochastic process $\{Y(t)\}$ obtained from $X(t)$ via a smooth function $f(x,t)$, i.e.

$$Y(t) = f(X(t),t), \quad 0 \le t \le T. \quad (12.42)$$

For example, the time t price of a derivative security written on the underlying security price $X(t)$ is given by $Y(t)$ in (12.42) for a certain function $f(x,t)$. The next theorem is due to Ito (1944). Here and hereafter, we shall assume that the partial derivatives of $f(x,t)$ are continuous. The proof of the next result is similar to the proof of Theorem 1.1 and left to the reader.

Theorem 12.1 (Ito's Formula) *Let $X(t)$ be a stochastic process that satisfies the SDE (12.36), i.e.*

$$\mathrm{d}X = \mu(X,t)\mathrm{d}t + \sigma(X,t)\mathrm{d}z, \quad 0 \le t \le T.$$

Then, for a smooth function $f(x,t)$, $Y(t) = f(X(t),t)$ satisfies the SDE

$$\mathrm{d}Y = \mu_Y(t)\mathrm{d}t + \sigma_Y(t)\mathrm{d}z, \quad 0 \le t \le T,$$

where
$$\mu_Y(t) = f_t(X,t) + f_x(X,t)\mu(X,t) + \frac{f_{xx}(X,t)}{2}\sigma^2(X,t)$$
and
$$\sigma_Y(t) = f_x(X,t)\sigma(X,t).$$

Martingale techniques are the main tool for working with diffusions, and Ito's formula provides a source for construction of martingales. Let $\{X(t)\}$ solve the SDE (12.36), and define the operator L_t by

$$L_t f(x,t) = \mu(x,t) f_x(x,t) + \frac{1}{2}\sigma^2(x,t) f_{xx}(x,t) \tag{12.43}$$

for a sufficiently smooth function $f(x,t)$. The operator L_t is often called the *generator* of $\{X(t)\}$. Using the generator L_t, it is readily seen that Ito's formula (Theorem 12.1) can be stated as

$$df(X,t) = (L_t f(X,t) + f_t(X,t))\,dt + f_x(X,t)\sigma(X,t)dz.$$

Also, since the Ito integral is a martingale under some regularity conditions (see Proposition 12.5), we can obtain a martingale by isolating the Ito integral. That is, the process $\{M_f(t)\}$ defined by

$$M_f(t) = f(X,t) - \int_0^t (L_u f + f_t)(X,u)du, \quad 0 \le t \le T,$$

is a martingale under regularity conditions.

Combining the result with the optional sampling theorem (Proposition 5.5), we then have the following important result. See Exercises 12.15 and 12.16 for some applications.

Proposition 12.8 (Dynkin's Formula) *Let τ be any stopping time such that $\tau \le T$. Then, for a sufficiently smooth function $f(x,t)$, we have*

$$E[f(X(\tau),\tau)] = f(X(0),0) + E\left[\int_0^\tau (L_u f + f_t)(X(u),u)du\right].$$

Another application of Ito's formula is given in the next example.

Example 12.5 (Feynman–Kac Formula) Let $\{X(t)\}$ be a diffusion satisfying the SDE (12.36), i.e.

$$dX = \mu(X,t)dt + \sigma(X,t)dz, \quad 0 \le t \le T,$$

and suppose that there exists a solution to the PDE

$$f_t(x,t) + L_t f(x,t) = r(x,t) f(x,t), \quad 0 \le t \le T, \tag{12.44}$$

with the boundary condition $f(x,T) = g(x)$. Let $f(x,t)$ be its solution, and apply Ito's formula for $f(X,t)$ to obtain

$$df(X,t) = (f_t(X,t) + L_t f(X,t))\,dt + dM, \quad 0 \le t \le T,$$

where $dM = f_x(X,t)\sigma(X,t)dz$. It follows from (12.44) that

$$df(X,t) = r(X,t)f(X,t)dt + dM, \quad 0 \le t \le T.$$

Since this is a linear equation, it is solved as

$$f(X,T) = e^{\int_t^T r(X,u)du} \left[f(X,t) + \int_t^T e^{-\int_t^s r(X,u)du} dM(s) \right].$$

Note that the last term in the right-hand side is an Ito integral, which is a martingale under regularity conditions. Hence, taking the expectation and using the boundary condition, it follows that

$$f(X(t),t) = E\left[g(X(T))e^{-\int_t^T r(X,u)du} \Big| X(t) = x \right],$$

which has been obtained before; see (12.11).

Finally, we state Ito's formula in higher dimensions. For independent standard Brownian motions $\{z_i(t)\}$, $i = 1, 2, \ldots, d$, define

$$\mathbf{z}(t) = (z_1(t), z_2(t), \ldots, z_d(t))^\top, \quad t \geq 0,$$

where $\top$ denotes the transpose. The process $\{\mathbf{z}(t)\}$ is called the d-dimensional standard Brownian motion.

Let $\mathcal{F}_t$ be the σ-field generated from $\{\mathbf{z}(s), s \leq t\}$. Recall that, for any predictable process $\{h(t)\}$ such that $\int_0^t h^2(s)ds < \infty$, the Ito integral process $\int_0^t h(s)dz_i(s)$ is well defined (see Proposition 12.4). More generally, we can define the Ito processes by

$$dX_i = b_i(t)dt + \sum_{j=1}^d \sigma_{ij}(t)dz_j, \quad i = 1, 2, \ldots, n, \qquad (12.45)$$

where $0 \leq t \leq T$. Here, we have assumed that, while $\{b_i(t)\}$ are adapted to the filtration $\{\mathcal{F}_t\}$ and $\int_0^T |b_i(t)|dt < \infty$ almost surely, $\{\sigma_{ij}(t)\}$ are predictable and $\int_0^T \sigma_{ij}^2(t)dt < \infty$ almost surely. The SDEs (12.45) can be compactly written in vector form as

$$d\mathbf{X} = \mathbf{b}(t)dt + \boldsymbol{\sigma}(t)d\mathbf{z}, \quad 0 \leq t \leq T,$$

where $\mathbf{X}(t) = (X_1(t), X_2(t), \ldots, X_n(t))^\top$ and $\mathbf{b}(t) = (b_1(t), b_2(t), \ldots, b_n(t))^\top$ are n-dimensional vectors, and where $\boldsymbol{\sigma}(t) = (\sigma_{ij}(t))$ is an $n \times d$ matrix. In particular, if $\mathbf{b}(t) = \mathbf{b}(\mathbf{X},t)$ and $\boldsymbol{\sigma}(t) = \boldsymbol{\sigma}(\mathbf{X},t)$, then we have the multivariate SDE of the form

$$d\mathbf{X} = \mathbf{b}(\mathbf{X},t)dt + \boldsymbol{\sigma}(\mathbf{X},t)d\mathbf{z}, \quad 0 \leq t \leq T. \qquad (12.46)$$

It is easy to see that the solution to (12.46), if it exists, is an n-dimensional diffusion.

Let $\mathbf{X}(t)$ be a solution to the SDE (12.46), and consider a sufficiently smooth function $f(x_1, x_2, \ldots, x_n)$. In the following, we shall denote

$$f_i(x_1, x_2, \ldots, x_n) = \frac{\partial}{\partial x_i} f(x_1, x_2, \ldots, x_n)$$

and
$$f_{ij}(x_1, x_2, \ldots, x_n) = \frac{\partial^2}{\partial x_j \partial x_i} f(x_1, x_2, \ldots, x_n).$$

Also, let
$$\mathbf{A}(t) = (a_{ij}(t)) \equiv \boldsymbol{\sigma}(t) \boldsymbol{\sigma}^\top(t).$$

It is easy to see from (12.45) that
$$\mathrm{d}X_i(t)\mathrm{d}X_j(t) = a_{ij}(t)\mathrm{d}t, \quad i, j = 1, 2, \ldots, n,$$

since $\{z_i(t)\}$ are independent. The next result is Ito's formula in the multivariate setting.

$$\mathrm{d}f(X_1, \ldots, X_n) = \sum_{i=1}^n f_i(X_1, \ldots, X_n)\mathrm{d}X_i + \frac{1}{2}\sum_{i=1}^n \sum_{j=1}^n f_{ij}(X_1, \ldots, X_n)a_{ij}(t)\mathrm{d}t. \tag{12.47}$$

The proof is left in Exercise 12.17.

12.7 Exercises

Exercise 12.1 For a Brownian motion $\{X(t)\}$ with drift μ and diffusion coefficient σ, show that (12.3) holds for any $\delta > 0$. Also, confirm the condition (12.4) for the cases $r = 3$ and $r = 4$.

Exercise 12.2 Suppose that a diffusion process $\{X(t)\}$ with drift function $\mu(x)$ and diffusion function $\sigma(x)$ has absorbing boundaries a and b, where $a < X(0) < b$, and consider the first passage time T_z that the process reaches the boundary z, $z = a, b$, for the first time. We are interested in the probability

$$u(x) = P\{T_b < T_a | X(0) = x\}, \quad a < x < b.$$

Let $h > 0$ be sufficiently small. Prove that

$$u(x) = E[u(X(h)) | X(0) = x] + o(h).$$

Also, assuming that $u(x)$ is sufficiently smooth, prove that

$$u(X(h)) = u(x) + u'(x)\mu(x)h + \frac{\sigma^2(x)}{2}u''(x)h + o(h),$$

using Taylor's expansion. Finally, prove that $u(x)$ satisfies the ordinary differential equation (ODE)

$$0 = \mu(x)u'(x) + \frac{\sigma^2(x)}{2}u''(x), \quad a < x < b,$$

with boundary conditions $u(a) = 0$ and $u(b) = 1$.

Exercise 12.3 In the same setting as Exercise 12.2, let

$$w(x) = E\left[\int_0^T g(X(t))\mathrm{d}t \,\Big|\, X(0) = x\right], \quad a < x < b,$$

where $T = \min\{T_a, T_b\}$. Prove that $w(x)$ satisfies the ODE

$$-g(x) = \mu(x)w'(x) + \frac{1}{2}\sigma^2(x)w''(x), \quad a < x < b,$$

with boundary conditions $w(a) = w(b) = 0$.

Exercise 12.4 Prove that the PDE (12.10) holds for any time-homogeneous diffusion $\{X(t)\}$ with sufficiently smooth drift function $\mu(x)$, diffusion function $\sigma(x)$ and killing rate $k(x)$.

Exercise 12.5 For two standard Brownian motions $\{z_1(t)\}$ and $\{z_2(t)\}$, suppose that the correlation coefficient between $\Delta z_1(t)$ and $\Delta z_2(t)$ is ρ. Let $\Delta z(t)$ be the increment of another standard Brownian motion $\{z(t)\}$, independent of $\Delta z_1(t)$. Then, from Proposition 10.4, we have $\Delta z_2(t) \stackrel{d}{=} \rho \Delta z_1(t) + \sqrt{1-\rho^2}\Delta z(t)$, where $\stackrel{d}{=}$ stands for equality in law. Using this and (12.16), prove that (12.17) holds.

Exercise 12.6 For a standard Brownian motion $\{z(t)\}$ and any $\beta \in \mathbf{R}$, let

$$V(t) = \exp\left\{\beta z(t) - \frac{\beta^2 t}{2}\right\}, \quad t \geq 0.$$

Prove that $\{V(t)\}$ is a martingale with respect to the Brownian filtration. Also, let $\{N(t)\}$ be a Poisson process with intensity $\lambda > 0$. Show that the following are martingales with respect to $\{\mathcal{F}_t^N\}$, where $\mathcal{F}_t^N = \sigma(N(s), s \leq t)$.

(a) $N(t) - \lambda t$ (b) $e^{\log(1-u)N(t) + u\lambda t}$, $0 < u < 1$.

Exercise 12.7 Prove that Proposition 12.2 holds.

Exercise 12.8 Let $M(t) = \int_0^t e^{z^2(s)} dz(s)$, where $\{z(t)\}$ is a standard Brownian motion. Let $\tau_n = \inf\{t > 0 : e^{z^2(t)} = n\}$. Prove that $\tau_n \to \infty$ as $n \to \infty$ and that $\{M(t)\}$ is a local martingale with the localizing sequence $\{\tau_n\}$. Also, prove that $E\left[e^{z^2(t)}\right] = \infty$ for $t \geq 1/2$, whence $\{M(t)\}$ is *not* a martingale.

Exercise 12.9 Prove that (12.34) holds for any simple processes $\{f(t)\}$ and $\{g(t)\}$, for which the expectations exist.

Exercise 12.10 In the SDE (12.36), suppose that $\mu(x,t) = \mu x$ and $\sigma(x,t) = \sigma x$, where μ and σ are constants. Prove that this model satisfies both the growth condition (C1) and the Lipschitz condition (C2). How about the case that $\sigma(x,t) = \min\{\sigma x^2, M\}$ for some $M > 0$?

Exercise 12.11 In the SDE (12.36), suppose that

$$\mu(x,t) = a(\kappa - x), \quad \sigma(x,t) = \sigma x^\alpha, \quad x \geq 0,$$

where a, κ, σ, and α are constants with $0.5 \leq \alpha < 1$. Confirm that the model does not satisfy the Lipschitz condition (C2); but, show that the SDE has a solution. *Hint*: According to Yamada and Watanabe (1971), if there is a constant $L > 0$ such that

$$|\sigma(x,t) - \sigma(y,t)| \leq L|x-y|^\beta, \quad \beta \geq \frac{1}{2},$$

EXERCISES

then there exists a solution to the SDE. Take $\beta = \alpha$ here.

Exercise 12.12 Consider the linear SDE

$$dX = (\alpha(t) + \beta(t)X)dt + (\gamma(t) + \delta(t)X)dz, \quad 0 \le t \le T,$$

where $\alpha(t)$, $\beta(t)$, $\gamma(t)$, and $\delta(t)$ are deterministic functions of time t. Suppose that $X(t) = U(t)V(t)$, where $U(0) = 1$, $V(0) = X(0)$ and

$$dU = \beta(t)U dt + \delta(t)U dz, \quad dV = a(t)dt + b(t)dz,$$

for some $a(t)$ and $b(t)$. Determine $U(t)$ and $V(t)$ and show that the solution to the SDE is given by

$$X(t) = U(t)\left(X(0) + \int_0^t \frac{\alpha(s) - \delta(s)\gamma(s)}{U(s)}ds + \int_0^t \frac{\gamma(s)}{U(s)}dz(s)\right).$$

Confirm that the solution to the SDE (12.40) is given by (12.41).

Exercise 12.13 Consider the SDE

$$dX = \mu(X)dt + \sigma(X)dz, \quad 0 \le t \le T,$$

where $\sigma(x) \ge \varepsilon > 0$, and let $Y(t) = f(X(t))$ for a sufficiently smooth function $f(x)$. Suppose

$$dY = \mu_Y(t)dt + dz, \quad 0 \le t \le T.$$

Prove that the function $f(x)$ must be of the form

$$f(x) = \int_0^x \frac{1}{\sigma(u)}du + C$$

for some constant C. On the other hand, let $Z(t) = g(X(t))$ for a sufficiently smooth function $g(x)$, and suppose

$$dZ = \sigma_Z(t)dz, \quad 0 \le t \le T.$$

Determine the diffusion function $\sigma_Z(t)$.

Exercise 12.14 In the time-homogeneous case, the generator (12.43) is defined by

$$Lf(x) = \mu(x)f_x(x) + \frac{1}{2}\sigma^2(x)f_{xx}(x),$$

where $\sigma^2(x) \ge \varepsilon > 0$. Any solution to the equation $Lf = 0$ is called a *harmonic* function for L. Show that a positive harmonic function is given by

$$f(x) = \int_c^x \exp\left\{-\int_c^u \frac{2\mu(y)}{\sigma^2(y)}dy\right\}du$$

for some c. Using this result, obtain a solution for Exercise 12.2.

Exercise 12.15 In the context of Exercise 12.3, let $g(x) = 1$ so that $w(x)$ is a solution to the equation $Lw = -1$ with $w(a) = w(b) = 0$. Using Dynkin's formula, show that $E[T] = w(x) < \infty$.

Exercise 12.16 Consider a Brownian motion $\{X(t)\}$ with drift $\mu > 0$ and diffusion coefficient $\sigma \neq 0$ starting from $X(0) = 0$. For $z > 0$, let T_z denote the first passage time to the state z. Prove that the Laplace transform of T_z is given by
$$E\left[e^{-\theta T_z}\right] = \exp\left\{-\frac{z}{\sigma^2}\left(\sqrt{\mu^2 + 2\sigma^2\theta} - \mu\right)\right\}.$$
Hint: Take $f(x,t) = \exp\left\{\lambda x - (\lambda\mu + \lambda^2\sigma^2/2)t\right\}$ and use Dynkin's formula.

Exercise 12.17 Prove that Ito's formula in the multivariate setting is given by (12.47).

CHAPTER 13

A Continuous-Time Model for Securities Market

The first half of this chapter describes a general continuous-time model for securities market, parallel to the discrete-time counterpart discussed in Chapter 5. Similar limiting argument to the ones given so far can be applied for those results. The Black–Scholes model is discussed in detail as an application of the no-arbitrage pricing theorem. The remaining part of this chapter concerns important pricing methods, risk-neutral and forward-neutral methods, together with their applications for the pricing of contingent claims. The reader should consult any textbooks of finance such as Merton (1990), Neftci (1996), Björk (1998), and Mikosch (1998) for discussions of continuous-time securities market models.

13.1 Self-Financing Portfolio and No-Arbitrage

Suppose that the set of time epochs consists of the closed interval $[0, T]$, where $T < \infty$ and the time $t = 0$ denotes the current time. As in Chapter 5, we consider a financial market in which there are available $n + 1$ financial securities, the one numbered 0 being a risk-free security (e.g. default-free money-market account) and the others numbered $1, 2, \ldots, n$ being risky securities (e.g. stocks), where $n \geq 1$.

Let $S_i(t)$ denote the time t price of security i, where the current prices $S_i(0)$ are known by all the investors in the market. The security price process denoted by $\{\mathbf{S}(t); 0 \leq t \leq T\}$
(or $\{\mathbf{S}(t)\}$ for short), where

$$\mathbf{S}(t) = (S_0(t), S_1(t), \ldots, S_n(t))^\top, \quad 0 \leq t \leq T,$$

is a vector-valued *stochastic process* in continuous time. Other terminology is the same as Chapter 5. For example, $\theta_i(t)$ denotes the number of security i possessed at time t, $0 \leq t \leq T$, and $\boldsymbol{\theta}(t) = (\theta_0(t), \theta_1(t), \ldots, \theta_n(t))^\top$ is the *portfolio* at that time. Also, $d_i(t)$ denotes the dividend *rate* paid at time t by security i, while the cumulative dividend paid by security i until time t is denoted by $D_i(t) = \int_0^t d_i(s)\mathrm{d}s$.

Throughout this chapter, we fix a probability space $(\Omega, \mathcal{F}, P)$ equipped with filtration $\{\mathcal{F}_t; 0 \leq t \leq T\}$, where $\mathcal{F}_t$ denotes the information about security prices available in the market at time t. For example, $\mathcal{F}_t$ is the smallest σ-field generated from $\{\mathbf{S}(u); u \leq t\}$. However, the information can be any as far as the time t prices of the securities can be known based on the information. That is, the prices $S_i(t)$ are *measurable* with respect to $\mathcal{F}_t$.

Recall that, when determining the portfolio $\theta(t)$, we cannot use the future information about the price processes. This restriction has been formulated by predictability of the portfolio process in the discrete-time setting. As we saw in Definition 12.7, an adapted process $\{X(t)\}$ is *predictable* if $X(t)$ is left-continuous in time t. Hence, as for the discrete-time case, we assume that, while the price process $\{\mathbf{S}(t)\}$ as well as the dividend processes $\{d_i(t)\}$ in the continuous-time securities market are adapted to the filtration $\{\mathcal{F}_t\}$, the portfolio process $\{\boldsymbol{\theta}(t)\}$ is predictable with respect to $\{\mathcal{F}_t\}$.

The value process $\{V(t)\}$ in the continuous-time setting is similar to the one given in the discrete-time case. However, it is not appropriate to start with (5.3) in order to define self-financing portfolios. Instead, we take (5.6) as an equivalent definition. Recall that, in order to derive the desired continuous-time model from the discrete-time counterpart, it is enough to replace the difference by a differential and the sum by an integral.

For a portfolio process $\{\boldsymbol{\theta}(t)\}$, let

$$V(t) = \sum_{i=0}^{n} \theta_i(t) S_i(t), \quad 0 \le t \le T. \tag{13.1}$$

The process $\{V(t)\}$ is called the *value process*. Note the difference between (13.1) and (5.2). In particular, in the continuous-time case, the dividend *rates* do not appear in the portfolio *value* (13.1).

Now, note that Equation (5.6) can be rewritten as

$$dV(t) = \sum_{i=0}^{n} \theta_i(t) \{dS_i(t) + d_i(t) dt\}, \quad 0 \le t \le T. \tag{13.2}$$

As in Section 5.2, let us define the time t *gain* obtained from security i by

$$G_i(t) = S_i(t) + D_i(t), \quad i = 0, 1, \ldots, n,$$

or, in the differential form,

$$dG_i(t) = dS_i(t) + d_i(t) dt,$$

whence (13.2) is represented as

$$dV(t) = \sum_{i=0}^{n} \theta_i(t) dG_i(t), \quad 0 \le t \le T,$$

or, in the integral form, as

$$V(t) = V(0) + \sum_{i=0}^{n} \int_0^t \theta_i(s) dG_i(s), \quad 0 \le t \le T. \tag{13.3}$$

In particular, when the securities pay no dividends, the portfolio value is reduced to the stochastic integral with respect to the price processes. That is,

we have
$$V(t) = V(0) + \sum_{i=0}^{n} \int_0^t \theta_i(s) \mathrm{d}S_i(s), \quad 0 \le t \le T.$$

See Exercise 13.1 for the case of the denominated value process $\{V^*(t)\}$.

Referring to Theorem 5.1, we can now define the self-financing portfolio formally.

Definition 13.1 (Self-Financing Portfolio) A portfolio process $\{\boldsymbol{\theta}(t)\}$ is said to be *self-financing* if the time t portfolio value $V(t)$ is represented by (13.3).

A contingent claim X is said to be *attainable* if there exists some self-financing trading strategy $\{\boldsymbol{\theta}(t); 0 \le t \le T\}$, called the *replicating portfolio*, such that $V(T) = X$. Hence, the attainable claim X is represented as

$$X = V(0) + \sum_{i=0}^{n} \int_0^T \theta_i(t) \mathrm{d}G_i(t) \tag{13.4}$$

for some self-financing portfolio process $\{\boldsymbol{\theta}(t)\}$. In this case, the portfolio process is said to *generate* the contingent claim X.

The definition of arbitrage opportunities is unchanged in the continuous-time setting (see Definition 5.5). That is, an *arbitrage opportunity* is the existence of some self-financing trading strategy $\{\boldsymbol{\theta}(t)\}$ such that (a) $V(0) = 0$, and (b) $V(T) \ge 0$ and $V(T) > 0$ with positive probability. Recall that an arbitrage opportunity is a risk-free way of making profit. For the securities market model to be sensible from the economic standpoint, there cannot exist any arbitrage opportunities.

The no-arbitrage pricing theorem (see Theorem 5.2) is also unchanged in the continuous-time case. That is, for a given contingent claim X, suppose that there exists a replicating trading strategy $\{\boldsymbol{\theta}(t)\}$ given by (13.4). If there are no arbitrage opportunities in the market, $V(0)$ is the correct value of the contingent claim X.

13.2 Price Process Models

So far, we have not mentioned how to construct a security price process in the continuous-time setting. In this section, we provide basic price process models that have been considered in finance literature.

In the finance literature, it is common to consider the rate of return of a security rather than its price process directly. Recall that the rate of return in the discrete-time case is defined by (5.4). In the continuous-time setting, since the difference is replaced by a differential, we need to consider an instantaneous rate of return. Namely, let $R(t)$ be the time t instantaneous rate of return of a security $S(t)$. From (5.4), we have

$$R(t)\mathrm{d}t = \frac{\mathrm{d}S(t) + d(t)\mathrm{d}t}{S(t)}, \quad 0 \le t \le T, \tag{13.5}$$

where $d(t)$ is the dividend rate paid at time t from the security.

Now, suppose that the instantaneous rate of return of a security is expressed by a sum of a deterministic trend and unanticipated vibrations (noise). The vibrations are, for example, due to a temporary imbalance between supply and demand, new information that causes marginal changes in the stock's value, etc. Hence, it is plausible to model the impact of such noise by a standard Brownian motion, because the impact is to produce a marginal change in the price. See, e.g., Samuelson (1965) and Merton (1973, 1976) for justifications of the use of Brownian motions in security price modeling. The instantaneous rate of return of the security during the time interval $[t, t+\mathrm{d}t]$ is then expressed as

$$R(t)\mathrm{d}t = \mu(S(t),t)\mathrm{d}t + \sigma(S(t),t)\mathrm{d}z(t), \qquad (13.6)$$

where $\{z(t)\}$ is a standard Brownian motion; cf. Example 10.2. Equation (13.6) says that the instantaneous rate of return is a sum of the deterministic trend $\mu(S(t),t)$ and the noise $\sigma(S(t),t)\mathrm{d}z(t)$. The trend $\mu(S,t)$ is called the *mean rate of return*, while the term $\sigma(S,t)$ is called the *volatility*.

From (13.5) and (13.6), we obtain

$$\mathrm{d}S = [\mu(S,t)S - d(t)]\mathrm{d}t + \sigma(S,t)S\mathrm{d}z, \quad 0 \le t \le T, \qquad (13.7)$$

where we use the simplified notation, as before. In particular, if the security pays no dividends, the SDE (13.7) for the price process becomes

$$\frac{\mathrm{d}S}{S} = \mu(S,t)\mathrm{d}t + \sigma(S,t)\mathrm{d}z, \quad 0 \le t \le T. \qquad (13.8)$$

It is noted that we have implicitly assumed the existence of a solution in the SDE (13.7) or (13.8).

Example 13.1 In the model (13.8), we are very often asked to obtain the SDE satisfied by the log-price $Y(t) = \log S(t)$. To this end, we need to apply Ito's formula (Theorem 12.1). Then, our first task is to identify what the function $f(x,t)$ is. In this case, the function transforms the price x into $\log x$, so that the transformation is given by $f(x,t) = \log x$. Hence, the necessary partial derivatives are obtained as $f_x(x,t) = 1/x$, $f_{xx}(x,t) = -1/x^2$, and $f_t(x,t) = 0$. We can now apply Ito's formula to obtain

$$\mathrm{d}\log S = \left\{\mu(S,t) - \frac{1}{2}\sigma^2(S,t)\right\}\mathrm{d}t + \sigma(S,t)\mathrm{d}z. \qquad (13.9)$$

In particular, if the mean rate of return as well as the volatility are constants, $\mu(x,t) = \mu$ and $\sigma(x,t) = \sigma$ say, then

$$\mathrm{d}\log S = \nu\mathrm{d}t + \sigma\mathrm{d}z, \quad \nu \equiv \mu - \frac{\sigma^2}{2}.$$

Since the right-hand side of this equation does not depend on S, we can integrate it over $[0,t]$ so that

$$\log S(t) - \log S(0) = \nu t + \sigma z(t), \quad 0 \le t \le T.$$

It follows that

$$S(t) = S(0)\mathrm{e}^{\nu t + \sigma z(t)}, \quad 0 \le t \le T. \qquad (13.10)$$

PRICE PROCESS MODELS

This algebra leads to the Black–Scholes formula (4.21) for European call options. Conversely, suppose that the SDE for the log-price is given by

$$d \log S = \mu(S,t)dt + \sigma(S,t)dz, \quad 0 \le t \le T.$$

Then, it is readily seen that

$$\frac{dS}{S} = \left\{\mu(S,t) + \frac{\sigma^2(S,t)}{2}\right\} dt + \sigma(S,t)dz, \quad 0 \le t \le T. \tag{13.11}$$

These results are quite useful in financial engineering.

The next result is obtained by combining (13.9) and (13.11) and quite useful for the pricing of derivative securities in a stochastic interest-rate economy. See Exercise 13.2 for a slight generalization of this result.

Proposition 13.1 (Division Rule) *Consider the two SDEs*

$$\frac{dX}{X} = \mu_X(t)dt + \sigma_X(t)dz, \quad t \ge 0,$$

and

$$\frac{dY}{Y} = \mu_Y(t)dt + \sigma_Y(t)dz, \quad t \ge 0,$$

and let $D(t) = X(t)/Y(t)$ where $Y(t) > 0$. Then, the process $\{D(t)\}$, if it exists, follows the SDE

$$\frac{dD}{D} = \mu_D(t)dt + \sigma_D(t)dz, \quad t \ge 0,$$

where

$$\mu_D(t) = \mu_X(t) - \mu_Y(t) - \sigma_Y(t)(\sigma_X(t) - \sigma_Y(t))$$

and $\sigma_D(t) = \sigma_X(t) - \sigma_Y(t)$.

Proof. Note that we assume the same Brownian motion $\{z(t)\}$ for the two SDEs. Take the logarithm to obtain

$$\log D(t) = \log X(t) - \log Y(t).$$

It follows from (13.9) that

$$d \log D = \left\{\mu_X(t) - \frac{\sigma_X^2(t)}{2} - \mu_Y(t) + \frac{\sigma_Y^2(t)}{2}\right\} dt + (\sigma_X(t) - \sigma_Y(t)) \, dz.$$

The proposition follows at once from (13.11). □

Next, consider the multivariate case where $S_0(t) = B(t)$, the money-market account, and

$$\mathbf{S}(t) = (S_1(t), S_2(t), \ldots S_n(t))^\top, \quad 0 \le t \le T.$$

We assume for simplicity that each security pays no dividends and, as in (13.8), each security price follows the SDE

$$dS_i = S_i[\mu_i(\mathbf{S},t)dt + \sigma_i(\mathbf{S},t)dz_i], \quad i = 1, 2, \ldots, n, \tag{13.12}$$

with correlation structure
$$dz_i dz_j = \rho_{ij}(\mathbf{S},t)dt, \quad i \neq j,$$
where $\mu_i(\mathbf{S},t)$ is the mean rate of return and $\sigma_i(\mathbf{S},t)$ denotes the volatility of $S_i(t)$. Note that these parameters may depend on the other security prices.

Define the n-dimensional mean vector and the $n \times n$ volatility matrix by
$$\boldsymbol{\mu}(\mathbf{S},t) = (\mu_1(\mathbf{S},t),\ldots,\mu_n(\mathbf{S},t))^\top, \quad \widetilde{\boldsymbol{\sigma}}(\mathbf{S},t) = (\widetilde{\sigma}_{ij}(\mathbf{S},t)),$$
respectively, where
$$\sum_{k=1}^n \widetilde{\sigma}_{ik}(\mathbf{S},t)\widetilde{\sigma}_{jk}(\mathbf{S},t) = \sigma_i(\mathbf{S},t)\sigma_j(\mathbf{S},t)\rho_{ij}(\mathbf{S},t). \tag{13.13}$$
with $\rho_{ii}(\mathbf{S},t) = 1$. Let $\{\widetilde{\mathbf{z}}(t)\}$ be the n-dimensional standard Brownian motion, and define
$$d\mathbf{S} = \mathbf{S}_D \left[\boldsymbol{\mu}(\mathbf{S},t)dt + \widetilde{\boldsymbol{\sigma}}(\mathbf{S},t)d\widetilde{\mathbf{z}} \right], \quad 0 \leq t \leq T, \tag{13.14}$$
where $\mathbf{S}_D(t)$ denotes the $n \times n$ diagonal matrix with diagonal components $S_i(t)$. It is not hard to show that the solutions to (13.12) and (13.14), if they exist, are the same in law. The proof is left in Exercise 13.3. Also, note the difference between the SDE (13.14) and the SDE (12.46).

Example 13.2 (Product Rule) Let $n = 2$ and suppose that the price vector $\{(S_1(t), S_2(t))\}$ follows the SDE (13.12), i.e.
$$dS_i = S_i[\mu_i(t)dt + \sigma_i(t)dz_i], \quad i = 1, 2,$$
with correlation $dz_1 dz_2 = \rho(t)dt$. Let $K(t) = S_1(t)S_2(t)$. We want to obtain the SDE for $\{K(t)\}$. To this end, we apply Ito's formula (12.47). Let $\{\widetilde{\mathbf{z}}(t)\}$ be a two-dimensional standard Brownian motion, and define
$$\begin{aligned} dS_1 &= S_1[\mu_1(t)dt + \sigma_1(t)d\widetilde{z}_1], \\ dS_2 &= S_2[\mu_2(t)dt + \rho(t)\sigma_2(t)d\widetilde{z}_1 + \sqrt{1-\rho^2(t)}\sigma_2(t)d\widetilde{z}_2]. \end{aligned}$$

That is, the volatility matrix is given by
$$\widetilde{\boldsymbol{\sigma}}(t) = \begin{pmatrix} \sigma_1(t) & 0 \\ \rho(t)\sigma_2(t) & \sqrt{1-\rho^2(t)}\sigma_2(t) \end{pmatrix}$$
in the above notation; cf. the Cholesky decomposition (Proposition 10.3). Now, let
$$f(x_1,x_2,t) = x_1 x_2.$$
Then, $f_t = 0$, $f_1 = x_2$, $f_2 = x_1$ and $f_{12} = 1$. It follows that
$$\frac{dK}{K} = \frac{dS_1}{S_1} + \frac{dS_2}{S_2} + \rho(t)\sigma_1(t)\sigma_2(t)dt.$$
Hence, the mean rate of return of the process $\{K(t)\}$ is given by
$$\mu_K(t) = \mu_1(t) + \mu_2(t) + \rho(t)\sigma_1(t)\sigma_2(t),$$
and its volatility is $\sigma_K(t) = [\sigma_1^2(t) + \sigma_2^2(t) + 2\rho(t)\sigma_1(t)\sigma_2(t)]^{1/2}$.

PRICE PROCESS MODELS

Finally, we consider a *jump-diffusion model* first introduced by Merton (1976). Suppose that the noise in the security price is composed of two types of changes, the 'normal' vibrations and the 'abnormal' vibrations. While the former is modeled by a standard Brownian motion, the latter may be due to the arrival of important information that has more than a marginal effect on price changes. Since important information arrives only at discrete points in time, this component is modeled by a jump process reflecting the non-marginal impact of the information.

Let $\{N(t)\}$ be a Poisson process with intensity $\{\lambda(t)\}$, and let $\eta(t)$ be a random variable representing the size of time t jump, if any, with finite mean $m(t) = E[\eta(t)]$. It is assumed that the jump process is independent of the normal vibrations (the Brownian motion). Note that, during the time interval $(t, t + \mathrm{d}t]$, the increment in the Poisson process takes values of either 0 or 1. That is,

$$P_t\{\mathrm{d}N = 1\} = 1 - P_t\{\mathrm{d}N = 0\} = \lambda(t)\mathrm{d}t,$$

where $\mathrm{d}N = N(t + \mathrm{d}t) - N(t)$ and P_t denotes the conditional probability given the information $\mathcal{F}_t$. The jump size $\eta(t)$ is determined independently given the information. In this model, the filtration $\{\mathcal{F}_t\}$ is the natural one generated by the Brownian motion $\{z(t)\}$ and the Poisson process $\{N(t)\}$. The instantaneous rate of return of the security during the time interval $(t, t + \mathrm{d}t]$ is then expressed as

$$\frac{\mathrm{d}S}{S} = (\mu(t) - \lambda(t)m(t))\mathrm{d}t + \sigma(t)\mathrm{d}z + \eta(t)\mathrm{d}N, \quad 0 \leq t \leq T, \qquad (13.15)$$

if it pays no dividends. Note that, under the assumptions, we have $\mathrm{d}z\mathrm{d}N = 0$ since they are independent and $(\mathrm{d}N)^n = \mathrm{d}N$, $n \geq 1$, since $\mathrm{d}N$ takes values of either 0 or 1.

Example 13.3 Merton (1976) considered a special case of (13.15) to derive the European call option premium written on the stock. Namely, consider the SDE

$$\frac{\mathrm{d}S}{S} = (\mu - \lambda m)\mathrm{d}t + \sigma \mathrm{d}z + \left(\mathrm{e}^Y - 1\right)\mathrm{d}N, \quad 0 \leq t \leq T, \qquad (13.16)$$

where μ, σ, and λ are positive constants, and where Y is a normally distributed random variable with

$$m = E\left[\mathrm{e}^Y - 1\right] = \mathrm{e}^{\mu_y + \sigma_y^2/2} - 1.$$

The Brownian motion $\{z(t)\}$, the Poisson process $\{N(t)\}$, and the random variable Y are assumed to be independent.

In this example, we obtain the SDE for $\log S(t)$. To this end, from Taylor's expansion (1.16), we have

$$\mathrm{d}\log S = \sum_{n=1}^{\infty} \frac{(-1)^{n-1}}{n} \left(\frac{\mathrm{d}S}{S}\right)^n.$$

Note that $\mathrm{d}z\mathrm{d}N = 0$ and $(\mathrm{d}N)^n = \mathrm{d}N$, $n \geq 1$. Also, from (13.16), we obtain

$$\left(\frac{\mathrm{d}S}{S}\right)^n = \begin{cases} \sigma^2\mathrm{d}t + \left(\mathrm{e}^Y - 1\right)^2 \mathrm{d}N, & n = 2, \\ \left(\mathrm{e}^Y - 1\right)^n \mathrm{d}N, & n \geq 3. \end{cases}$$

It follows that

$$\mathrm{d}\log S = \nu\mathrm{d}t + \sigma\mathrm{d}z + Y\mathrm{d}N, \quad \nu \equiv \mu - \frac{\sigma^2}{2} - \lambda m,$$

where we have used the identity

$$\log(1 + x) = \sum_{n=1}^{\infty} \frac{(-1)^{n-1}}{n} x^n.$$

Hence, the log-price is obtained as

$$\log \frac{S(t)}{S(0)} = \nu t + \sigma z(t) + \sum_{i=1}^{N(t)} Y_i, \quad 0 \leq t \leq T,$$

where Y_i denote independent replica of Y and $\sum_{i=1}^{N(t)} Y_i = 0$ if $N(t) = 0$. See Exercise 13.12 for the pricing of a European call option in the jump-diffusion setting.

13.3 The Black–Scholes Model

Suppose that $S_0(t) = B(t)$, the money-market account, and $B(t)$ follows the ordinary differential equation (ODE)

$$\mathrm{d}B(t) = rB(t)\mathrm{d}t, \quad 0 \leq t \leq T,$$

where r is a positive constant. Also, in our securities market model, let $n = 1$ and suppose that $S_1(t) = S(t)$ pays no dividends and follows the SDE

$$\mathrm{d}S = S[\mu\mathrm{d}t + \sigma\mathrm{d}z], \quad 0 \leq t \leq T, \tag{13.17}$$

where μ and σ are positive constants and where $\{z(t)\}$ is a standard Brownian motion. This is the setting that Black and Scholes (1973) considered to derive the European call option premium (4.21).

Let $S(0) = S$, and consider a European call option with strike price K and maturity T written on $\{S(t)\}$. Suppose that the time t price of the option is given by $C(t) = f(S(t), t)$ for some smooth function $f(S, t)$. Then, from Ito's formula, we obtain

$$\frac{\mathrm{d}C}{C} = \mu_C(t)\mathrm{d}t + \sigma_C(t)\mathrm{d}z, \quad 0 \leq t \leq T, \tag{13.18}$$

where

$$\mu_C(t) = \frac{f_t(S,t) + f_S(S,t)\mu S + f_{SS}(S,t)\sigma^2 S^2/2}{C}$$

is the mean rate of return of the option, and

$$\sigma_C(t) = \frac{f_S(S,t)\sigma S}{C}$$

THE BLACK–SCHOLES MODEL

is the volatility. The proof is left to the reader.

Suppose that we invest the fraction $w(t)$ of the wealth $W(t)$ into the security $S(t)$ and the remaining into the option $C(t)$. Then, the rate of return of the portfolio is given by

$$\frac{dW}{W} = w\frac{dS}{S} + (1-w)\frac{dC}{C} \qquad (13.19)$$

under the self-financing assumption. The proof is left in Exercise 13.4. Substitution of (13.17) and (13.18) into (13.19) yields

$$\frac{dW}{W} = (w\mu + (1-w)\mu_C(t))dt + (w\sigma + (1-w)\sigma_C(t))dz.$$

Now, suppose $\sigma \neq \sigma_C(t)$, and let

$$w(t) = -\frac{\sigma_C(t)}{\sigma - \sigma_C(t)}. \qquad (13.20)$$

We then obtain

$$w(t)\sigma + (1-w(t))\sigma_C(t) = 0,$$

whence the portfolio is risk-free.

By the no-arbitrage condition, the rate of return of the portfolio must be equal to that of the risk-free security. That is, if the portfolio is constructed as in (13.20), we *must* have

$$w(t)\mu + (1-w(t))\mu_C(t) = -\frac{\mu\sigma_C(t)}{\sigma - \sigma_C(t)} + \frac{\mu_C(t)\sigma}{\sigma - \sigma_C(t)} = r.$$

It follows that

$$\frac{\mu - r}{\sigma} = \frac{\mu_C(t) - r}{\sigma_C(t)}. \qquad (13.21)$$

Equation (13.21) shows that the mean excess return per unit of risk for the stock is equal to that of the derivative, the same conclusion obtained as in the discrete model (see Theorem 7.1). Note that the *market price of risk* in this case is determined uniquely, since the stock is a traded asset and the parameters in the left-hand side in (13.21) are observed in the market.

Substitution of the mean rate of return $\mu_C(t)$ and the volatility $\sigma_C(t)$ into (13.21) yields the partial differential equation (PDE)

$$rSf_S(S,t) + f_t(S,t) + \frac{\sigma^2 S^2}{2}f_{SS}(S,t) - rf(S,t) = 0.$$

It should be noted that the PDE does not involve the mean rate of return μ of the underlying stock. Also, the same PDE can be derived even when the volatility as well as the risk-free interest rate are dependent on the underlying stock price and time. Namely, we obtain

$$r(S,t)Sf_S(S,t) + f_t(S,t) + \frac{\sigma^2(S,t)S^2}{2}f_{SS}(S,t) = r(S,t)f(S,t) \qquad (13.22)$$

in the general setting (cf. the PDE (12.10)).

The PDE (13.22) does not depend on the payoff function $h(x)$ of the derivative. The price of the derivative $C(t) = f(S(t),t)$ is obtained by solving the

PDE (13.22) with the boundary condition $C(T) = h(T)$. In particular, in the case of the European call option with strike price K, we have

$$f(S,T) = \{S - K\}_+, \quad f(0,t) = 0. \tag{13.23}$$

It can be readily verified that the Black–Scholes equation (4.21), i.e.

$$f(S,t) = S\Phi(d) - Ke^{-r(T-t)}\Phi(d - \sigma\sqrt{T-t}),$$

where $S(t) = S$ and

$$d = \frac{\log[S/K] + r(T-t)}{\sigma\sqrt{T-t}} + \frac{\sigma\sqrt{T-t}}{2},$$

satisfies the PDE (13.22) and the boundary condition (13.23). The proof is left in Exercise 13.5.

Alternatively, from the Feynman–Kac formula (see Example 12.5), the solution to the PDE (13.22) with boundary condition (13.23) is given, under regularity conditions, by

$$f(S,t) = E^* \left[\{S(T) - K\}_+ e^{-\int_t^T r(S(u),u)du} \Big| S(t) = S \right], \tag{13.24}$$

where the price process $\{S(t)\}$ follows the SDE

$$\frac{dS}{S} = r(S,t)dt + \sigma(S,t)dz, \quad 0 \le t \le T. \tag{13.25}$$

Here, E^* indicates that the expectation is taken for the process defined by (13.25), not for the original process (13.17). Note that the mean rate of return in the SDE (13.25) is equal to the risk-free interest rate. The reason for this will become clear later.

When $r(S,t) = r$ and $\sigma(S,t) = \sigma$, the SDE (13.25) can be solved as in (13.10) and given by

$$S(T) = S(t)e^{\nu(T-t) + \sigma[z(T) - z(t)]}, \quad \nu \equiv r - \frac{\sigma^2}{2}.$$

Evaluation of the expectation in (13.24) then leads to the Black–Scholes formula (4.21). The calculation is similar to the one given in Example 11.2 and omitted. The other method to derive the Black–Scholes formula will be discussed in the next section.

Finally, we point out that the partial derivatives of the price function $f(S,t)$ are used as risk indices of the derivative security. From (13.18) and (13.22), it is readily seen that

$$dC = f_S dS + \frac{C - f_S S}{B} dB, \tag{13.26}$$

where $f_S(S,t)$ is denoted by f_S for notational simplicity. Note that, under the self-financing condition, (13.26) is equivalent to

$$C(t) = f_S S(t) + \frac{C(t) - f_S S(t)}{B(t)} B(t).$$

Hence, the derivative (the call option in this case) is replicated by taking f_S units of the stock and $C(t) - f_S S(t)$ units of cash. The position f_S is called the *delta* of the derivative.

Formally, the delta of a derivative security is defined as the rate of change in its price with respect to the price of the underlying security. In particular, the delta of the call option in the Black–Scholes setting is given by

$$f_S(S,t) = \Phi(d), \quad d = \frac{\log[S/K] + r(T-t)}{\sigma\sqrt{T-t}} + \frac{\sigma\sqrt{T-t}}{2},$$

where $\Phi(d)$ denotes the standard normal distribution function; see Exercise 4.8. It is readily seen that the delta is decreasing in K (see Exercise 13.17 for a more general result).

Also, from Ito's formula, we have

$$\mathrm{d}f(S,t) = f_S \mathrm{d}S + \frac{1}{2} f_{SS} (\mathrm{d}S)^2 + f_t \mathrm{d}t, \qquad (13.27)$$

where $C = f(S,t)$. Hence, the change in the derivative price is explained by the three factors f_S, f_t, and f_{SS}. The rate of change in the value of the derivative with respect to time, f_t, is called the *theta* of the derivative. The term f_{SS} is called the *gamma*, which represents the rate of change in the derivative's delta f_S with respect to the price of the underlying security. See Exercise 4.8 for the theta and the gamma of a call option in the Black–Scholes setting.

13.4 The Risk-Neutral Method

In order to simplify the exposition, we continue to assume that $n = 1$ and $S_1(t) = S(t)$ is the time t price of a risky stock. Also, the risk-free security $S_0(t)$ is the money-market account $B(t)$.

Let $\{z(t)\}$ be a standard Brownian motion on a given probability space $(\Omega, \mathcal{F}, P)$, and assume that the filtration $\{\mathcal{F}_t; 0 \le t \le T\}$ is generated by the Brownian motion $\{z(t)\}$.

Let $r(t)$ be the time t instantaneous spot rate, and assume that $\{r(t)\}$ is a non-negative process, adapted to the filtration. The money-market account $B(t)$ is defined by

$$\mathrm{d}B(t) = r(t)B(t)\mathrm{d}t, \quad 0 \le t \le T, \qquad (13.28)$$

with $B(0) = 1$.

On the other hand, the stock price process $\{S(t)\}$ under the physical probability measure P evolves according to the SDE

$$\mathrm{d}S = [\mu(t)S - \delta(t)]\mathrm{d}t + \sigma(t)S\mathrm{d}z, \quad 0 \le t \le T, \qquad (13.29)$$

where $\{\sigma(t)\}$ is a positive stochastic process, adapted to the filtration, and satisfies $\int_0^T E[\sigma^2(t)]\mathrm{d}t < \infty$ for the sake of simplicity. The term $\delta(t)$ represents the dividend rate at time t that the stock pays continuously, and we assume that the process $\{\delta(t)\}$ is adapted to the filtration and satisfies

$\int_0^T E[|\delta(t)|]dt < \infty$. Note that, as before, the mean rate of return $\mu(t)$ plays no role for the pricing of derivatives. It is assumed throughout this section that the SDE (13.29) has a non-negative, square integrable solution.

Consider a European contingent claim written on the stock with payoff function $h(x)$ and maturity T. The time t price of the claim is denoted by $C(t)$. Suppose that the contingent claim is replicated through a *self-financing* dynamic strategy of trading $\theta(t)$ units of the underlying stock and $b(t)$ units of the money market. That is,

$$\begin{aligned} C(t) &\equiv b(t)B(t) + \theta(t)S(t) \\ &= C(0) + \int_0^t b(u)dB(u) + \int_0^t \theta(u)dG(u), \end{aligned} \quad (13.30)$$

where $G(t) = S(t) + \int_0^t \delta(u)du$, $0 \le t \le T$, and $C(T) = h(S(T))$ at the maturity T; see (13.3) and (13.26). Note from (13.29) that

$$dG = S[\mu(t)dt + \sigma(t)dz], \quad 0 \le t \le T, \quad (13.31)$$

under the physical probability measure P.

Let us define the process $\{z^*(t)\}$ by

$$\sigma(t)dz^* = (\mu(t) - r(t))dt + \sigma(t)dz, \quad 0 \le t \le T. \quad (13.32)$$

We then have

$$dG = S[r(t)dt + \sigma(t)dz^*], \quad 0 \le t \le T. \quad (13.33)$$

Let P^* be a probability measure that makes the process $\{z^*(t)\}$ a standard Brownian motion. The existence of such a probability measure is guaranteed by Girsanov's theorem stated below.

As before, we denote the denominated price of $S(t)$ with numeraire $B(t)$ by $S^*(t)$, i.e. $S^*(t) \equiv S(t)/B(t)$. By Ito's division rule (Proposition 13.1), we have from (13.29) and (13.32) that

$$dS^* = -\delta^*(t)dt + \sigma(t)S^*dz^*, \quad (13.34)$$

where $\delta^*(t) = \delta(t)/B(t)$. Hence, under the regularity conditions, the denominated process $\{S^*(t)\}$ is a martingale under the new probability measure P^* if and only if the stock pays no dividends (see Proposition 12.6).

Now, from (13.28) and (13.30), we obtain

$$dC = r(t)Cdt + \theta(t)\sigma(t)Sdz^*, \quad 0 \le t \le T.$$

It follows from Ito's formula that

$$C^*(t) = C^*(0) + \int_0^t \theta(u)\sigma(u)S^*(u)dz^*(u), \quad 0 \le t \le T. \quad (13.35)$$

Hence, the denominated price process $\{C^*(t)\}$ is always a martingale with respect to the new probability measure P^* under technical conditions. Note that the European contingent claim pays $C(T) = h(S(T))$ at the maturity T. Hence, the time t price $C(t)$ of the claim is given by

$$C(t) = B(t)E_t^*\left[\frac{h(S(T))}{B(T)}\right], \quad 0 \le t \le T, \quad (13.36)$$

THE RISK-NEUTRAL METHOD 227

under the no-arbitrage condition, where E_t^* denotes the conditional expectation operator under the new probability measure P^* given the information $\mathcal{F}_t$.

Summarizing, in order to calculate the time t price $C(t)$ of a contingent claim;

(RN1) Find a probability measure P^* under which the process $\{z^*(t)\}$ defined by (13.32) is a standard Brownian motion, and

(RN2) Calculate the expected value (13.36) under P^*.

This method for the pricing of derivatives is called the *risk-neutral method* by the reason explained below. See Theorem 5.3 for the discrete-time case.

Before proceeding, we state Girsanov's theorem formally. For a process $\{\beta(t)\}$ satisfying $E\left[\exp\left\{\frac{1}{2}\int_0^T \beta^2(u)\mathrm{d}u\right\}\right] < \infty$, let

$$Y(T) = \exp\left\{\int_0^T \beta(t)\mathrm{d}z(t) - \frac{1}{2}\int_0^T \beta^2(t)\mathrm{d}t\right\}, \qquad (13.37)$$

and define

$$\widetilde{P}(A) = E[1_A Y(T)], \quad A \in \mathcal{F}_T. \qquad (13.38)$$

We note that $\widetilde{P}$ defines a probability measure and the two probability measures $\widetilde{P}$ and P are equivalent, since $Y(T) > 0$. The proof of the next result is beyond the scope of this book and omitted. See Proposition 6.5 and its proof for the random walk case (see also Section 11.4).

Proposition 13.2 (Girsanov) *The process $\{\widetilde{z}(t)\}$ defined by*

$$\widetilde{z}(t) = z(t) - \int_0^t \beta(u)\mathrm{d}u, \quad 0 \le t \le T,$$

is a standard Brownian motion under $\widetilde{P}$.*

It remains to calculate the expectation in (13.36) to determine the price $C(t)$ of the contingent claim. To this end, we need to investigate the stock's price process $\{S(t)\}$ under the new probability measure P^*. By substituting (13.32) into the SDE (13.29), we obtain

$$\mathrm{d}S = [r(t)S - \delta(t)]\mathrm{d}t + \sigma(t)S\mathrm{d}z^*, \quad 0 \le t \le T, \qquad (13.39)$$

where $\{z^*(t)\}$ is a standard Brownian motion under P^*. That is, the mean rate of return of the stock $S(t)$ under P^* is equal to that of the risk-free security; see (13.28). Hence, under the new probability measure P^*, two securities have the same mean rate of return while the volatilities (i.e. risks) are different.

If there were such a securities market, investors would be risk-neutral, since otherwise the risky stock cannot exist. Because of this reason, the measure P^* is often called the *risk-neutral probability measure* (see Definition 5.8). Note that P^* is *not* a physical measure and used only for the pricing of derivative securities. In other words, under the physical probability measure P associated

---
* The Girsanov theorem can be extended to a higher dimensional case as it stands.

with a risk-averse market, the risky stock can never have $r(t)$ as its mean rate of return. Moreover, since the risk-neutral probability measure is unique, the market is complete (see Theorem 5.4) in this setting.

In general, evaluation of the expected value in (13.36) requires some numerical techniques (see Chapter 10). However, in the Black–Scholes setting, where the risk-free interest rate r and the volatility σ are positive constants, we obtain

$$S(t) = S \exp\left\{\left(r - \frac{\sigma^2}{2}\right)t + \sigma z^*(t)\right\}, \quad 0 \le t \le T, \qquad (13.40)$$

with $S(0) = S$. Hence, the stock price $S(T)$ follows a log-normal distribution under P^*. The next result is useful for evaluating the expectation. The proof is left in Exercise 13.6.

Proposition 13.3 *Let $\sigma(t)$ be a deterministic function of time t such that $\int_0^T \sigma^2(t)\mathrm{d}t < \infty$, and suppose that the price process $\{S(t)\}$ follows the SDE*

$$\mathrm{d}S = \sigma(t)S\mathrm{d}z, \quad t \ge 0,$$

where $\{z(t)\}$ is a standard Brownian motion. Then,

$$S(t) = S \exp\left\{-\frac{\psi^2}{2} + \int_0^t \sigma(u)\mathrm{d}z(u)\right\}, \quad \psi^2 = \int_0^t \sigma^2(u)\mathrm{d}u,$$

and

$$E[\{S(t) - K\}_+] = S\Phi(d) - K\Phi(d - \psi),$$

where $S(0) = S$ and

$$d = \frac{\log[S/K]}{\psi} + \frac{\psi}{2}, \quad \psi > 0.$$

We are now in a position to apply the risk-neutral method for evaluating various derivative securities. See Exercises 13.8–13.13 for other applications of the risk-neutral method.

Example 13.4 (Compound Option) For a stock price process $\{S(t)\}$, consider a derivative security with maturity T written on the stock. Under technical conditions, the time t price of the derivative is given by $C(t) = f(S(t), t)$ for some smooth function $f(S, t)$. But, since the underlying stock price $S(t)$ fluctuates in time, the derivative price also fluctuates in time. Hence, the derivative can be an underlying asset for a new derivative. Let $h(x)$ be the payoff function of the new derivative with maturity $\tau < T$. According to the risk-neutral method, the premium of the new derivative is given by

$$C_{\mathrm{new}} = E^*\left[\frac{h(f(S(\tau), \tau))}{B(\tau)}\right], \quad \tau < T.$$

In particular, if the underlying asset is the Black–Scholes call option and the new derivative is again a call option, we have

$$C_{\mathrm{new}} = \mathrm{e}^{-r\tau} E^*\left[\{f(S(\tau), \tau) - L\}_+\right],$$

THE RISK-NEUTRAL METHOD

where $f(S,t)$ is given by (4.21) and $\{S(t)\}$ follows (13.40) under the risk-neutral probability measure P^*. An explicit solution of the *compound option* is available in Geske (1979).

Example 13.5 (Look-Back Option) For a price process $\{S(t)\}$, let

$$M(T) = \max_{0 \le t \le T} S(t), \quad T > 0.$$

Hence, $M(T)$ is the maximum of the price process during the time interval $[0,T]$. Suppose that the payoff at the maturity T is given by $\{M(T) - K\}_+$. This European option is called a *look-back* call option. We denote the option premium by L.

According to the risk-neutral method, we need to consider the price process $\{S(t)\}$ given by (13.40). Let $Y(t) = \log[S(t)/S]$, where $S(0) = S$, and define $M_Y(T) = \max_{0 \le t \le T} Y(t)$. Then, we calculate

$$L = e^{-rT} E^* \left[\left\{ S e^{M_Y(T)} - K \right\}_+ \right].$$

Since the process $\{Y(t)\}$ is a Brownian motion with drift $\nu \equiv r - \sigma^2/2$ and diffusion coefficient σ, the density function $g(m)$ and the distribution function $G(m)$ of $M_Y(T)$ are given, respectively, by

$$g(m) = \frac{2}{\sqrt{2\pi T}\sigma} \exp\left\{-\frac{(m-\nu T)^2}{2\sigma^2 T}\right\} - \frac{2\nu}{\sigma^2} e^{2\nu m/\sigma^2} \Phi\left(\frac{-m-\nu T}{\sigma\sqrt{T}}\right)$$

and

$$G(m) = \Phi\left(\frac{m-\nu T}{\sigma\sqrt{T}}\right) - e^{2\nu m/\sigma^2} \Phi\left(\frac{-m-\nu T}{\sigma\sqrt{T}}\right).$$

The proof is left in Exercise 13.7. It follows that

$$L = S e^{-rT} \int_{\log[K/S]}^{\infty} e^m g(m) dm - K e^{-rT}(1 - G(\log[K/S])).$$

Now, since

$$\int_{\log[K/S]}^{\infty} e^m \frac{2}{\sqrt{2\pi T}\sigma} \exp\left\{-\frac{(m-\nu T)^2}{2\sigma^2 T}\right\} dm$$
$$= 2 e^{rT} \Phi\left(\frac{\log[S/K] + (r + \sigma^2/2)T}{\sigma\sqrt{T}}\right),$$

it suffices to calculate

$$M \equiv \frac{2r - \sigma^2}{\sigma^2} \int_{\log[K/S]}^{\infty} e^{2rm/\sigma^2} \Phi\left(\frac{-m-\nu T}{\sigma\sqrt{T}}\right) dm.$$

Using the symmetry of the standard normal density, we obtain

$$M = \frac{2r - \sigma^2}{\sigma^2} \int_{\log[K/S]}^{\infty} e^{2rm/\sigma^2} \left[1 - \Phi\left(\frac{m+\nu T}{\sigma\sqrt{T}}\right)\right] dm.$$

By integration by parts, we then have after some algebra that

$$M = -\left(1 - \frac{\sigma^2}{2r}\right)\left\{\left(\frac{K}{S}\right)^{2r/\sigma^2}\Phi\left(\frac{\log[S/K] - \nu T}{\sigma\sqrt{T}}\right)\right.$$
$$\left. - e^{rT}\Phi\left(\frac{\log[S/K] + (\nu + \sigma^2)T}{\sigma\sqrt{T}}\right)\right\}.$$

It follows that the premium of the look-back call option is given by

$$L = S\Phi(\xi) - Ke^{-rT}\Phi\left(\xi - \sigma\sqrt{T}\right)$$
$$+ S\frac{\sigma^2}{2r}\left[\Phi(\xi) - e^{-rT}\left(\frac{K}{S}\right)^{2r/\sigma^2}\Phi(\zeta)\right],$$

where

$$\xi = \frac{\log[S/K] + (r + \sigma^2/2)T}{\sigma\sqrt{T}}, \quad \zeta = \frac{\log[S/K] - (r - \sigma^2/2)T}{\sigma\sqrt{T}}.$$

We note that the first term of the premium is just the Black–Scholes equation (4.21). See Goldman, Sosin, and Gatto (1979) for details of the pricing of look-back options.

Example 13.6 (Forward and Futures Prices) A futures price resembles a forward price in that they are based on the price of a security at a fixed, future time. In particular, at the delivery date, they are equal to the underlying security price. However, there are subtle differences that can cause the two prices to be different.

Let $\{S(t)\}$ be a security price process, and consider a forward contract with delivery date T. At time $t < T$, the *forward* price $F_T(t)$ is determined to enter the contract. The payoff of the contract at the delivery date T is given by

$$X = S(T) - F_T(t).$$

According to the risk-neutral method, the time t value of the contract is given by

$$B(t)E_t^*\left[\frac{S(T) - F_T(t)}{B(T)}\right], \quad t < T,$$

which must be equal to 0, since nothing is required to enter the contract at time t. Since $F_T(t)$ is determined at time t (so that it is $\mathcal{F}_t$-measurable), it follows that

$$F_T(t) = \frac{B(t)}{v(t,T)}E_t^*\left[\frac{S(T)}{B(T)}\right] = \frac{S(t)}{v(t,T)}.$$

Here, we have used the fact that the denominated price process $\{S^*(t)\}$ is a martingale under the risk-neutral probability measure P^*. Hence, we recovered the result (4.17) for the forward price.

On the other hand, the *futures* price, $f_T(t)$ say, is determined so that the value of the contract during its life is equal to 0. It follows that the process $\{f_T(t)\}$ itself is a martingale under P^*. Since $f_T(T) = S(T)$ at the delivery date, we must have

$$f_T(t) = E_t^*[S(T)], \quad 0 \leq t \leq T. \tag{13.41}$$

In particular, when interest rates are deterministic, we have $f_T(t) = F_T(t)$. This is so, because $v(t,T) = B(t)/B(T)$ is deterministic in this case. See Exercise 13.15 for the relationship between the forward and futures prices in a stochastic interest rate economy. See also Exercise 13.9 for Black's formula on a futures option.

13.5 The Forward-Neutral Method

In this section, we take the default-free discount bond as the numeraire. For this purpose, we denote the time t price of the discount bond with maturity T by $v(t,T)$, and assume that the discount bond price follows the SDE

$$\frac{\mathrm{d}v(t,T)}{v(t,T)} = r(t)\mathrm{d}t + \sigma_v(t)\mathrm{d}z^*, \quad 0 \leq t \leq T, \tag{13.42}$$

under the risk-neutral probability measure P^*, where $\sigma_v(t)$ denotes the volatility of the discount bond and $r(t)$ is the risk-free interest rate.

Suppose that the stock pays no dividends and its price process $\{S(t)\}$ follows the SDE

$$\frac{\mathrm{d}S}{S} = r(t)\mathrm{d}t + \sigma(t)\mathrm{d}z^*, \quad 0 \leq t \leq T, \tag{13.43}$$

under P^*, where $\sigma(t)$ denotes the volatility of the stock price.

Recall that, when the stock pays no dividends, the value defined by

$$S^T(t) \equiv \frac{S(t)}{v(t,T)}, \quad 0 \leq t \leq T,$$

is the *forward* price. It follows from Ito's division rule (Proposition 13.1) that

$$\frac{\mathrm{d}S^T}{S^T} = -\sigma_v(t)(\sigma(t) - \sigma_v(t))\mathrm{d}t + (\sigma(t) - \sigma_v(t))\mathrm{d}z^*.$$

Define the process $\{z^T(t)\}$ so that

$$(\sigma(t) - \sigma_v(t))\mathrm{d}z^T = -\sigma_v(t)(\sigma(t) - \sigma_v(t))\mathrm{d}t + (\sigma(t) - \sigma_v(t))\mathrm{d}z^*.$$

From Girsanov's theorem (Proposition 13.2), there exists, under technical conditions, a probability measure P^T under which the process $\{z^T(t)\}$ is a standard Brownian motion. Note that

$$\frac{\mathrm{d}S^T}{S^T} = \sigma^T(t)\mathrm{d}z^T, \quad \sigma^T(t) \equiv \sigma(t) - \sigma_v(t). \tag{13.44}$$

Hence, the new probability measure P^T is called *forward-neutral*, because the forward price process $\{S^T(t)\}$ is a martingale under P^T. See El Karoui, Jeanblanc, and Shreve (1998) for the formal definition of the forward-neutral probability measure. The relationship between the risk-neutral and forward-neutral probability measures is discussed in Section 5.6.

Consider now a European contingent claim written on the stock with payoff function $h(x)$ and maturity T. The time t price of the claim is denoted by $C(t)$. Suppose that the contingent claim is replicated through a self-financing dynamic strategy of trading $\theta(t)$ units of the underlying stock and $b(t)$ units of the discount bond. That is,

$$\begin{aligned} C(t) &\equiv b(t)v(t,T) + \theta(t)S(t) \\ &= C(0) + \int_0^t b(u)\mathrm{d}v(u,T) + \int_0^t \theta(u)\mathrm{d}S(u) \end{aligned}$$

and $C(T) = h(S(T))$ at the maturity T; cf. (13.30). It follows from Ito's formula that

$$C^T(t) = C^T(0) + \int_0^t \theta(u)\sigma^T(u)S^T(u)\mathrm{d}z^T(u), \quad 0 \leq t \leq T, \qquad (13.45)$$

where $C^T(t) = C(t)/v(t,T)$. Hence, the forward price process $\{C^T(t)\}$ is a martingale under the forward-neutral probability measure P^T under technical conditions. Note that the European contingent claim pays $C(T) = h(S(T))$ at the maturity T. Hence, the time t price $C(t)$ of the claim is given by

$$C(t) = v(t,T)E_t^T\left[h(S(T))\right], \quad 0 \leq t \leq T, \qquad (13.46)$$

where E_t^T denotes the conditional expectation operator under P^T given the information $\mathcal{F}_t$.

Summarizing, in order to calculate the time t price $C(t)$ of a contingent claim;

(FN1) Find a probability measure P^T under which the denominated price process $\{S^T(t)\}$ is a martingale, and

(FN2) Calculate the expected value (13.46) under P^T.

This method for the pricing of derivatives is called the *forward-neutral method* by the obvious reasons explained above. It should be noted that the forward-neutral method is applicable even for the case that the underlying security pays dividends, although the denominated price process $\{S^T(t)\}$ is no longer a martingale under P^T. The proof is left in Exercise 13.16.

The advantage of the forward-neutral method over the risk-neutral method seems obvious in a stochastic interest-rate economy. To see this, from (13.36) and (13.46), we have

$$C(t) = B(t)E_t^*\left[\frac{X}{B(T)}\right] = v(t,T)E_t^T[X] \qquad (13.47)$$

where X is the $\mathcal{F}_T$-measurable random variable representing the payoff of a contingent claim. Hence, while the knowledge about the two-dimensional distribution of $(X, B(T))$ is required for the risk-neutral method, the marginal distribution of X suffices to price the contingent claim in the forward-neutral method. This advantage becomes significant in a Gaussian interest-rate economy, as the next example reveals.

Example 13.7 In this example, we assume that there exists a risk-neutral probability measure P^*, as given, and suppose that the stock price process $\{S(t)\}$ follows the SDE

$$\frac{dS}{S} = r(t)dt + \sigma_1 dz_1 + \sigma_2 dz_2, \quad 0 \leq t \leq T,$$

under P^*, where σ_i, $i = 1, 2$, are constants and $r(t)$ denotes the risk-free interest rate. Here, the standard Brownian motions $\{z_1(t)\}$ and $\{z_2(t)\}$ are independent. Suppose also that the default-free discount bond price $\{v(t,T)\}$ follows the SDE

$$\frac{dv(t,T)}{v(t,T)} = r(t)dt + \sigma_v(t)dz_1, \quad 0 \leq t \leq T,$$

where the volatility $\sigma_v(t)$ is a deterministic function of time t. Note that the instantaneous covariance between them is given by

$$\frac{dS(t)}{S(t)} \frac{dv(t,T)}{v(t,T)} = \sigma_1 \sigma_v(t)dt,$$

and the correlation is $\sigma_1/\sqrt{\sigma_1^2 + \sigma_2^2}$.

From above discussions, there exists a forward-neutral probability measure P^T under which we have

$$\frac{dS^T}{S^T} = (\sigma_1 - \sigma_v(t))dz_1^T + \sigma_2 dz_2^T, \quad 0 \leq t \leq T.$$

Here $\{z_i^T(t)\}$, $i = 1, 2$ are independent, standard Brownian motions under P^T. Now, define another standard Brownian motion $\{z^T(t)\}$ under P^T such that

$$\sigma(t)dz^T \stackrel{d}{=} (\sigma_1 - \sigma_v(t))dz_1^T + \sigma_2 dz_2^T, \quad 0 \leq t \leq T,$$

where $\stackrel{d}{=}$ stands for equality in law and where

$$\sigma^2(t) = \sigma_1^2 + \sigma_2^2 - 2\sigma_1 \sigma_v(t) + \sigma_v^2(t).$$

Also, consider the process $\{\hat{S}^T(t)\}$ defined by

$$\frac{d\hat{S}^T}{\hat{S}^T} = \sigma(t)dz^T, \quad 0 \leq t \leq T.$$

Then, the two processes $\{\hat{S}^T(t)\}$ are $\{S^T(t)\}$ are equal in law. Recall that the distribution of $\{S^T(t)\}$ is only relevant for the pricing of derivative securities. Hence, for example, the European call option with maturity T and strike price K written on the stock is given as follows: In Proposition 13.3, put $t \leftarrow T$ and $S \leftarrow S^T(0) = S/v(0,T)$. Then,

$$E^T[\{S^T(T) - K\}_+] = \frac{S}{v(0,T)}\Phi(d) - K\Phi(d - \psi),$$

where $S(0) = S$ and

$$d = \frac{\log[S/Kv(0,T)]}{\psi} + \frac{\psi}{2}.$$

It follows that the call option premium is given by

$$C(0) = S\Phi(d) - Kv(0,T)\Phi(d-\psi).$$

Notice that our main assumption is that the volatilities σ_i, $i = 1, 2$, and $\sigma_v(t)$ are deterministic, i.e. the stock and the discount bond prices are log-normally distributed with a certain correlation.

13.6 The Interest-Rate Term Structure

This section considers the pricing of default-free discount bonds. The time t discount bond price maturing at time T is denoted by $v(t,T)$. When viewed as a function of maturity T, the discount bond price $v(t,T)$ is called the *term structure* of interest rates. Since the term structure changes stochastically in a complicated fashion, it has been a great interest how to model the term structure in both academy and industry. The reader is referred to, e.g., Brigo and Mercurio (2001) and Musiela and Rutkowski (1997) for more detailed discussions of interest-rate term structure models. Recall that we have already discussed in Section 7.5 about some ideas of term structure models in the binomial setting.

13.6.1 Spot-Rate Models

Let $r(t)$ be the time t instantaneous spot rate under the physical probability measure P, and assume that it follows the SDE

$$\mathrm{d}r = a(m-r)\mathrm{d}t + \sigma r^\gamma \mathrm{d}z, \quad t \geq 0, \tag{13.48}$$

where a, m, σ, and γ are constants. Recall that, unless $\gamma = 0$ or $1/2 \leq \gamma \leq 1$, the (path-by-path) solution may not exist (see Exercises 12.11 and 12.12).

The drift function in the SDE (13.48) reflects a restoring force directed towards the level m, whose magnitude is proportional to the distance. This model is called the *mean-reverting model*. See Exercise 13.18 for the analytical expression of the mean spot rate.

The SDE (13.48) has four parameters (a, m, γ, σ), and there are some important special cases according to the parameter values.

The Vasicek Model

Suppose that $a, m > 0$ and $\gamma = 0$. That is, we assume that

$$\mathrm{d}r = a(m-r)\mathrm{d}t + \sigma \mathrm{d}z, \quad t \geq 0. \tag{13.49}$$

This model was first considered by Vasicek (1977) and has been proved very useful in finance applications. The SDE (13.49) is linear and can be solved in closed form as

$$r(t) = m + (r(0) - m)\mathrm{e}^{-at} + \sigma \int_0^t \mathrm{e}^{-a(t-s)} \mathrm{d}z(s);$$

see (12.41). The process $\{r(t)\}$ is known as the *Ornstein–Uhlenbeck* process (OU process for short). It is readily seen that $r(t)$ is normally distributed, whence the spot rates in the Vasicek model become negative with positive probability. However, the probability is often negligible, and the merit from its analytical tractability makes the model have a practical interest.

The CIR Model

Next, suppose that a, $m > 0$ and $\gamma = 1/2$ in the SDE (13.48). That is, we assume that
$$dr = a(m - r)dt + \sigma\sqrt{r}\,dz, \quad t \geq 0. \tag{13.50}$$
This model was first considered by Cox, Ingersoll, and Ross (1985) and has been used by many authors for the pricing of interest-rate derivatives. The very reason is that, in contrast to the Vasicek model, the CIR model does not allow the spot rate to become negative. That is, according to Feller (1951), the spot rate $r(t)$ stays positive when $\sigma^2 \leq 2a$, while $r(t) \geq 0$ when $\sigma^2 > 2a$. Moreover, the probability distribution of $r(t)$ is known and its density function is given by
$$f_t(r) = c\,\mathrm{e}^{-c(u+r)}\left(\frac{r}{u}\right)^{q/2} I_q(2c\sqrt{ur}), \quad r \geq 0, \tag{13.51}$$
where
$$c = \frac{2a}{\sigma^2(1 - \mathrm{e}^{-at})}, \quad u = r(0)\mathrm{e}^{-at}, \quad q = \frac{2am}{\sigma^2} - 1,$$
and the function
$$I_q(x) = \sum_{k=0}^{\infty} \frac{(x/2)^{2k+q}}{k!\Gamma(k+q+1)}$$
is the modified Bessel function of the first kind and order q. Here, $\Gamma(x)$ denotes the gamma function defined in (3.23).

The density function (13.51) is the non-central, chi-square distribution with $2(q+1)$ degrees of freedom and non-centrality $2cu$. The moment generating function (MGF) is given by
$$m(\theta) = \left(\frac{c}{c-\theta}\right)^{q+1} \exp\left\{\frac{cu\theta}{c-\theta}\right\}. \tag{13.52}$$
The proof is left in Exercise 13.19. Moreover, the density function (13.51) can be expressed as
$$f_t(r) = \sum_{k=0}^{\infty} \frac{(cu)^k}{k!}\mathrm{e}^{-cu} g_{k+q+1,c}(r), \tag{13.53}$$
where the function
$$g_{\alpha,\lambda}(r) = \frac{1}{\Gamma(\alpha)}\lambda^\alpha r^{\alpha-1}\mathrm{e}^{-\lambda r}, \quad r > 0,$$
denotes the density function of the gamma distribution with mean α/λ and variance α/λ^2 (see Exercise 3.13). Hence, the non-central, chi-square distribution is a Poisson mixture of gamma distributions (cf. Equation (3.25)). See,

The Geometric Brownian Motion Model

Suppose that $\gamma = 1$ and $m = 0$ in the SDE (13.48). That is, we assume that
$$dr = \beta r dt + \sigma r dz, \quad t \geq 0,$$
where $\beta = -a$, which is the SDE for a familiar geometric Brownian motion. The advantage of this model is that the spot rates always stay positive. The model is a continuous limit of the BDT model in discrete time developed by Black, Derman, and Toy (1990). The disadvantage of this model is that the mean of the money-market account diverges. More specifically, define
$$B(t, t+h) = \exp\left\{\int_t^{t+h} r(s) ds\right\}.$$
For sufficiently small $h > 0$, we have $E[B(t, t+h)] \approx E[e^{hr(t)}]$; but, as we have already seen in (3.19) (see also Exercise 3.11), the mean $E[e^{hr(t)}]$ does not exist.

13.6.2 Pricing of Discount Bonds

We now turn our attention to the pricing of default-free discount bonds. For this purpose, suppose that the spot rate process $\{r(t)\}$ follows the SDE
$$dr = \mu(r,t) dt + \sigma(r,t) dz, \quad t \geq 0, \tag{13.54}$$
under the physical probability measure P, and assume that the default-free discount bond is a contingent claim written on the spot rate $r(t)$.

Recall that the spot rates cannot be traded in the market. Hence, in this framework, the discount bond is not replicated in terms of the underlying variable (the spot rate in this case). This is the major difference between the following arguments and those given for the Black–Scholes setting.

Consider the default-free discount bond with maturity T, and let $F(t)$ be its time t price. It is assumed that there exists a smooth function $f(r,t)$ such that $F(t) = f(r(t), t)$, $0 \leq t \leq T$. By Ito's formula, we then obtain
$$\frac{dF}{F} = \mu_F(r(t), t) dt + \sigma_F(r(t), t) dz, \quad 0 \leq t \leq T,$$
where
$$\mu_F(r,t) = \frac{1}{f(r,t)}\left(f_r(r,t)\mu(r,t) + f_t(r,t) + \frac{f_{rr}(r,t)}{2}\sigma^2(r,t)\right) \tag{13.55}$$
and
$$\sigma_F(r,t) = \frac{f_r(r,t)\sigma(r,t)}{f(r,t)}. \tag{13.56}$$
Here, as before, $f_r(r,t)$ denotes the partial derivative of $f(r,t)$ with respect to r, etc.

THE INTEREST-RATE TERM STRUCTURE

As for the Black–Scholes model (see Section 13.3), we construct a risk-free portfolio by using two discount bonds with different maturities. If there are no arbitrage opportunities, the excess rate of return per unit of risk must be the same for the two bonds. This implies that the *market price of risk* defined by

$$\lambda(t) = \frac{\mu_F(r(t),t) - r(t)}{\sigma_F(r(t),t)} \qquad (13.57)$$

does not depend on the maturity.

Suppose that the market price of risk $\lambda(t)$ is known in the market. Then, from the Girsanov theorem (Proposition 13.2), we know that there exists a probability measure P^*, equivalent to the physical measure P, under which the process $\{z^*(t)\}$ defined by

$$dz^* = dz + \lambda(t)dt, \quad 0 \le t \le T, \qquad (13.58)$$

is a standard Brownian motion.

For the money-market account $B(t) = \exp\{\int_0^t r(s)ds\}$, define the denominated price $F^*(t) = F(t)/B(t)$ of the discount bond with maturity T. Then, Ito's division rule yields

$$\frac{dF^*}{F^*} = (\mu_F(t) - r(t))dt + \sigma_F(t)dz, \quad 0 \le t \le T.$$

If the market price of risk $\lambda(t)$ is known, we obtain from (13.57) and (13.58) that

$$\frac{dF^*}{F^*} = \sigma_F(t)dz^*, \quad 0 \le t \le T.$$

Hence, the denominated price process $\{F^*(t)\}$ is a martingale under P^* and so, P^* is the risk-neutral probability measure. According to the risk-neutral method (13.36), we then have

$$F^*(t) = E_t^*\left[\frac{1}{B(T)}\right], \quad 0 \le t \le T,$$

since the discount bond pays 1 dollar for sure at the maturity T. It follows that

$$v(t,T) = E_t^*\left[e^{-\int_t^T r(s)ds}\right], \quad 0 \le t \le T, \qquad (13.59)$$

provided that the market price of risk $\lambda(t)$ is known.

As before, we need to obtain the SDE for the spot-rate process under the risk-neutral probability measure P^* in order to calculate the bond price (13.59). That is, from (13.58) and (13.54), we have

$$dr = (\mu(r,t) - \sigma(r,t)\lambda(t))dt + \sigma(r,t)dz^*, \quad 0 \le t \le T. \qquad (13.60)$$

The SDE (13.60) differs from (13.54) only by the drift. Hence, in order to value the discount bond in this setting, we need to adjust the drift $\mu(r,t)$ such that

$$m(t) \equiv \mu(r,t) - \sigma(r,t)\lambda(t).$$

If the 'risk-adjusted' drift $m(t)$ depends only on the state $r(t)$ and time t, then the discount bond price is given by

$$v(t,T) = E^* \left[e^{-\int_t^T r(s)ds} \middle| r(t) \right], \quad 0 \le t \le T,$$

since the resulting spot-rate process $\{r(t)\}$ under P^* is Markovian.

Consider the Markovian case, i.e. $m(t) = m(r(t), t)$ say. Then, by substituting the mean rate of return $\mu_F(r,t)$ in (13.55) and volatility $\sigma_F(r,t)$ in (13.56) into (13.57), it is readily shown that the bond price function $f(r,t)$ satisfies the partial differential equation (PDE)

$$f_t(r,t) + m(r,t)f_r(r,t) + \frac{\sigma^2(r,t)}{2}f_{rr}(r,t) = rf(r,t). \tag{13.61}$$

The proof is left to the reader. The boundary condition is given by $f(r,T) = 1$, since the discount bond pays 1 dollar for sure at the maturity T.

The Affine Model

There is a class of spot-rate models that the PDE (13.61) can be solved with ease. That is, suppose that the risk-adjusted drift $m(r,t)$ and the diffusion coefficient $\sigma(r,t)$ are of the form

$$m(r,t) = \alpha_1(t) + \alpha_2(t)r, \quad \sigma^2(r,t) = \beta_1(t) + \beta_2(t)r, \tag{13.62}$$

respectively, where $\alpha_i(t)$ and $\beta_i(t)$, $i=1,2$, are deterministic functions of time t. Then, substituting (13.62) into (13.61), we obtain

$$f_t(r,t) + (\alpha_1(t) + \alpha_2(t)r)f_r(r,t) + \frac{\beta_1(t) + \beta_2(t)r}{2}f_{rr}(r,t) = rf(r,t). \tag{13.63}$$

Suppose that the discount bond price $v(t,T) = f(r(t),t)$ is given by

$$f(r,t) = e^{a_T(t) + b_T(t)r}, \quad 0 \le t \le T,$$

for some smooth, deterministic functions $a_T(t)$ and $b_T(t)$. Since $f(r,T) = 1$, they must satisfy

$$a_T(T) = b_T(T) = 0. \tag{13.64}$$

Furthermore, we have

$$f_t = [a'_T(t) + b'_T(t)r]f, \quad f_r = b_T(t)f, \quad f_{rr} = b_T^2(t)f.$$

Hence, substituting these equations into (13.63) and rearranging the terms, we obtain

$$\left(b'_T(t) + \alpha_2(t)b_T(t) + \frac{\beta_2(t)b_T^2(t)}{2} - 1 \right) r$$
$$+ \left(a'_T(t) + \alpha_1(t)b_T(t) + \frac{\beta_1(t)b_T^2(t)}{2} \right) = 0.$$

It follows that, in order for the PDE (13.63) to hold, the functions $a_T(t)$ and

THE INTEREST-RATE TERM STRUCTURE

$b_T(t)$ must satisfy the simultaneous ordinary differential equation (ODE)

$$\begin{cases} b_T'(t) = -\alpha_2(t)b_T(t) - \dfrac{\beta_2(t)b_T^2(t)}{2} + 1, \\ a_T'(t) = -\alpha_1(t)b_T(t) - \dfrac{\beta_1(t)b_T^2(t)}{2}, \end{cases} \quad (13.65)$$

together with the boundary condition (13.64). Note that solving the ODE (13.65) numerically is much easier than solving the PDE (13.63). In summary, we have the following.

Theorem 13.1 (Affine Model) *Under the risk-neutral probability measure P^*, suppose that the risk-adjusted drift $m(r,t)$ and the diffusion coefficient $\sigma(r,t)$ of the spot rate $r(t)$ are given by (13.62). Then, the default-free discount bond price $v(t,T)$ is obtained as*

$$v(t,T) = e^{a_T(t) + b_T(t)r(t)}, \quad 0 \le t \le T, \quad (13.66)$$

where $a_T(t)$ and $b_T(t)$ satisfy the simultaneous ODE (13.65) and the boundary condition (13.64).

From (13.66) and (4.7), the yield of the discount bond is given by

$$Y(t,T) = -\frac{b_T(t)}{T-t} r(t) - \frac{a_T(t)}{T-t}.$$

Since the yield is an affine function of the spot rate, the model resulting from (13.62) is called the *affine model*. Note that both $a_T(t)$ and $b_T(t)$ are deterministic functions of time t. Hence, the randomness in the discount bond price (13.66) is due to the spot rate $r(t)$ only. See Exercise 13.21 for a two-factor affine model.

Suppose now that the diffusion coefficient is a deterministic function of time, i.e. $\sigma(r,t) = \sigma(t)$ say, so that, in the notation of (13.62), we have

$$\beta_1(t) = \sigma^2(t), \quad \beta_2(t) = 0.$$

Then, from (13.65), $b_T(t)$ satisfies the ODE

$$b_T'(t) = -\alpha_2(t)b_T(t) + 1, \quad 0 \le t \le T,$$

with $b_T(T) = 0$, which can be solved as

$$b_T(t) = -\int_t^T e^{\int_t^s \alpha_2(u)du} ds, \quad 0 \le t \le T. \quad (13.67)$$

The function $a_T(t)$ is given by

$$a_T(t) = \int_t^T \alpha_1(u)b_T(u)du + \frac{1}{2}\int_t^T \sigma^2(u)b_T^2(u)du.$$

Example 13.8 In the Vasicek model (13.49), suppose that the market price of risk $\lambda(t)$ is a constant, $\lambda(t) = \lambda$ say. Then, from (13.60), we obtain the SDE

$$dr = a(\bar{r} - r)dt + \sigma dz^*, \quad 0 \le t \le T, \quad (13.68)$$

under the risk-neutral probability measure P^*, where
$$\bar{r} = m - \frac{\sigma}{a}\lambda$$
is a positive constant that represents the risk-adjusted, mean-reverting level. The SDE (13.68) is an affine model with
$$m(r,t) = a\bar{r} - ar, \quad \sigma^2(r,t) = \sigma^2.$$
Since the diffusion coefficient is constant, it follows from (13.67) that
$$b_T(t) = -\frac{1 - e^{-a(T-t)}}{a}$$
and
$$a_T(t) = -(b_T(t) + T - t)\left(\bar{r} - \frac{\sigma^2}{2a^2}\right) - \frac{\sigma^2}{4a}b_T^2(t).$$
Hence, the default-free discount bond price in the Vasicek model is given by
$$v(t,T) = H_1(T-t)e^{-H_2(T-t)r(t)}, \quad 0 \le t \le T,$$
where $H_2(t) = (1 - e^{-at})/a$ and
$$H_1(t) = \exp\left\{\frac{(H_2(t) - t)(a^2\bar{r} - \sigma^2/2)}{a^2} - \frac{\sigma^2 H_2^2(t)}{4a}\right\}.$$
See Exercise 13.20 for a related model.

The Quadratic Model

Let $y(t)$ represent a latent variable that affects the spot rate $r(t)$, and assume that $\{y(t)\}$ follows the linear SDE
$$dy = (ay + b)dt + \sigma dz^*, \quad 0 \le t \le T, \tag{13.69}$$
where a, b, and σ are constants and $\{z^*(t)\}$ denotes a standard Brownian motion under a given risk-neutral probability measure P^*.

Suppose that the spot rate is given by a quadratic form of y, $r(t) = y^2(t)$ say. Such a quadratic interest-rate model has been considered by, e.g., Jamshidian (1996) and Rogers (1995).

The advantage of this model is, among others, that the interest rates stay non-negative by the very definition and the transition probability density of $y(t)$ is known in closed form. Also, by Ito's formula, the spot-rate process $\{r(t)\}$ follows the SDE
$$dr = \left[2y(ay+b) + \sigma^2\right]dt + 2\sigma y dz^*, \quad 0 \le t \le T. \tag{13.70}$$
Since $y(t) = \text{sign}[y(t)]\sqrt{r(t)}$, the SDE (13.70) looks similar to the CIR model (13.50) with a non-linear drift.

Let $v(t,T)$ denote the time t price of the default-free discount bond with maturity T, and suppose that it is given by
$$v(t,T) = \exp\left\{\alpha(t) + \beta(t)y(t) + \gamma(t)y^2(t)\right\}, \quad 0 \le t \le T,$$

where $\alpha(t)$, $\beta(t)$, and $\gamma(t)$ are (unknown) deterministic functions of time with the boundary condition $\alpha(T) = \beta(T) = \gamma(T) = 0$. By Ito's formula, we then obtain

$$\frac{dv(t,T)}{v(t,T)} = \left\{\alpha'(t) + b\beta(t) + \sigma^2\gamma(t) + \frac{\sigma^2}{2}\beta^2(t)\right.$$
$$+ \left[\beta'(t) + a\beta(t) + 2b\gamma(t) + 2\sigma^2\beta(t)\gamma(t)\right]y(t)$$
$$+ \left.\left[\gamma'(t) + 2a\gamma(t) + 2\sigma^2\gamma^2(t)\right]y^2(t)\right\}dt$$
$$+ \sigma\left[\beta(t) + 2\gamma(t)y(t)\right]dz^*.$$

Since the discounted bond price is a martingale under P^*, we conclude that the drift term is equal to the spot rate $r(t) = y^2(t)$ for all $t \leq T$. It follows that the unknown functions must satisfy the simultaneous ODE

$$\begin{cases} \alpha'(t) + b\beta(t) + \sigma^2\gamma(t) + \dfrac{\sigma^2}{2}\beta^2(t) = 0, \\ \beta'(t) + a\beta(t) + 2b\gamma(t) + 2\sigma^2\beta(t)\gamma(t) = 0, \\ \gamma'(t) + 2a\gamma(t) + 2\sigma^2\gamma^2(t) = 1, \end{cases}$$

together with the boundary condition $\alpha(T) = \beta(T) = \gamma(T) = 0$. See Exercise 13.22 for a related problem.

13.7 Pricing of Interest-Rate Derivatives

In this section, we explain the pricing of default-free interest-rate derivatives such as bond options, caps, and swaps. For this purpose, the forward-neutral method plays a crucial role, since the interest rates are stochastic. In this section, we consider European-type derivatives only. The reader is referred to, e.g., Rebonato (1996, 1999), Pelsser (2000), and Brigo and Mercurio (2001) for more detailed discussions of the pricing of interest-rate derivatives.

13.7.1 Bond Options

Let $v(t,\tau)$ denote the time t price of the default-free discount bond maturing at time τ, and consider a contingent claim written on $v(t,\tau)$ with maturity $T < \tau$. Let $h(x)$ denote the payoff function of the claim. According to the forward-neutral method (13.46), the time t price of the claim is given by

$$C(t) = v(t,T)E_t^T\left[h(v(T,\tau))\right], \quad t \leq T < \tau. \tag{13.71}$$

Hence, we only need the univariate distribution of $v(T,\tau)$ under the forward-neutral probability measure P^T.

Suppose that there are no arbitrage opportunities in the market and that the discount bond price follows the SDE

$$\frac{dv(t,\tau)}{v(t,\tau)} = r(t)dt + \sigma(t,\tau)dz^*, \quad 0 \leq t \leq \tau, \tag{13.72}$$

under a given risk-neutral probability measure P^*, where $r(t)$ denotes the time t (default-free) spot rate and $\{z^*(t)\}$ is a standard Brownian motion under P^*.

Let us denote the T forward price of the discount bond by

$$v_T(t,\tau) = \frac{v(t,\tau)}{v(t,T)}, \quad t \leq T < \tau.$$

Since $v(t,T)$ also follows the SDE (13.72) with the volatility $\sigma(t,\tau)$ being replaced by $\sigma(t,T)$, Ito's division rule shows that

$$\frac{\mathrm{d}v_T(t,\tau)}{v_T(t,\tau)} = \mu_T(t)\mathrm{d}t + [\sigma(t,\tau) - \sigma(t,T)]\mathrm{d}z^*, \quad 0 \leq t \leq T,$$

for some mean rate of return $\mu_T(t)$.

Now, the forward-neutral probability measure P^T makes the forward price process $\{v_T(t,\tau)\}$ a martingale. Hence, the change of measure is such that

$$\mu_T(t)\mathrm{d}t + [\sigma(t,\tau) - \sigma(t,T)]\mathrm{d}z^* = [\sigma(t,\tau) - \sigma(t,T)]\mathrm{d}z^T,$$

where $\{z^T(t)\}$ is a standard Brownian motion under P^T. Then, the T forward discount bond price follows the SDE

$$\frac{\mathrm{d}v_T(t,\tau)}{v_T(t,\tau)} = [\sigma(t,\tau) - \sigma(t,T)]\mathrm{d}z^T, \quad 0 \leq t \leq T, \qquad (13.73)$$

and, under some technical conditions, the process $\{v_T(t,\tau)\}$ is indeed a martingale under P^T. Since

$$v(T,\tau) = v_T(T,\tau)$$

by the assumption, we can in principle determine the distribution of $v(T,\tau)$ under P^T from (13.73).

Example 13.9 (Affine Model) In the case of the affine model (13.62), we have from (13.66) that

$$v_T(t,\tau) = \frac{v(t,\tau)}{v(t,T)} = \mathrm{e}^{a_\tau(t) - a_T(t) + (b_\tau(t) - b_T(t))r(t)}.$$

It follows, after some algebra, that

$$\frac{\mathrm{d}v_T(t,\tau)}{v_T(t,\tau)} = (b_\tau(t) - b_T(t))\sigma(r(t),t)\left\{-b_T(t)\sigma(r(t),t)\mathrm{d}t + \mathrm{d}z^*\right\},$$

where

$$\sigma^2(r,t) = \beta_1(t) + \beta_2(t)r.$$

Hence, the change of measure is such that

$$-b_T(t)\sigma(r(t),t)\mathrm{d}t + \mathrm{d}z^* = \mathrm{d}z^T,$$

and we obtain

$$\frac{\mathrm{d}v_T(t,\tau)}{v_T(t,\tau)} = (b_\tau(t) - b_T(t))\sigma(r(t),t)\mathrm{d}z^T, \quad 0 \leq t \leq T. \qquad (13.74)$$

In particular, suppose that the diffusion coefficient $\sigma(t)$ is a deterministic

PRICING OF INTEREST-RATE DERIVATIVES

function of time t, i.e. $\beta_1(t) = \sigma^2(t)$ and $\beta_2(t) = 0$. Then, from (13.67), we obtain

$$b_\tau(t) - b_T(t) = -\int_T^\tau e^{\int_t^s \alpha_2(u)du} ds.$$

It follows from (13.74) that

$$\frac{dv_T(t,\tau)}{v_T(t,\tau)} = \theta(t)dz^T, \quad 0 \le t \le T,$$

where

$$\theta(t) = -\sigma(t)\int_T^\tau e^{\int_t^s \alpha_2(u)du} ds,$$

which is a deterministic function of time t. Therefore, in this setting, the forward price $v_T(t,\tau)$ follows a log-normal distribution under the forward-neutral probability measure P^T. Hence, by setting

$$\sigma_F^2 = \int_0^T \theta^2(t)dt,$$

it follows from Proposition 13.3 that

$$E_t^T\left[\{v_T(T,\tau) - K\}_+\right] = v_T(t,\tau)\Phi(d) - K\Phi(d - \sigma_F),$$

where

$$d = \frac{\log[v_T(t,\tau)/K]}{\sigma_F} + \frac{\sigma_F}{2}.$$

Since $v_T(T,\tau) = v(T,\tau)$ and $v_T(t,\tau) = v(t,\tau)/v(t,T)$, we obtain the call option premium as

$$C(t) = v(t,\tau)\Phi(d) - Kv(t,T)\Phi(d - \sigma_F), \quad t \le T < \tau. \tag{13.75}$$

Note that the premium (13.75) does not include the term $\alpha_1(t)$. In fact, it can be shown that, in the affine model with deterministic volatility, prices of interest-rate derivatives are irrelevant to the risk-adjusted, mean-reverting level.

13.7.2 The Forward-LIBOR and Valuation of Swaps

Suppose that

$$0 \le T_0 < \cdots < T_i < T_{i+1} < \cdots,$$

and denote $\delta_i \equiv T_{i+1} - T_i > 0$. The interest rate defined by

$$L_i(t) = \frac{v(t,T_i) - v(t,T_{i+1})}{\delta_i v(t,T_{i+1})}, \quad 0 \le t \le T_i, \tag{13.76}$$

is called the T_i *forward-LIBOR* at time t. Note that

$$1 + \delta_i L_i(t) = \exp\left\{\int_{T_i}^{T_{i+1}} f(t,s)ds\right\} = \frac{v(t,T_i)}{v(t,T_{i+1})}.$$

Hence, $L_i(T_i) = L(T_i, T_{i+1})$ is the LIBOR rate (4.15) at time T_i.

Recall that, for any security $C(t)$ paying no dividends, the forward price defined by $C^T(t) = C(t)/v(t,T)$ is a martingale under the forward-neutral probability measure P^T. It follows that

$$\frac{C(t)}{v(t,T)} = E_t^T\left[\frac{C(\tau)}{v(\tau,T)}\right], \quad t \le \tau \le T. \tag{13.77}$$

Substitution of $T = T_{i+1}$ and $C(t) = v(t,T_i) - v(t,T_{i+1})$ into (13.77) yields

$$\frac{v(t,T_i) - v(t,T_{i+1})}{v(t,T_{i+1})} = E_t^{T_{i+1}}\left[\frac{v(\tau,T_i) - v(\tau,T_{i+1})}{v(\tau,T_{i+1})}\right].$$

It follows from (13.76) that

$$L_i(t) = E_t^{T_{i+1}}[L_i(\tau)], \quad t \le \tau \le T_i,$$

whence the T_i forward-LIBOR $L_i(t)$ is a martingale under the forward-neutral probability measure $P^{T_{i+1}}$.

Example 13.10 In this example, we recover the present value (4.25) of the cashflows paid by the floating side of the interest-rate swap considered in Example 5.4 via the forward-neutral method. Suppose that the interest paid by the floating side at time T_{i+1} is given by $\delta_i L_i(T_i)$. Then, the time t value of the total payments for the floating side is equal to

$$V_{\text{FL}}(t) = B(t)E_t^*\left[\sum_{i=0}^{n-1} \frac{\delta_i L_i(T_i)}{B(T_{i+1})}\right] = B(t)\sum_{i=0}^{n-1}\delta_i E_t^*\left[\frac{L_i(T_i)}{B(T_{i+1})}\right].$$

From (13.47), we have

$$B(t)E_t^*\left[\frac{L_i(T_i)}{B(T_{i+1})}\right] = v(t,T_{i+1})E_t^{T_{i+1}}[L_i(T_i)].$$

It follows that

$$V_{\text{FL}}(t) = \sum_{i=0}^{n-1} v(t,T_{i+1})\delta_i L_i(t),$$

since $\{L_i(t)\}$ is a martingale under $P^{T_{i+1}}$. Noting that

$$v(t,T_{i+1})\delta_i L_i(t) = v(t,T_i) - v(t,T_{i+1}),$$

we obtain (4.25).

Since the T_i forward-LIBOR $L_i(t)$ is a martingale under $P^{T_{i+1}}$, it may be plausible to assume that the process $\{L_i(t)\}$ follows the SDE

$$\frac{\mathrm{d}L_i}{L_i} = \sigma_i(t)\mathrm{d}z^{i+1}, \quad 0 \le t \le T_i, \tag{13.78}$$

for some volatility $\sigma_i(t)$, where $\{z^{i+1}(t)\}$ is a standard Brownian motion under the forward-neutral probability measure $P^{T_{i+1}}$.

Now, suppose that the volatility $\sigma_i(t)$ in (13.78) is a deterministic function of time t. Then, the LIBOR rate $L_i(T_i)$ is log-normally distributed under

$P^{T_{i+1}}$. On the other hand, the time t price of the associated caplet is given by

$$\mathrm{Cpl}_i(t) = B(t)E_t^* \left[\frac{\delta_i\{L_i(T_i) - K\}_+}{B(T_{i+1})} \right], \quad 0 \le t \le T_i;$$

see (4.28). According to the forward-neutral method, we have

$$\mathrm{Cpl}_i(t) = \delta_i v(t, T_{i+1}) E_t^{T_{i+1}} [\{L_i(T_i) - K\}_+], \quad 0 \le t \le T_i.$$

Hence, defining

$$\nu_i^2 = \int_t^{T_i} \sigma_i^2(s)\mathrm{d}s,$$

it follows from Proposition 13.3 that

$$\mathrm{Cpl}_i(t) = \delta_i v(t, T_{i+1})\{L_i \Phi(d_i) - K\Phi(d_i - \nu_i)\}, \tag{13.79}$$

where $L_i = L_i(t)$ and

$$d_i = \frac{\log[L_i/K]}{\nu_i} + \frac{\nu_i}{2}, \quad \nu_i > 0.$$

Note that, when $\delta_i = 1$, Equation (13.79) coincides with Black's formula given in Exercise 13.9.

Since a cap is a portfolio of the caplets, if $t < T_0$ then the time t price of the cap with cap rate K is given by

$$\mathrm{Cap}(t) = \sum_{i=0}^{n-1} \delta_i v(t, T_{i+1})\{L_i \Phi(d_i) - K\Phi(d_i - \nu_i)\}.$$

The valuation of floors is similar and omitted (see Exercise 13.23).

Example 13.11 Suppose that the discount bond price $v(t,T)$ is log-normally distributed as in the Vasicek model (see Example 13.8). By the definition (13.76), however, the T_i forward LIBOR $L_i(t)$ cannot be log-normally distributed, since the log-normal distribution is not closed under a sum of random variables. In fact, consider the Vasicek model (13.68). After some algebra, it can be shown that the forward LIBOR follows the SDE

$$\frac{\mathrm{d}L_i}{L_i} = \mu(t)\mathrm{d}t + \sigma \left(1 + \frac{1}{\delta_i L_i}\right) \frac{e^{-a(T_i - t)} + e^{-a(T_{i+1} - t)}}{a}\mathrm{d}z^*$$

for some drift $\mu(t)$, where $\{z^*(t)\}$ is a standard Brownian motion under the risk-neutral probability measure P^*. Since the volatility depends on the forward LIBOR itself, it cannot be log-normally distributed. Hence, the assumption that the volatility $\sigma_i(t)$ in the SDE (13.78) is deterministic is *not* consistent to the Vasicek model.

13.8 Pricing of Corporate Debts

For the pricing of corporate debts, there are two major approaches in finance literature. The first approach, called the *structural approach*, considers a firm's asset value and defines default as occurring either at maturity or when the firm value process reaches a prespecified default boundary for the first time. On the other hand, the second approach, called the *reduced-form approach*, assumes an exogenous hazard rate process as given, which represents the likelihood of unexpected default of the firm, and calculates the expected, discounted payoffs under the risk-neutral probability measure. This section reviews the two approaches for modeling the term structure of credit spreads. See, e.g., Ammann (2001) and Bielecki and Rutkowski (2002) for rigorous mathematical treatments of this topic.

13.8.1 The Structural Approach

Consider a corporate firm, and let $A(t)$ denote its asset value at time t. We assume that the firm's default occurs when the asset value becomes less than or equal to some prespecified boundary $d(t)$ for the first time. That is, the default epoch is modeled by the *first passage time* defined by

$$\tau = \inf\{t \geq 0 : A(t) \leq d(t)\}, \qquad (13.80)$$

where $\tau = \infty$ if the event is empty.

Suppose that the firm issues a debt of face value F with maturity T. When there is no default until the maturity, it is assumed that claimholders receive the face value F if the firm can repay the whole amount of the debt, while they will receive only a fraction of the firm's asset value, $\alpha_1 A(T)$ say, if $A(T) < F$. On the other hand, when default occurs at time τ before maturity T, we assume that the claimholders receive only $\alpha_2 A(\tau)$ at the default epoch; see Figure 13.1. The structural approach describes the firm's asset value $A(t)$ in terms of a stochastic differential equation (SDE) and evaluates the recovery value of the debt in the risk-neutral paradigm.

An attractive feature of this approach is that we can analyze how firm-specific variables such as debt ratio influence debt values. Also, it can treat complex contingent claims written on the firm's asset value. However, if the firm value is assumed to follow a diffusion process for the sake of analytical tractability, it takes time to reach the default boundary so that the firm never defaults unexpectedly (i.e., by surprise). See, e.g., Madan and Unal (2000), Duffie and Lando (2001), and Kijima and Suzuki (2001) for models to overcome this flaw.

Let $r(t)$ be the default-free spot rate at time t, and suppose that there is a risk-neutral probability measure P^* as given. Suppose further that the firm's asset value process $\{A(t)\}$ follows the SDE

$$\frac{\mathrm{d}A}{A} = r(t)\mathrm{d}t + \sigma(t)\mathrm{d}z^*, \quad 0 \leq t \leq T, \qquad (13.81)$$

where $\sigma(t)$ denotes the volatility of the asset value and $\{z^*(t)\}$ is a standard Brownian motion under P^*.

Suppose for simplicity that, when default occurs before the maturity T, the recovery value, if any, is invested into the default-free discount bond until the maturity T. Then, in the framework described above, the payoff X received

PRICING OF CORPORATE DEBTS

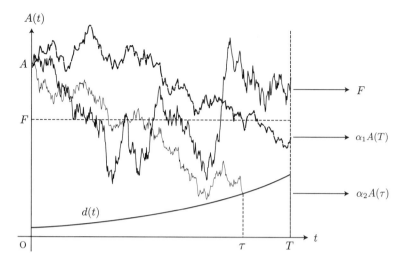

Figure 13.1 *The structure of default and recovery value*

at maturity T is equal to F if $\tau > T$ and $A(T) \geq F$, $X = \alpha_1 A(T)$ if $\tau > T$ and $A(T) < F$, or $X = \alpha_2 A(\tau)/v(\tau, T)$ if $\tau \leq T$, where $v(t, T)$ denotes the time t price of the default-free discount bond maturing at time T. Using the indicator function 1_A, i.e. $1_A = 1$ if A is true and $1_A = 0$ otherwise, the payoff X is represented as

$$X = F 1_{\{\tau > T,\, A(T) \geq F\}} + \alpha_1 A(T) 1_{\{\tau > T,\, A(T) < F\}} + \frac{\alpha_2 A(\tau)}{v(\tau, T)} 1_{\{\tau \leq T\}}. \quad (13.82)$$

It is assumed that $0 \leq \alpha_1, \alpha_2 < 1$.

According to the risk-neutral method, the time t price of the debt, $v^c(t, T)$ say, is given by

$$v^c(t, T) = E_t^* \left[e^{-M(t,T)} X \right], \quad 0 \leq t \leq T, \quad (13.83)$$

where

$$M(t, T) = \int_t^T r(u) \mathrm{d}u.$$

By specifying the SDE for the firm's asset value $\{A(t)\}$, the default boundary $d(t)$, and the default-free spot rate $\{r(t)\}$, there have been proposed various structural models in the literature.

In particular, Black and Cox (1976) assumed that the firm's asset value $\{A(t)\}$ follows the geometric Brownian motion

$$A(t) = A(0) e^{\nu t + \sigma z^*(t)}, \quad 0 \leq t \leq T; \quad \nu \equiv r - \frac{\sigma^2}{2}, \quad (13.84)$$

under the risk-neutral probability measure P^*, while the default boundary is given by

$$d(t) = d e^{-\alpha_0 (T-t)}, \quad 0 \leq t \leq T,$$

for some positive constant α_0, where $d < F$. Since the default-free discount bond price is given by

$$v(t,T) = e^{-r(T-t)}, \quad 0 \le t \le T,$$

the payoff X at the maturity is obtained from (13.82) as

$$X = F1_{\{\tau>T,\, A(T)\ge F\}} + \alpha_1 A(T) 1_{\{\tau>T,\, A(T)<F\}} + \alpha_2 d\, e^{(r-\alpha_0)(T-\tau)} 1_{\{\tau \le T\}}. \tag{13.85}$$

In order to derive the first-passage-time distribution, we define

$$X(t) = \log \frac{A(t)}{A} - \alpha_0 t, \quad 0 \le t \le T,$$

where $A(0) = A$. From (13.84), we obtain

$$\log \frac{A(t)}{A} = \nu t + \sigma z^*(t).$$

Note that
$$A(t) > d(t) \iff X(t) > K,$$

where $K \equiv \log[d/A] - \alpha_0 T$.

Let
$$M(t) = \min_{0 \le u \le t} X(u), \quad 0 \le t \le T,$$

so that
$$\{\tau > T,\, A(T) \ge F\} = \{M(T) > K,\, X(T) \ge \log[F/A] - \alpha_0 T\}.$$

Hence, denoting
$$G_t(x,m) = P^*\{X(t) > x,\, M(t) > m\}, \quad m \le 0,\, x \ge m,$$

we have
$$E^*\left[1_{\{\tau>T,\, A(T)\ge F\}}\right] = G_T\left(\log[F/A] - \alpha_0 T,\, K\right),$$

where, from the result in Exercise 13.7,

$$G_t(x,m) = \Phi\left(\frac{-x + (\nu - \alpha_0)t}{\sigma\sqrt{t}}\right) - e^{2(\nu-\alpha_0)m/\sigma^2} \Phi\left(\frac{-x + 2m + (\nu - \alpha_0)t}{\sigma\sqrt{t}}\right).$$

The other terms in the payoff (13.85) can be evaluated similarly.

In summary, the corporate debt in the Black–Cox model is evaluated as

$$\begin{aligned} v^c(0,T) &= F e^{-rT} \left[\Phi(z_1) - y^{2(\theta-1)} \Phi(z_2)\right] \\ &\quad + A\alpha_1 \left[\Phi(z_3) + y^{2\theta} \Phi(z_4)\right] \\ &\quad + A(\alpha_2 - \alpha_1) \left[\Phi(z_5) + y^{2\theta} \Phi(z_6)\right], \end{aligned} \tag{13.86}$$

where $A(0) = A$, $y = e^K = d e^{-\alpha_0 T}/A$, $\theta = 1 + (\nu - \alpha_0)/\sigma^2$, and

$$z_1 = \frac{\log[A/F] + \nu T}{\sigma\sqrt{T}}, \quad z_2 = \frac{\log[A/F] + 2K + \nu T}{\sigma\sqrt{T}},$$

$$z_3 = -\frac{\log[A/F] + (\nu + \sigma^2)T}{\sigma\sqrt{T}},$$

$$z_4 = \frac{\log[A/F] + 2K + (\nu + \sigma^2)T}{\sigma\sqrt{T}},$$

$$z_5 = \frac{K - \theta\sigma^2 T}{\sigma\sqrt{T}}, \quad z_6 = \frac{K + \theta\sigma^2 T}{\sigma\sqrt{T}}.$$

The proof is left in Exercises 13.24 and 13.25.

13.8.2 Continuous-Time Hazard Models

Let X be a continuous, non-negative random variable representing the lifetime of a corporate bond. It is assumed that X has a density function $f(x)$, $x \geq 0$, for simplicity.

As in the discrete-time case (see Figure 8.1), suppose that the current time is $t > 0$ and the bond is currently alive. That is, we are given the information that the lifetime of the bond is not less than t. The information alters the probability distribution of the lifetime, since

$$P\{X > t + x | X > t\} = \frac{P\{X > t + x\}}{P\{X > t\}}, \quad x \geq 0.$$

Recall that $P\{X > t + x | X > t\} = P\{X > x\}$, i.e. X has the *memoryless* property if and only if X is exponentially distributed. Hence, in general, the lifetime distribution changes in time according to the information that the bond is still alive.

Definition 13.2 (Hazard Rate) The *hazard rate* is the intensity that default occurs at the next instance given no default before that time. Mathematically, the hazard rate at time t for the continuous, non-negative random variable X is defined by

$$h(t) \equiv \lim_{\Delta t \to 0} \frac{1}{\Delta t} P\{t < X \leq t + \Delta t | X > t\}, \tag{13.87}$$

provided that $P\{X > x\} > 0$. The function $h(t)$ with respect to time t is called the *hazard function*.

Exponential distributions are characterized by constant hazard rates. That is, if $X \sim Exp(\lambda)$, then we have from (13.87) that

$$h(x) = \frac{\lambda e^{-\lambda x}}{e^{-\lambda x}} = \lambda, \quad x \geq 0.$$

Conversely, it is readily shown that if $h(x) = \lambda$ then $X \sim Exp(\lambda)$.

The hazard function $h(t)$ can be calculated from either the density function $f(t)$, the cumulative default probability $F(t) = P\{X \leq t\}$ or the survival probability $S(t) = P\{X > t\}$. Conversely, any of these can be deduced from the hazard function, since

$$S(t) = \exp\left\{-\int_0^t h(u)du\right\}, \quad t \geq 0.$$

The proof is left to the reader.

Example 13.12 As in Example 8.1, the lifetime X can be generated as follows. Let Y be an exponentially distributed random variable with parameter 1, i.e.

$$P\{Y > x\} = e^{-x}, \quad x \geq 0,$$

and let $H(t) = \int_0^t h(u)du$ be the *cumulative* hazard rate. Suppose $H(t) \to \infty$ as $t \to \infty$, and let

$$\tau = \min\{t : H(t) \geq Y\}. \tag{13.88}$$

The random variable τ is well defined, since $H(t)$ is non-decreasing in t and diverges as $t \to \infty$. Also, it is easily seen that the event $\{\tau > t\}$ is equivalent to the event $\{Y > H(t)\}$. It follows that

$$P\{\tau > t\} = P\{Y > H(t)\} = e^{-H(t)} = S(t),$$

and so, τ and X are equal in law. Hence, the lifetime X can be generated from the hazard function $h(t)$ and the exponentially distributed random variable Y through (13.88). Cf. Example 3.3 for a special case.

We next define (continuous) Cox processes. Let $\{h(t)\}$ be a non-negative stochastic process in continuous time, and define $H(t) = \int_0^t h(u)du$, $t \geq 0$. Suppose that we are given a realization of $h(t)$, $t \geq 0$, and define the non-negative random variable τ as in (13.88). In other words, we define τ to be any random variable distributed by

$$P\{\tau > t | h(u), u \geq 0\} = e^{-H(t)}, \quad t \geq 0,$$

for the given realization $h(t)$, $t \geq 0$. By the law of total probability (2.31), it follows that

$$P\{\tau > t\} = E\left[e^{-H(t)}\right], \quad t \geq 0. \tag{13.89}$$

Note that the construction of the default time τ is a continuous limit of the discrete Cox process described in Section 8.2. We call the process $\{N(t)\}$ a (continuous) *Cox process* if $N(t) = 1_{\{\tau \leq t\}}$, where 1_A denotes the indicator function meaning that $1_A = 1$ if A is true and $1_A = 0$ otherwise. See Kijima and Muromachi (2000) for details. We note that the Cox process will be discretized and the conditional independence assumption is employed for the purpose of Monte Carlo simulation.

A Cox process is formally defined as follows. Suppose for simplicity that the hazard rate process $\{h(t)\}$ follows the SDE

$$dh = \mu_h(t)dt + \sigma_h(t)dz, \quad 0 \leq t \leq T, \tag{13.90}$$

where $\{z(t)\}$ is a standard Brownian motion defined on a given probability space. We denote the natural Brownian filtration by $\{\mathcal{F}_t^z\}$, where $\mathcal{F}_t^z = \sigma(z(s), s \leq t)$.

Let $\{K(t)\}$ be a Poisson process with the intensity process $\{h(t)\}$ defined on the same probability space. We assume that $K(0) = 0$. Let τ denote the default epoch of a firm defined by the first occurrence of events in the Poisson process, i.e.

$$\tau = \inf\{t > 0; K(t) = 1\}, \tag{13.91}$$

PRICING OF CORPORATE DEBTS 251

where $\tau = \infty$ if the event is empty. Let $\mathcal{F}_t = \mathcal{F}_t^z \vee \mathcal{F}_t^K$, where $\mathcal{F}_t^K = \sigma\left(1_{\{\tau \leq s\}}, s \leq t\right)$. Clearly, τ is a stopping time with respect to $\{\mathcal{F}_t\}$, but not with respect to $\{\mathcal{F}_t^z\}$. Here, it is assumed that all the filtrations satisfy the usual conditions regarding the continuity and completeness. The conditional expectation operator given $\mathcal{F}_t$ is denoted by E_t with $E = E_0$.

Let $t < u \leq T$, and consider the conditional probability $P\{\tau > u | \mathcal{F}_T^z\}$ on the event $\{\tau > t\}$. From (13.91), we obtain

$$P\{\tau > u | \mathcal{F}_T^z\} = P\{K(u) = 0 | \mathcal{F}_T^z\} = \exp\left\{-\int_t^u h(y) dy\right\} \quad (13.92)$$

on the event $\{\tau > t\}$. It follows that

$$P_t\{\tau > u\} = E_t\left[\exp\left\{-\int_t^u h(y) dy\right\}\right] \quad \text{on } \{\tau > t\}, \quad (13.93)$$

which agrees with (13.89). See Exercise 13.26 for a related problem.

13.8.3 The Reduced-Form Approach

Consider a corporate firm, and suppose that the firm issues a debt with face value F and maturity T. Let $\delta(\tau)$, $\delta(\tau) < F$, denote the *recovery rate* when default occurs at time τ before the maturity. Assuming for simplicity that $F = 1$, this means that claimholders will receive only $\delta(\tau)$ at default epoch τ, $\tau \leq T$, if any. When there is no default until the maturity, they will receive the face value $F = 1$ at the maturity for sure. Supposing that, when default occurs before maturity T, the recovery fraction is invested into the default-free discount bond with the same maturity, the payoff X is given by

$$X = \frac{\delta(\tau)}{v(\tau, T)} 1_{\{\tau \leq T\}} + 1_{\{\tau > T\}},$$

where $v(t, T)$ denotes the time t price of the default-free discount bond with maturity T. The payoff X should be compared with that in (13.82).

Suppose that there exists a risk-neutral probability measure P^* as given. Then, according to the risk-neutral method, the time t price of the debt is given from (13.83) as

$$v^c(t, T) = E_t^*\left[e^{-M(t,T)}\left(\frac{\delta(\tau)}{v(\tau, T)} 1_{\{\tau \leq T\}} + 1_{\{\tau > T\}}\right)\right] \quad \text{on } \{\tau > t\}. \quad (13.94)$$

It can be shown under some regularity conditions that

$$E_t^*\left[e^{-M(t,T)}\frac{\delta(\tau)}{v(\tau, T)} 1_{\{\tau \leq T\}}\right] = E_t^*\left[e^{-M(t,\tau)}\delta(\tau) 1_{\{\tau \leq T\}}\right].$$

Hence, Equation (13.94) can be rewritten as

$$v^c(t, T) = E_t^*\left[e^{-M(t,\tau)}\delta(\tau) 1_{\{\tau \leq T\}} + e^{-M(t,T)} 1_{\{\tau > T\}}\right] \quad \text{on } \{\tau > t\}. \quad (13.95)$$

As we have already discussed in Chapter 8, there are three treatments for the recovery rate. That is, the recovery formulation is RMV (recovery of market value) if $\delta(u) = (1-L(u))v^c(u,T)$, RT (recovery of treasury) if $\delta(u) = (1-L(u))v(u,T)$, and RFV (recovery of face value) if $\delta(u) = (1-L(u))$. Here, $L(u)$ denotes the *fractional loss* at time u under the risk-neutral probability measure P^*.

Let $h(t)$ represent the hazard rate for τ under the risk-neutral probability measure P^*, and assume that the process $N(t) = 1_{\{\tau \le t\}}$ is a Cox process. In what follows, we assume that the hazard-rate process $\{h(t)\}$ follows the SDE (13.90), and the cumulative hazard rate is denoted by

$$H(t,T) = \int_t^T h(u)du, \quad 0 \le t \le T.$$

Then, from (13.93), we obtain

$$P_t^*\{\tau > T\} = E_t^*\left[1_{\{\tau > T\}}\right] = E_t^*\left[e^{-H(t,T)}\right] \quad (13.96)$$

on the event $\{\tau > t\}$.

The major advantage of this approach is its analytical tractability and ability of generating a flexible and realistic term structure of credit spreads. However, nobody knows what determines the hazard-rate process. In particular, since the default mechanism is not related to the firm value, we cannot examine the impact of firm-specific variables on debt values.

The Duffie–Singleton Model

In the Duffie–Singleton model, the recovery formulation is RMV. Then, since

$$P_t^*\{u < \tau \le u + dt | \mathcal{F}_T^z\} = h(u)e^{-H(t,u)}dt, \quad t \le u < T,$$

we obtain from (13.95) that

$$v^c(t,T) = E_t^*\left[\int_t^T e^{-\hat{H}(t,u)}h(u)(1-L(u))v^c(u,T)du + e^{-\hat{H}(t,T)}\right], \quad (13.97)$$

where

$$\hat{H}(t,T) = \int_t^T [r(u) + h(u)]du.$$

Let us define the *risk-adjusted spot rate* by

$$R(t) = r(t) + L(t)h(t), \quad 0 \le t \le T.$$

Then, the integral equation (13.97) can be solved as

$$v^c(t,T) = E_t^*\left[e^{-\int_t^T R(u)du}\right], \quad 0 \le t \le T,$$

on the event $\{\tau > t\}$; cf. Exercise 8.7. The proof is left to the reader. See Duffie, Schroder, and Skiadas (1996) for detailed discussions of the existence of the unique solution to the integral equation (13.97).

The Jarrow–Turnbull Model

In the pricing formula (13.94), if we assume the RT recovery formulation, we obtain

$$v^c(t,T) = E_t^* \left[e^{-M(t,T)} \{(1-L(\tau))1_{\{\tau \leq T\}} + 1_{\{\tau > T\}}\} \right] \quad \text{on } \{\tau > t\}.$$

In particular, when the fractional loss is constant, $L(t) = 1 - \delta$ say, we obtain

$$\begin{aligned}
v^c(t,T) &= E_t^* \left[e^{-M(t,T)} \{\delta + (1-\delta)1_{\{\tau > T\}}\} \right] \\
&= \delta v(t,T) + (1-\delta) E_t^* \left[e^{-M(t,T)} 1_{\{\tau > T\}} \right].
\end{aligned}$$

From (13.92), we have

$$E^* \left[1_{\{\tau > T\}} | \mathcal{F}_T^z \right] = e^{-H(t,T)} \quad \text{on } \{\tau > t\}.$$

It follows that

$$\begin{aligned}
E_t^* \left[e^{-M(t,T)} 1_{\{\tau > T\}} \right] &= E_t^* \left[e^{-M(t,T)} E_t^* \left[1_{\{\tau > T\}} | \mathcal{F}_T^z \right] \right] \\
&= E_t^* \left[e^{-M(t,T) - H(t,T)} \right],
\end{aligned}$$

whence we obtain

$$v^c(t,T) = \delta v(t,T) + (1-\delta) E_t^* \left[e^{-\hat{H}(t,T)} \right] \quad \text{on } \{\tau > t\},$$

where $\hat{H}(t,T) = M(t,T) + H(t,T)$.

13.9 Exercises

Exercise 13.1 Suppose that $S_0(t) > 0$ with $S_0(0) = 1$, and it pays no dividends. Let $S_i^*(t) = S_i(t)/S_0(t)$, $i = 1, 2, \ldots, n$, etc. Confirm that, for any self-financing portfolio $\{\boldsymbol{\theta}(t)\}$, the denominated value process is given by

$$V^*(t) = V^*(0) + \sum_{i=1}^n \int_0^t \theta_i(s) \mathrm{d}G_i^*(s), \quad 0 \leq t \leq T,$$

where $\mathrm{d}G_i^*(t) = \mathrm{d}S_i^*(t) + d_i^*(t)\mathrm{d}t$; cf. (5.14).

Exercise 13.2 Consider the two SDEs

$$\frac{\mathrm{d}X}{X} = \mu_X(t)\mathrm{d}t + \sigma_X(t)\mathrm{d}z_1, \quad t \geq 0,$$

and

$$\frac{\mathrm{d}Y}{Y} = \mu_Y(t)\mathrm{d}t + \sigma_Y(t)\mathrm{d}z_2, \quad t \geq 0,$$

where $\mathrm{d}z_1 \mathrm{d}z_2 = \rho(t)\mathrm{d}t$. Mimicking the arguments given in Example 13.2, show that the process $\{D(t)\}$ defined by $D(t) = X(t)/Y(t)$ has the mean rate of return

$$\mu_D(t) = \mu_X(t) - \mu_Y(t) - \sigma_Y(t)(\rho(t)\sigma_X(t) - \sigma_Y(t))$$

and the volatility $\sigma_D(t) = [\sigma_X^2(t) + \sigma_Y^2(t) - 2\rho(t)\sigma_X(t)\sigma_Y(t)]^{1/2}$.

Exercise 13.3 Calculate the instantaneous variance/covariance $dS_i dS_j$ in the SDE (13.12) as well as those in the SDE (13.14) to confirm that they are identical.

Exercise 13.4 Let $\alpha(t)$ and $\beta(t)$ be the number of security $S(t)$ and derivative $C(t)$, respectively, possessed at time t. Under the self-financing assumption (see Definition 13.1), we have
$$dW = \alpha(t)dS + \beta(t)dC.$$
Prove that Equation (13.19) holds.

Exercise 13.5 Prove that the Black–Scholes formula $f(S,t)$ given by (4.21) satisfies the PDE (13.22) and the boundary condition (13.23). *Hint*: Use the identity proved in Exercise 4.8.

Exercise 13.6 By mimicking the calculation given in Example 11.2, prove that Proposition 13.3 holds.

Exercise 13.7 For a standard Brownian motion $\{z(t)\}$, let
$$M(T) = \max_{0 \le t \le T} z(t).$$
The joint distribution of $(z(T), M(T))$ is given, from the result in Exercise 11.9, as
$$P\{z(T) \le x,\ M(T) \le m\} = \Phi\left(x/\sqrt{T}\right) - \Phi\left((x-2m)/\sqrt{T}\right).$$
Consider now $X(t) = z(t) + \mu t$, and let $M_\mu(T) = \max_{0 \le t \le T} X(t)$. Let $Y(T) = \exp\{\mu z(T) - \mu^2 T/2\}$. Applying Girsanov's theorem (Proposition 13.2), show that
$$P\{M_\mu(T) \le m\} = \Phi\left(\frac{m - \mu T}{\sqrt{T}}\right) - e^{2\mu m}\Phi\left(\frac{-m - \mu T}{\sqrt{T}}\right), \quad m > 0.$$
In general, for a Brownian motion with drift μ and diffusion coefficient σ, the distribution function is given by
$$P\{M(t) \le m\} = \Phi\left(\frac{m - \mu t}{\sigma\sqrt{t}}\right) - e^{2\mu m/\sigma^2}\Phi\left(\frac{-m - \mu t}{\sigma\sqrt{t}}\right).$$

Exercise 13.8 (Digital Option) Suppose that the price process $\{S(t)\}$ under the risk-neutral probability measure P^* is given by (13.40), and consider a contingent claim that pays 1 dollar at maturity T only when the event $\{S(T) \ge K\}$ occurs. Obtain the premium of this derivative.

Exercise 13.9 (Black's Formula) In the same setting as Exercise 13.8, the futures price is given by
$$f_T(t) = e^{r(T-t)}S(t), \quad 0 \le t \le T.$$
Show that the premium of the European call (futures option) with strike price K and maturity $s < T$ written on the futures price is given by
$$c_f(t) = e^{-r(s-t)}\left\{f\Phi(d) - K\Phi\left(d - \sigma\sqrt{s-t}\right)\right\}, \quad 0 \le t \le s,$$

where $f_T(t) = f$ and
$$d = \frac{\log[f/K]}{\sigma\sqrt{s-t}} + \frac{\sigma\sqrt{s-t}}{2}.$$
Also, show that the put–call parity of the futures options is given by
$$c_f(t) + v(t,s)K = p_f(t) + v(t,s)F_s(t),$$
where $p_f(t)$ is the premium of the futures put option and $F_s(t)$ is the s forward price of the underlying security $S(t)$.

Exercise 13.10 (Knock-Out Option) In the same setting as Exercise 13.8, consider a knock-out option that expires worthless if the underlying price ever hits the pre-specified level H. That is, let τ be the first passage (hitting) time to the level H. Then, the premium of this option is given by
$$C_1 = e^{-rT} E^* \left[\{S(T) - K\}_+ 1_{\{\tau > T\}} \right],$$
where $S(0), K > H$. Prove that
$$C_1 = S\Phi(\xi) - Ke^{-rT}\Phi\left(\xi - \sigma\sqrt{T}\right)$$
$$- \left[S\left(\frac{S}{H}\right)^{\gamma-2} \Phi(\eta) - Ke^{-rT}\left(\frac{S}{H}\right)^{\gamma} \Phi\left(\eta - \sigma\sqrt{T}\right) \right],$$
where $S(0) = S$, $\gamma = 1 - 2r/\sigma^2$ and
$$\xi = \frac{\log[S/K] + (r + \sigma^2/2)T}{\sigma\sqrt{T}}, \quad \eta = \frac{\log[H^2/SK] + (r + \sigma^2/2)T}{\sigma\sqrt{T}}.$$
Note that the first term of the premium is the Black–Scholes equation. See Cox and Rubinstein (1985, p.411) for the interpretation of the second term.

Exercise 13.11 In the same setting as Exercise 13.10, consider in turn a contingent claim that pays R dollars whenever the underlying price hits the level H. That is, the premium of this option is given by
$$C_2 = E^* \left[R e^{-r\tau} 1_{\{\tau \leq T\}} \right].$$
Prove that
$$C_2 = R e^{|z|} \left[1 - \Phi\left(\frac{|z| + (r + \sigma^2/2)t}{\sigma\sqrt{t}}\right) \right.$$
$$\left. + e^{-(2r+\sigma^2)|z|/\sigma^2} \Phi\left(\frac{-|z| + (r + \sigma^2/2)t}{\sigma\sqrt{t}}\right) \right],$$
where $z = \log[H/S]$.

Exercise 13.12 (Jump-Diffusion) Suppose that the underlying stock price follows the SDE (13.16) with μ being replaced by the risk-free interest rate r under the risk-neutral probability measure P^*. Assuming that the risk-neutral method is applicable,[†] show that the price of the call option with strike price

---
[†] See Merton (1976) for justifications of this assumption.

K and maturity T written on the stock is given by
$$c = \sum_{n=0}^{\infty} \frac{(\widehat{\lambda}T)^n}{n!} e^{-\widehat{\lambda}T} \mathrm{BS}(S,T,K;r_n,\sigma_n),$$
where $S(0) = S$, $\widehat{\lambda} = \lambda(1+m)$,
$$r_n = r - \lambda m + \frac{n}{T}\gamma, \quad \sigma_n^2 = \sigma^2 + \frac{n}{T}\sigma_\eta^2,$$
with $\gamma \equiv \log(1+m) = \mu_y + \sigma_y^2/2$, and
$\mathrm{BS}(S,T,K;r,\sigma)$ denotes the Black–Scholes price function of the call option with volatility σ and instantaneous interest rate r.

Exercise 13.13 Suppose that, under the risk-neutral probability measure P^*, the underlying stock price follows the SDE
$$\mathrm{d}\log S = (\nu - q\log S)\mathrm{d}t + \sigma \mathrm{d}z(t), \quad 0 \le t \le T,$$
where $\nu = r - \sigma^2/2$ and where r, q, and σ are positive constants. Note that the model assumes the dividend rate to be $\delta(S,t) = qS\log S$. Show that
$$S(T) = SA\exp\left\{-\frac{\psi^2}{2} + \sigma\int_0^T e^{-q(T-t)}\mathrm{d}z(t)\right\},$$
where $S(0) = S$, $\psi^2 = \sigma^2(1 - e^{-2qT})/2q$, and
$$\log A = \left(e^{-qT} - 1\right)\log S + \frac{\nu}{q}\left(1 - e^{-qT}\right) + \frac{\psi^2}{2}.$$
Also, show that the premium of the European call option with strike price K and maturity T written on the stock is given by
$$c = Se^{-rT}A\Phi(d) - Ke^{-rT}\Phi(d - \psi), \quad \psi > 0.$$
Finally, show that the price c is neither convex with respect to S nor increasing in σ. See Kijima (2002) for details.

Exercise 13.14 Suppose that the process $\{X(t)\}$ follows the SDE
$$\frac{\mathrm{d}X}{X} = \mu_X(t)\mathrm{d}t + \sigma_X(t)\mathrm{d}z, \quad 0 \le t \le T,$$
where $\mu_X(t)$ and $\sigma_X(t)$ are sufficiently smooth. Prove that
$$X(t) = E_t\left[e^{-\int_t^T \mu_X(s)\mathrm{d}s}X(T)\right], \quad 0 \le t \le T.$$
Hint: Define $Y(t) = e^{-\int_0^t \mu_X(s)\mathrm{d}s}X(t)$, and obtain the SDE for $\{Y(t)\}$.

Exercise 13.15 Suppose that the stock price $\{S(t)\}$ follows the SDE (13.43) and the discount bond price follows (13.42) under the risk-neutral probability measure P^*. Using the result in Exercise 13.14, show that
$$F_T(t) = E_t^*\left[e^{-\int_t^T \mu_F(s)\mathrm{d}s}S(T)\right],$$

EXERCISES

where $\mu_F(t) = -\sigma_v(t)[\sigma(t) - \sigma_v(t)]$. Also, assuming that the volatilities $\sigma(t)$ and $\sigma_v(t)$ are deterministic functions of time t, obtain the relationship between the futures price and the forward price. See Jamshidian (1993) for more general results.

Exercise 13.16 Suppose in turn that the stock price follows the SDE (13.39) and the discount bond price follows (13.42) under the risk-neutral probability measure P^*. Obtain the SDE for the denominated price process $\{S^T(t)\}$, and show that Equation (13.46) still holds in this setting.

Exercise 13.17 Consider a European call option with strike price K and maturity T. From (13.46), the premium of the option is given by

$$c(K) = v(0,T) E^T \left[\{S(T) - K\}_+ \right].$$

Let $F(S, x) = P^T \{S(T) \leq x | S(0) = S\}$. Show that

$$c'(K) = -v(0,T)\{1 - F(S, K)\}.$$

Note: It is known that the distribution function $F(S, x)$ is non-increasing in S for all $x > 0$ if the stock price process is a diffusion. Hence, $c'(K)$ is non-increasing in S, so that the delta c_S is non-increasing in the strike price K in the case of diffusion. See Kijima (2002) for details.

Exercise 13.18 Suppose that a solution to the SDE (13.48) exists and is integrable. Show that the mean spot rate, $M(t) = E[r(t)]$, is given by

$$M(t) = m + (r(0) - m) e^{-at}, \quad t \geq 0,$$

provided that $a \neq 0$. *Hint*: First, obtain the ODE for $M(t)$.

Exercise 13.19 Show that the density function (13.51) can be expressed as (13.53). Using this, prove that the MGF of the spot rate $r(t)$ in the CIR model is given by (13.52).

Exercise 13.20 (Hull–White) Suppose that the spot-rate process $\{r(t)\}$ under the risk-neutral probability measure P^* follows the SDE

$$dr = (\phi(t) - ar)dt + \sigma dz^*, \quad 0 \leq t \leq T,$$

where a and σ are positive constants, $\phi(t)$ is a deterministic function of time t, and $\{z^*(t)\}$ is a standard Brownian motion under P^*. This affine model is called the Hull–White model (1990) or the *extended* Vasicek model. Show that the discount bond price is given by

$$v(t,T) = H_1(t,T) e^{-H_2(t,T)r(t)}, \quad 0 \leq t \leq T,$$

where $H_2(t,T) = \left(1 - e^{-a(T-t)}\right)/a$ and

$$H_1(t,T) = \exp\left\{\frac{1}{2} \int_t^T \sigma^2 H_2^2(u,T) du - \int_t^T \phi(u) H_2(u,T) du\right\}.$$

Also, obtain the forward rate $f(t,T)$ and its derivative with respect to t so as to prove that

$$\phi(t) = a f(0,t) + \frac{\partial}{\partial t} f(0,t) + \frac{\sigma^2}{2a} \left(1 - e^{-2at}\right).$$

See Hull and White (1994a) and Kijima and Nagayama (1994) for a related result.

Exercise 13.21 Hull and White (1994b) considered the following two-factor affine model: Under a given risk-neutral probability measure P^*, suppose that

$$\begin{cases} dr = (\phi(t) + u - ar)dt + \sigma_r dz_1^*, \\ du = -budt + \sigma_u dz_2^*, \end{cases}$$

where a, b, σ_r, and σ_u are positive constants, $\phi(t)$ is a deterministic function of time, and $\{z_1^*(t)\}$ and $\{z_2^*(t)\}$ are standard Brownian motions with correlation coefficients ρ, i.e. $dz_1^* dz_2^* = \rho dt$. The model assumes that the mean-reverting level $\phi(t) + u(t)$ of the spot rate under P^* is stochastically varying in time. Let $v(t,T) = f(r(t), u(t), t)$, and suppose that

$$f(r,u,t) = e^{A_T(t) + B_T(t)r + C_T(t)u}, \quad 0 \le t \le T,$$

with the boundary condition $A_T(T) = B_T(T) = C_T(T) = 0$, where $A_T(t)$, $B_T(t)$ and $C_T(t)$ are deterministic functions of time. Show that the following simultaneous ODE holds:

$$\begin{cases} B_T'(t) = aB_T(t) + 1, \\ C_T'(t) = bC_T(t) - B_T(t), \\ A_T'(t) = -\phi(t)B_T(t) - \dfrac{\sigma_r^2}{2}B_T^2(t) - \dfrac{\sigma_u^2}{2}C_T^2(t) - \rho B_T(t)C_T(t)\sigma_r\sigma_u. \end{cases}$$

See Hull and White (1994b) for the analytical solution to the ODE.

Exercise 13.22 (Cameron–Martin Formula) For a standard Brownian motion $\{z(t)\}$, suppose that $E\left[\exp\{-\int_0^t q(s)z^2(s)ds\}\right]$ exists, where $q(t)$ is a deterministic function of time t. Show that

$$E\left[\exp\left\{-\int_0^T q(s)z^2(s)ds\right\}\right] = \exp\left(\frac{1}{2}\int_0^T \gamma(s)ds\right)$$

where $\gamma(t)$ is a unique solution of the Riccati equation

$$\gamma'(t) = 2q(t) - \gamma^2(t)$$

with the boundary condition $\gamma(T) = 0$. *Hint*: Consider $d\left(\gamma(t)z^2(t)\right)$ and use the result that $E[Y(T)] = 1$ in (13.37) for $\beta(t) = \gamma(t)z(t)$.

Exercise 13.23 The time t price of a floorlet is given by

$$\text{Fll}_i(t) = \delta_i v(t, T_{i+1}) E_t^{T_{i+1}} \left[\{K - L_i(T_i)\}_+\right].$$

Using this, prove the following *parity* relationship between the caplet and the corresponding floorlet:

$$\text{Cpl}_i(t) - \text{Fll}_i(t) = \delta_i v(t, T_{i+1}) \left[L_i(t) - K\right].$$

Exercise 13.24 The second term in the right-hand side of (13.85) is evaluated as

$$E^*\left[A(T)1_{\{\tau > T,\, A(T) < F\}}\right] = Ae^{\alpha_0 T} E^*\left[e^{X(T)} 1_{\{X(T) < L,\, M(T) > K\}}\right],$$

EXERCISES

where $L = \log[F/A] - \alpha_0 T$. Show that

$$E^* \left[e^{X(T)} 1_{\{X(T)<L,\, M(T)>K\}} \right]$$
$$= \frac{e^{(r-\alpha_0)T}}{\sigma\sqrt{T}} \int_K^L \left[\phi\left(\frac{-x+\theta\sigma^2 T}{\sigma\sqrt{T}}\right) - y^{2\theta}\phi\left(\frac{-x+\theta\sigma^2 T + 2K}{\sigma\sqrt{T}}\right) \right] dx$$

and

$$e^{-rT} E^* \left[A(T) 1_{\{\tau>T,\, A(T)<F\}} \right] = A \left[\Phi(z_3) - \Phi(z_5) - y^{2\theta}\left[\Phi(z_6) - \Phi(z_4)\right] \right],$$

where $\phi(z)$ and $\Phi(z)$ denote the standard normal density and distribution functions, respectively, and z_i, $i = 3, \ldots, 6$, are defined below (13.86).

Exercise 13.25 In the Black–Cox model, show that

$$E^* \left[e^{-(r-\alpha_0)\tau} 1_{\{\tau \leq T\}} \right] = e^K \int_0^T \frac{-K}{\sigma\sqrt{2\pi t^3}} \exp\left\{ -\frac{(-K+\theta\sigma^2 t)^2}{2\sigma^2 t} \right\} dt$$
$$= \Phi\left(\frac{-K+\theta\sigma^2 T}{\sigma\sqrt{T}}\right) - e^{2\theta K} \Phi\left(\frac{K+\theta\sigma^2 T}{\sigma\sqrt{T}}\right)$$

and

$$e^{-rT} E^* \left[d\, e^{(r-\alpha_0)(T-\tau)} 1_{\{\tau \leq T\}} \right] = A \left[\Phi(z_5) + y^{2\theta} \Phi(z_6) \right],$$

where z_5 and z_6 are defined below (13.86).

Exercise 13.26 For the default time epoch τ defined in (13.91), let

$$M(t) = \int_0^t 1_{\{K(u-)=0\}}[dK(u) - h(u)du].$$

Show that the process $\{M(t)\}$ is a martingale with respect to the filtration $\{\mathcal{F}_t\}$. *Hint*: Since $\int_0^t 1_{\{K(u-)=0\}} dK(u) = 1_{\{\tau \leq t\}}$, we have

$$M(t) = 1_{\{\tau \leq t\}} - \int_0^t 1_{\{K(u-)=0\}} h(u) du.$$

Apply (13.92) to the case $\{\tau > t\}$. The case $\{\tau \leq t\}$ is obvious.

References

Ait-Sahalia, Y. (1996). "Testing continuous-time models of the spot interest rate." *Review of Financial Studies*, **9**, 385–426.

Ammann, M. (2001). *Credit Risk Valuation: Methods, Models, and Applications.* Second Edition. Springer, Berlin.

Anderson, W.J. (1991). *Continuous-Time Markov Chains: An Applications Oriented Approach.* Springer, New York.

Bartle, R.G. (1976). *The Elements of Real Analysis.* Second Edition. Wiley, New York.

Bielecki, T.R. and M. Rutkowski (2002). *Credit Risk: Modeling, Valuation, and Hedging.* Springer, Berlin.

Björk, T. (1998). *Arbitrage Theory in Continuous Time.* Oxford University Press, Oxford.

Black, F. and J. Cox (1976). "Valuing corporate securities: Some effects on bond indenture provisions." *Journal of Finance*, **31**, 351–367.

Black, F., E. Derman and W. Toy (1990). "A one-factor model of interest rates and its application to treasury bond options." *Financial Analysts Journal*, 33–39.

Black, F. and M. Scholes (1973). "The pricing of options and corporate liabilities." *Journal of Political Economy*, **81**, 637–654.

Brigo, D. and F. Mercurio (2001). *Interest Rate Models: Theory and Practice.* Springer, Berlin.

Chan, K.C., G.A. Karolyi, F.A. Longstaff and A.B. Sanders (1992). "An empirical comparison of alternative models of the short-term interest rate." *Journal of Finance*, **47**, 1209–1227.

Chance, D.M. and D. Rich (1998). "The pricing of equity swaps and swaptions." *Journal of Derivatives*, **5**, 19–31.

Çinlar, E. (1975). *Introduction to Stochastic Processes.* Prentice-Hall, New Jersey.

Cox, J.C., J.E. Ingersoll and S.A. Ross (1985). "A theory of the term structure of interest rates." *Econometrica*, **53**, 385–407.

Cox, J.C. and S.A. Ross (1976). "The valuation of options for alternative stochastic processes." *Journal of Financial Economics*, **3**, 145–166.

Cox, J.C., S.A. Ross and M. Rubinstein (1979). "Option pricing: A simplified approach." *Journal of Financial Economics*, **7**, 229–268.

Cox, J.C. and M. Rubinstein (1985). *Option Markets.* Prentice-Hall, New York.

Duffie, D. (1999). "Credit swap valuation." *Financial Analyst Journal*, 73–87.

Duffie, D. and D. Lando (2001). "Term structures of credit spreads with incomplete accounting information." *Econometrica*, **69**, 633–664.

Duffie, D., M. Schroder and C. Skiadas (1996). "Recursive valuation of defaultable securities and the timing of resolution of uncertainty." *Annals of Applied Probability*, **6**, 1075–1090.

Duffie, D. and K. Singleton (1998). "Simulating correlated defaults." Preprint.

Duffie, D. and K. Singleton (1999). "Modeling term structures of defaultable bonds." *Review of Financial Studies*, **12**, 687–720.

Duffie, D. and K. Singleton (2002). *Credit Risk Modeling for Financial Institutions*. Princeton University Press, Princeton.

Elliott, R.J. and P.E. Kopp (1999). *Mathematics of Financial Markets*. Springer, New York.

El Karoui, N., M. Jeanblanc and S.E. Shreve (1998). "Robustness of the Black and Scholes formula." *Mathematical Finance*, **8**, 93–126.

Feller, W. (1951). "Two singular diffusion problems." *Annals of Mathematics*, **54**, 173–182.

Feller, W. (1957). *An Introduction to Probability Theory and Its Applications, Volume I*. Second Edition. Wiley, New York.

Feller, W. (1971). *An Introduction to Probability Theory and Its Applications, Volume II*. Second Edition. Wiley, New York.

Freedman, D. (1971). *Markov Chains*. Holden-Day, San Francisco.

Geman, H., N. El Karoui and J.C. Rochet (1995). "Changes of numeraire, changes of probability measure and option pricing." *Journal of Applied Probability*, **32**, 443–458.

Gerber, H.U. and E.S.W. Shiu (1994). "Option pricing by Esscher transforms," *Transactions of the Society of Actuaries*, **46**, 99–140.

Geske, R. (1979). "The valuation of compound options." *Journal of Financial Economics*, **7**, 63–81.

Gnedenko, B.V. and A.N. Kolmogorov (1954). *Limit Distributions for Sums of Independent Random Variables*. Addison-Wesley, Cambridge, MA.

Goldman, B.M., H.B. Sosin and M.A. Gatto (1979). "Path dependent options: Buy at the low, sell at the high." *Journal of Finance*, **34**, 1111–1128.

Grandell, J. (1991). *Aspects of Risk Theory*. Springer, New York.

Ho, T.S. and S. Lee (1986). "Term structure movements and pricing interest rate contingent claims." *Journal of Finance*, **41**, 1011–1028.

Hull, J.C. (2000). *Options, Futures, and Other Derivative Securities*. Fourth Edition. Prentice Hall, New York.

Hull, J.C. and A. White (1990). "Pricing interest-rate-derivative securities." *Review of Financial Studies*, **3**, 573–592.

Hull, J.C. and A. White (1994a). "Numerical procedures for implementing term structure models I: Single-factor models." *Journal of Derivatives*, **2**, 7–16.

Hull, J.C. and A. White (1994b). "Numerical procedures for implementing term structure models II: Two-factor models." *Journal of Derivatives*, **2**, 37–48.

Israel, R.B., J.S. Rosenthal and J.Z. Wei (2001). "Finding generators for Markov chains via empirical transition matrices, with applications to credit ratings." *Mathematical Finance*, **11**, 245–265.

Ito, K. (1944). "Stochastic integral." *Proc. Imperial Acad. Tokyo*, **20**, 519–524.

Jamshidian, F. (1993). "Option and futures evaluation with deterministic volatilities." *Mathematical Finance*, **3**, 149–159.

Jamshidian, F. (1996). "Bond, futures and option evaluation in the quadratic interest rate model." *Applied Mathematical Finance*, **3**, 93–115.

Jarrow, R.A. (1996). *Modelling Fixed Income Securities and Interest Rate Options*. McGraw-Hill, New York.

Jarrow, R.A., D. Lando and S.M. Turnbull (1997). "A Markov model for the term structure of credit risk spread." *Review of Financial Studies*, **10**, 481–523.

REFERENCES

Jarrow, R.A. and S.M. Turnbull (1995). "Pricing derivatives on financial securities subject to credit risk." *Journal of Finance*, **50**, 53–86.

Johnson, N.L. and S. Kotz (1970). *Distributions in Statistics: Continuous Univariate Distributions*. Wiley, New York.

Johnson, N.L., S. Kotz and A.W. Kemp (1992). *Univariate Discrete Distribution*. Second Edition. Wiley, New York.

Karatzas, I. and S.E. Shreve (1988). *Brownian Motion and Stochastic Calculus*. Springer, New York.

Karlin, S. and H.M. Taylor (1975). *A First Course in Stochastic Processes*. Second Edition. Academic Press, New York.

Kijima, M. (1997). *Markov Processes for Stochastic Modeling*. Chapman & Hall, London.

Kijima, M. (2000). "Valuation of a credit swap of the basket type." *Review of Derivatives Research*, **4**, 79–95.

Kijima, M. (2002). "Monotonicity and convexity of option prices revisited." *Mathematical Finance*, forthcoming.

Kijima, M. and K. Komoribayashi (1998). "A Markov chain model for valuing credit risk derivatives." *Journal of Derivatives*, **6**, 97–108.

Kijima, M., K. Komoribayashi and E. Suzuki (2002). "A multivariate Markov model for simulating correlated defaults." *Journal of Risk*, forthcoming.

Kijima, M., H. Li and M. Shaked (2001). "Stochastic processes in reliability." *Stochastic Processes: Theory and Methods*. Rao, C.R. and D.N. Shanbhag Eds., *Handbook in Statistics*, **18**, 471–510.

Kijima, M. and Y. Muromachi (2000). "Credit events and the valuation of credit derivatives of basket type." *Review of Derivatives Research*, **4**, 53–77.

Kijima, M. and Y. Muromachi (2001). "Pricing of equity swaps in a stochastic interest rate economy." *Journal of Derivatives*, **8**, 19–35.

Kijima, M. and I. Nagayama (1994). "Efficient numerical procedures for the Hull-White extended Vasicek model." *Journal of Financial Engineering*, **3**, 275–292.

Kijima, M. and T. Suzuki (2001). "A jump-diffusion model for pricing corporate debt securities in a complex capital structure." *Quantitative Finance*, **1**, 611–620.

Klebaner, F.C. (1998). *Introduction to Stochastic Calculus with Applications*. Imperial College Press, London.

Lando, D. (1998). "On Cox processes and credit risky securities." *Review of Derivatives Research*, **2**, 99–120.

Liptser, R.S. and A.N. Shiryaev (1989). *Theory of Martingales*. Kluwer, New York.

Loeve, M. (1978). *Probability Theory, I and II*. Springer, New York.

Madan, D. and H. Unal (1998). "Pricing the risks of default." *Review of Derivatives Research*, **2**, 121–160.

Madan, D. and H. Unal (2000). "A two-factor hazard rate model for pricing risky debt and the term structure of credit spreads." *Journal of Financial and Quantitative Analysis*, **35**, 43–65.

Markov, A.A. (1906). "Extension of the law of large numbers to dependent events," (in Russian) *Bull. Soc. Phys. Math. Kazan*, **15**, 135–156.

Marshall, A.W. and I. Olkin (1988). "Families of multivariate distributions." *Journal of the American Statistical Association*, **83**, 834–841.

Merton, R.C. (1973). "The theory of rational option pricing." *Bell Journal of Economics and Management Science*, **4**, 141–183.

Merton, R.C. (1976). "Option pricing when underlying stock returns are discontinuous." *Journal of Financial Economics*, **3**, 125–144.

Merton, R.C. (1990). *Continuous-Time Finance*. Blackwell, Cambridge.

Mikosch, T.(1998). *Elementary Stochastic Calculus with Finance in View*. World Scientific, Singapore.

Moro, B. (1995). "The full Monte." *RISK*, February, 57–58.

Musiela, M. and M. Rutkowski (1997). *Martingale Methods in Financial Modelling*. Springer, Berlin.

Nakagawa, T. and S. Osaki (1975). "The discrete Weibull distribution." *IEEE Transactions on Reliability*, **24**, 300–301

Neftci, S.N. (1996). *An Introduction to the Mathematics of Financial Derivatives*. Academic Press, San Diego, California.

Nelsen, R.B. (1999). *An Introduction to Copulas*. Springer, New York.

Neuts, M.F. (1973). *Probability*. Allyn and Bacon, Boston.

Niederreiter, H. (1992). *Random Number Generation and Quasi-Monte Carlo Methods*. SIAM.

Øksendal, B. (1998). *Stochastic Differential Equations: An Introduction with Applications*. Fifth Edition. Springer, Berlin.

Peizer, D.B. and J.W. Pratt (1968). "A normal approximation for binomial, F, beta and other common, related tail probabilities." *Journal of the American Statistical Association*, **63**, 1417–1456.

Pelsser, A. (2000). *Efficient Methods for Valuing Interest Rate Derivatives*. Springer, London.

Pliska, S.R. (1997). *Introduction to Mathematical Finance: Discrete Time Models*. Blackwell, Cambridge.

Puterman, M.L. (1994). *Markov Decision Processes: Discrete Stochastic Dynamic Programming*. Wiley, New York.

Rebonato, R. (1996). *Interest-Rate Option Models*. Wiley, Chichester.

Rebonato, R. (1999). *Volatility and Correlation*. Wiley, Chichester.

Ripley, B.D. (1987). *Stochastic Simulation*. Wiley, New York.

Rogers, L.C.G. (1995). "Which model for the term structure of interest rates should one use?" In *Mathematical Finance*, M. Davis, D. Duffie, W. Fleming and S. Shreve Eds., Springer, 93–116.

Ross, S.M. (1983). *Stochastic Processes*. Wiley, New York.

Ross, S.M. (1990). *A Course in Simulation*. Macmillian Publishing Company, New York.

Samuelson, P.A. (1965). "Rational theory of warrant pricing." *Industrial Management Review*, **6**, 13–31.

Shiryaev, A.N. (1999). *Essentials of Stochastic Finance: Facts, Models, Theory*. World Scientific, Singapore.

Singpurwalla, N.D. (1995). "Survival in dynamic environments." *Statistical Science*, **10**, 86–103.

Tezuka, S. (1995). *Uniform Random Numbers: Theory and Practice*. Kluwer Academic Publishers, New York.

Tong, Y.L. (1990). *The Multivariate Normal Distribution*. Springer, New York.

Vasicek, O.A. (1977). "An equilibrium characterization of the term structure." *Journal of Financial Economics*, **5**, 177–188.

Williams, D. (1992). "An appendix to Durbin's paper." *Journal of Applied Probability*, **29**, 302–303.

Yamada, T. and S. Watanabe (1971), "On the uniqueness of solutions of stochastic differential equations." *Journal of Math. Kyoto University*, **11**, 155–167.

Index

σ-additivity, 19, 48
σ-field, 19

Absorbing
  Markov chain, 145
  state, 99, 145
Absorption time, 146
Adapted, 76
Affine model, 238–239, 242
  two-factor, 258
Almost surely, 32
American
  call, 89, 118, 125
  option, 68, 88, 118
  put, 119
Antithetic variable method, 170
Arbitrage opportunity, 81, 111, 217
Archimedean copula, 36
Asian option, 171
Asset pricing
  fundamental theorem of, 87
  no-arbitrage, 82
Attainable, 80, 89, 120, 217
Average option, 171

Backward equation, 98–99
  Kolmogorov, 177, 180
Barrier option, 150
Base of natural logarithm, 1
Bayes formula, 38
BDT (Black–Derman–Toy) model, 236
Bernoulli
  distribution, 41, 95
  random variable, 129, 135
  trials, 41
Bilinearity, 31
Binomial
  distribution, 41, 44, 96
  lattice, 115, 117

  model, 41–42, 109, 111, 131, 148, 166, 181, 188
  pricing method, 114
Black–Cox model, 248, 259
Black–Scholes
  formula, 70, 185
  model, 181–182
Black's formula, 245, 254
Bond
  corporate, 63
  coupon-bearing, 63
  default-free, 63
  discount, 63, 87, 121, 231, 234
Box–Müller method, 172
Bradford distribution, 25
Brownian filtration, 206
Brownian motion, 177, 193
  driftless, 185
  geometric, 178, 236
  path, 198
  standard, 175, 210, 227

Call option, 68, 243, 255–256
  American, 89, 118, 125
  European, 115
Cameron–Martin formula, 258
Cap, 72, 245
  rate, 72
Caplet, 72, 245
Cauchy mean value theorem, 16
Central limit theorem, 158, 179
Chain rule, 5, 38, 85
Change
  of base, 4
  of measure, 49, 90, 104, 107, 185
  of variable, 17
Chapman–Kolmogorov equation, 154
Chebyshev's inequality, 39, 172
Chi-square distribution, 50, 59
Cholesky decomposition, 164

CIR (Cox–Ingersoll–Ross) model, 235
Coefficient
  correlation, 31
  diffusion, 177
  of variation, 58
Collective risk process, 189
Combination, 41
  linear, 54
Complete, 88, 228
Completely monotone, 36
Compound
  option, 228–229
  Poisson process, 190
Compounded interest, 61
Conditional
  density, 27
  distribution, 26
  expectation, 32, 84
  hazard rate, 136
  probability, 20
Conditionally independent, 38, 142
Confidence interval, 159
Constant hazard rate (CHR), 128
Contingent claim, 80
  defaultable, 134
Continuous, 4, 24, 27
  compounding, 62
Control variate, 171
  method, 171
Convergence
  dominated, 15
  in law, 35
Convolution, 57–58
Copula, 36
  Archimedean, 36
Corporate bond, 63
Correlation
  coefficient, 31
  negative, 31
  positive, 31
Counting process, 187
Coupon, 63
Coupon-bearing bond, 63
Covariance, 30
  matrix, 54
Cox process, 130, 135, 139, 250
Credit
  rating, 145, 151, 167
  risk, 63
  swap, 137

Cumulative
  default probability, 127, 249
  dividend, 77, 215
  hazard rate, 250
  probability, 42

Decomposition
  Cholesky, 164
  Doob, 94
  spectral, 150
Decreasing hazard rate (DHR), 128
Default
  model, 45
  time epoch, 45, 51, 246
Defaultable contingent claim, 134
Default-free
  bond, 63
  discount bond, 197
Delivery price, 66
Delta, 225, 257
  Kronecker, 98
Denominated
  price, 79
  process, 91
Density function, 24
  conditional, 27
  joint, 54
  marginal, 27
Derivative, 4, 61
  higher order, 5
  partial, 7
  Radon–Nikodym, 91–92
  security, 61
  total, 8
Difference, 5
Differentiable, 4, 8
Differential, 5
  Ito, 205
Differential equation
  partial, 177, 195, 223, 238
  stochastic, 205
Diffusion, 194, 205
  coefficient, 177
  function, 194
  process, 145, 194
Digital option, 254
Discount bond, 63, 87, 121, 231, 234
  default-free, 197
  price, 238–240

Discount function, 63
Distribution
  Bernoulli, 41, 95
  binomial, 41, 44, 96
  Bradford, 25
  chi-square, 50, 59
  conditional, 26
  double exponential, 59
  Erlang, 52, 58
  exponential, 51, 58, 189, 249
  extreme-value, 59
  first-passage-time, 248
  function, 24
  gamma, 58
  geometric, 45
  hyper-exponential, 58
  initial, 144
  inverse gamma, 59
  joint, 135
  limiting, 155
  logarithmic, 57
  log-normal, 48, 245
  marginal, 26
  mixture, 27
  negative binomial, 57
  non-central chi-square, 51, 235
  normal, 46
  Pascal, 57
  Poisson, 44, 52, 188
  posterior, 28
  prior, 28
  probability, 23
  quasi-limiting, 156
  quasi-stationary, 156
  standard normal, 46
  standard uniform, 53
  state, 144
  stationary, 155
  uniform, 53
  Weibull, 138
Dividend, 77
  cumulative, 77, 215
  rate, 215, 256
Division rule, 219
Dominated convergence, 15
Doob
  decomposition, 94
  martingale, 85, 201
Double exponential distribution, 59
Down-factor, 106, 149

Downward transition probability, 102
Drift, 177
  function, 194
  risk-adjusted, 238
Driftless Brownian motion, 185
Duffie–Singleton model, 252
Dynamic programming, 90
Dynkin's formula, 209

Elementary event, 19
Equality in law, 23, 35
Equation
  backward, 98–99
  Chapman–Kolmogorov, 154
  Fokker–Planck, 191
  forward, 98
  Kolmogorov, 177, 180, 191
  partial differential, 177, 195, 223, 238
  Riccati, 258
  stochastic difference, 161, 167
  stochastic differential, 205, 208, 210, 213
  stochastic recursive, 162
Equity swap, 93
Erlang distribution, 52, 58
Esscher transform, 49
Estimate, 161
Estimator, 161
European option, 68
  call, 115
Event, 19
  elementary, 19
Expectation, 28–30
  conditional, 32, 84
  linearity of, 30
Exponential
  distribution, 51, 58, 189, 249
  function, 2, 6, 9
  martingale, 208
Extended Vasicek model, 257
Extreme-value distribution, 59

Face value, 63
Factorial, 6
Fair game, 85
Feynman–Kac formula, 209, 224
Filtration, 76
  Brownian, 206

First passage time, 100, 192, 211, 214, 246
　distribution, 248
First-to-default time, 136, 139
Floor, 72
Floorlet, 72, 258
Fokker–Planck equation, 191
Formula
　Bayes, 38
　Black, 245, 254
　Black–Scholes, 70, 185
　Cameron–Martin, 258
　Dynkin, 209
　Feynman–Kac, 209, 224
　Ito, 11, 17, 208, 218
　Stirling, 108
Forward
　contract, 66
　equation, 98, 191
　LIBOR, 243
　price, 65, 68, 82, 91, 230–231
　rate, 65
　yield, 65
Forward-neutral, 231
　method, 92, 232, 241
　probability measure, 91, 231
Fractional loss, 252
Frictionless, 75
Fubini's theorem, 15
Function
　density, 24
　diffusion, 194
　discount, 63
　distribution, 24
　drift, 194
　exponential, 2, 6, 9
　gamma, 50
　generating, 39
　harmonic, 213
　hazard, 128, 249
　indicator, 29
　joint density, 27
　logarithmic, 2, 6
　modified Bessel, 235
　moment generating, 35, 39
　payoff, 67
　simple, 14
　survival, 36
　transition density, 177, 180
　transition probability, 176–177

Fundamental theorem of asset pricing, 87
Futures
　contract, 68
　price, 68, 230

Gain, 78, 216
Gamma, 225
　distribution, 58
　function, 50
Gaussian
　interest-rate economy, 232
　process, 208
Generate, 80, 217
Generating function, 39
Generator, 209, 213
Geometric
　Brownian motion, 178, 236
　distribution, 45
Girsanov's theorem, 184, 227, 237
Growth condition, 206, 212

Harmonic function, 213
Hazard
　function, 128, 249
　process, 148
　rate, 128–129, 136, 249–250
Hedging portfolio, 116, 118
Higher order derivative, 5
Homogeneous, 97, 144, 194
　spatially, 97
Hull–White model, 257
Hyper-exponential distribution, 58

IID (Independent and identically distributed), 28
Importance sampling, 169
In law
　convergence, 35
　equality, 23, 35
Incomplete, 88, 120
Increasing hazard rate (IHR), 128
Increment, 95, 175
　independent, 95, 175, 187
　stationary, 175, 187
Independent, 26, 28, 31, 85
　and identically distributed, 28
　conditionally, 38, 142

INDEX 269

increment, 95, 175, 187
Indeterminant form, 6
Indicator function, 29
Induced probability, 23
Inequality
  Chebyshev, 39, 172
  Jensen, 33, 85
  Markov, 38
  Schwarz, 39
Infinitesimal, 179
  mean, 177
  variance, 177
Information, 76, 176, 215
Initial distribution, 144
Instantaneous
  interest rate, 65
  rate of return, 217
Integrable, 84
  square, 200
  uniformly, 200
Integral
  Ito, 203–204
  Riemann, 13
  Riemann–Stieltjes, 13
  stochastic, 79, 86, 202
  Stratonovich, 203
Integrand, 13
Integration by parts, 15, 17
Integrator, 13
Intensity, 187
Interest, 61
  compounded, 61
  rate, 61, 65, 106, 232
Interest-rate swap, 70, 83
In-the-money, 107
Inverse
  gamma distribution, 59
  transform method, 163
Ito
  differential, 205
  formula, 11, 17, 208, 218
  integral, 203–204
  process, 206

Jarrow–Turnbull model, 253
Jensen's inequality, 33, 85
JLT (Jarrow–Lando–Turnbull) model, 151
Joint

  density function, 27, 54
  distribution, 135
  probability, 20, 25
Jump-diffusion model, 221, 255

Kijima–Komoribayashi model, 153
Killing
  rate, 196
  time, 196–197
Knock-out option, 149, 255
Kolmogorov
  backward equation, 177, 180
  forward equation, 191
Kronecker's delta, 98

Laplace transform, 214
Law of large numbers
  strong, 157, 161
  weak, 172
Law of total probability, 21, 26–27, 35
Leverage, 93
L'Hospital's rule, 7, 16
LIBOR, 66
  rate, 66, 243
  forward, 243
Limiting distribution, 155
Linear
  combination, 54
  congruential method, 162
  stochastic differential equation, 208, 213
Linearity
  of conditional expectation, 33, 85
  of expectation, 30
Lipschitz condition, 16, 206, 212
Local martingale, 201, 204
Localizing sequence, 201
Logarithm
  base of natural, 1
  natural, 2
Logarithmic
  distribution, 57
  function, 2, 6
Log-normal distribution, 48, 245
Log-price, 218, 222
Look-back option, 229
Loss
  fraction, 134
  fractional, 252

Marginal
  density, 27
  distribution, 26
Market
  complete, 88, 228
  incomplete, 88
  price of risk, 113, 122, 223, 237
Markov
  chain, 141
  inequality, 38
  model, 151, 168
  property, 141–142
Martingale, 85, 109, 200, 207, 259
  Doob, 85, 201
  exponential, 208
  local, 201, 204
  measure, 87
  representation theorem, 205
  sub-, 94
  super-, 201
  transformation, 86
Matrix
  covariance, 54
  stochastic, 143
  transition, 143
  volatility, 220
Maturity, 63, 66, 68
  yield-to, 64
Mean, 29
  excess return per unit of risk, 113, 223
  infinitesimal, 177
  rate of return, 218
  sample, 50, 157
  square, 199
  vector, 54
Mean-reverting
  level, 258
  model, 234
Mean-value theorem, 6
  Cauchy, 16
Measurable, 76, 84
Measure
  change of, 49, 90, 104, 107, 185
  martingale, 87
  probability, 20
Memoryless property, 45, 51, 249
Merton's bound, 93
MGF (moment generating function), 35, 39, 43–44, 47–48

Mirror image, 99
Mixture distribution, 27
Mode, 42
Model
  affine, 238–239, 242
  BDT (Black–Derman–Toy), 236
  binomial, 41–42, 109, 111, 131, 148, 166, 181, 188
  Black–Cox, 248, 259
  Black–Scholes, 181–182
  CIR (Cox–Ingersoll–Ross), 235
  Duffie–Singleton, 252
  extended Vasicek, 257
  Hull–White, 257
  Jarrow–Lando–Turnbull (JLT), 151
  Jarrow–Turnbull, 253
  jump-diffusion, 221
  Kijima–Komoribayashi, 153
  mean-reverting, 234
  quadratic, 240
  trinomial, 119, 145
  Vasicek, 234, 239
Modified Bessel function, 235
Moment, 28, 135
  generating function, 35, 39, 43–44, 47–48
Money-market account, 61–62, 65, 106
Monotone
  completely, 36
Monotonicity
  of conditional expectation, 85
  of probability, 37
Monte Carlo simulation, 159
  quasi-, 169
Multivariate
  Markov model, 168
  normal distribution, 54
  stochastic differential equation, 210

Natural logarithm, 2
Negative
  binomial distribution, 57
  correlation, 31
  part, 12
No-arbitrage pricing, 82, 217
Non-central, chi-square distribution, 51, 235
Non-homogeneous, 144
Non-overlapping, 175

Non-stationary
  Poisson process, 192
Normal
  distribution, 46, 54
  random number, 163
Novikov condition, 207
Numeraire, 79, 231

Option
  American, 68, 88, 118
  Asian, 171
  average, 171
  barrier, 150
  call, 68, 243, 255–256
  compound, 228–229
  digital, 254
  European, 68
  knock-out, 149, 255
  look-back, 229
  put, 68
Optional sampling theorem, 88
Ornstein–Uhlenbeck (OU) process, 235

Partial
  derivative, 7
  differential equation, 177, 195, 223, 238
  sum, 95, 157
Partition, 12, 21
Pascal distribution, 57
Path-independent, 149
Payoff function, 67
PDE (partial differential equation), 177, 195, 223, 238
Percentile, 56
Perron–Frobenius theorem, 155
Poisson
  distribution, 44, 152, 188
  process, 53, 187–188, 190, 192, 221
Portfolio, 76, 215
  hedging, 116, 118
  replicating, 80, 112, 118, 121, 217
  self-financing, 77, 217
Positive
  correlation, 31
  definite, 54
  part, 12
  semi-definite, 54
Posterior distribution, 28

Predictable, 76, 204, 216
Premium, 70, 137
  risk, 114, 190
Present value (PV), 107
Price
  delivery, 66
  denominated, 79
  discount bond, 238–240
  forward, 65, 68, 82, 91, 230–231
  futures, 68, 230
  log-, 218, 222
  process, 76
  strike, 68
Primitive, 155
Prior distribution, 28
Probability, 19
  conditional, 20
  cumulative, 42
  distribution, 23
  forward-neutral, 91
  induced, 23
  joint, 20, 25
  law of total, 21, 26–27, 35
  measure, 20
  monotonicity of, 37
  of complementary event, 20
  of ruin, 190
  risk-neutral, 87, 112, 227, 237
  space, 20
  subjective, 22
  survival, 42, 127, 134, 249
  transition, 97, 100, 142–143
  vector, 144
Process
  collective risk, 189
  compound Poisson, 190
  counting, 187
  Cox, 130, 135, 139, 250
  denominated, 91
  diffusion, 145, 194
  Gaussian, 208
  hazard, 148
  Ito, 206
  Ornstein–Uhlenbeck (OU), 235
  Poisson, 53, 187–188, 221
  price, 76
  pure jump, 189
  renewal, 189
  return, 77
  simple, 203

stochastic, 75
stopped, 201
value, 77, 216
Product rule, 220
Protective put, 69
Pseudo random number, 162
Pure jump process, 189
Put
American, 119
option, 68
protective, 69
Put–call parity, 74, 82–83, 255

Quadratic
model, 240
variation, 198
Quasi-
limiting distribution, 156
Monte Carlo simulation, 169
stationary distribution, 156

Radon–Nikodym
derivative, 91–92
theorem, 91
Random
number, 161–163
variable, 22, 24, 129, 135
vector, 161
walk, 95–96, 185
Rate
cap, 72
dividend, 215, 256
forward, 65
hazard, 128–129, 249
interest, 61, 106
LIBOR, 66, 243
of return, 55, 64, 77–78, 217–218
recovery, 131–132, 251
spot, 65
swap, 71, 93
Realization, 22, 75, 197
Recombining, 117, 125, 149
Recovery
of face value (RFV), 139
of market value (RMV), 134, 139
of treasure (RT), 132
rate, 131–132, 251
value, 246
Reduced-form approach, 246

Refinement, 13
Reflecting, 109
Reflection principle, 99, 101, 191
Remainder, 9
Renewal process, 189
Replicating portfolio, 80, 112, 118, 121, 217
Return
process, 77
rate of, 55, 64, 77–78
Riccati equation, 258
Riemann
integral, 13
sum, 12
Riemann–Stieltjes
integral, 13
sum, 13, 202
Risk
credit, 63
index, 224
market price of, 113, 122, 223, 237
mean excess return per unit of, 113
premium, 114, 190
Risk-adjusted
drift, 238
spot rate, 252
Risk-averter, 114
Risk-neutral
method, 87, 227, 255
probability measure, 87, 112, 227, 237
Risk-premia adjustment, 152–153

Safety loading, 190
Sample
mean, 50, 157
path, 75, 194, 197
space, 19, 23
variance, 50, 160
Schwarz's inequality, 39
SDE (stochastic differential equation), 205, 208, 210, 213
Self-financing, 77–79
portfolio, 77, 217
Simple
function, 14
process, 203
Skip-free, 99
Small order, 8
Spatially homogeneous, 97

Spectral decomposition, 150
Spot rate, 65
  risk-adjusted, 252
Square integrable, 200
Standard
  Brownian motion, 175, 210, 227
  deviation, 29
  normal distribution, 46, 54
  uniform distribution, 53
Standardization, 47
State
  absorbing, 99, 145
  distribution, 144
  reflecting, 109
  space, 96
Stationary
  distribution, 155
  increment, 175, 187
Stirling's formula, 108
Stochastic
  difference equation, 161, 167
  differential equation, 205, 208, 210, 213
  integral, 79, 86, 202
  interest-rate economy, 232
  matrix, 143
  process, 75
  recursive equation, 162
Stopped process, 201
Stopping time, 88
Stratonovich integral, 203
Strike price, 68
Strong
  law of large numbers, 157, 161
  Markov property, 142
Structural approach, 245–246
Subinterval, 12
Subjective probability, 22
Submartingale, 94
Substochastic, 143
Sum
  partial, 95
  Riemann, 12
  Riemann–Stieltjes, 13, 202
Supermartingale, 201
Survival
  function, 36
  probability, 42, 127, 134, 249
Swap, 70
  credit, 137

equity, 93
interest-rate, 70, 83
rate, 71, 93
Swaption, 71
Symmetric, 96, 185

Taylor's expansion, 9
Term structure, 63, 123, 131, 234
Theorem
  asset pricing, 87
  central limit, 158, 179
  Fubini, 15
  Girsanov, 184, 227, 237
  martingale representation, 205
  mean value, 6, 16
  Optional sampling, 88
  Perron–Frobenius, 155
  Radon–Nikodym, 91
Theta, 225
Time-homogeneous, 97, 144, 194
Total
  derivative, 8
  variation, 199
Transform
  Esscher, 49
  inverse, 163
  Laplace, 214
Transition
  density function, 177, 180
  matrix, 143
  probability, 97, 100, 102, 142–143
  probability function, 176–177
Triangular array, 179
Trinomial model, 119, 145
Two-factor affine model, 258

Uncorrelated, 31
Uniform
  distribution, 53
  random number, 162
Uniformly integrable, 200
Unimodal, 42
Up-factor, 106, 149
Upward transition probability, 102

Value
  face, 63
  present, 107

process, 77, 216
recovery, 246
Value-at-risk (VaR), 55
Variance, 29
  infinitesimal, 177
  reduction method, 169
  sample, 50, 160
Variation
  coefficient of, 58
  quadratic, 198
  total, 199
Vasicek model, 234, 239
  extended,
Volatility, 70, 167, 218
  matrix, 220

Wald's identity, 109, 190
Weak law of large numbers, 172
Weibull distribution, 138
Weight, 55, 78

Yield, 64
  curve, 123
  forward, 65
  to-maturity, 64